Fortschritte der Chemie organischer Naturstoffe

Progress in the Chemistry of Organic Natural Products

48

Founded by L. Zechmeister
Edited by W. Herz, H. Grisebach, G. W. Kirby,
and Ch. Tamm

Authors:
P. Crews, R. E. Moore, S. Naylor,
P. S. Steyn, R. Vleggaar

Springer-Verlag
Wien New York 1985

Dr. W. Herz, Professor of Chemistry, Department of Chemistry,
The Florida State University, Tallahassee, Florida, U.S.A.

Prof. Dr. H. Grisebach, Biologisches Institut II, Lehrstuhl für Biochemie der Pflanzen,
Albert-Ludwigs-Universität, Freiburg i. Br., Federal Republic of Germany

G. W. Kirby, Sc. D., Regius Professor of Chemistry, Chemistry Department,
The University, Glasgow, Scotland

Prof. Dr. Ch. Tamm, Institut für Organische Chemie der Universität Basel,
Basel, Switzerland

With 33 Figures

© 1985 by Springer-Verlag/Wien

Softcover reprint of the hardcover 1st edition 1985

Library of Congress Catalog Card Number AC 39-1015

ISSN 0071-7886

ISBN-13: 978-3-7091-8817-0   e-ISBN-13: 978-3-7091-8815-6
DOI: 10.1007/978-3-7091-8815-6

Contents

List of Contributors

CREWS, Dr. P., Thimann Laboratories, Department of Chemistry, University of California, Santa Cruz, CA 95064, U.S.A.

MOORE, Professor R. E., Department of Chemistry, University of Hawaii, 2545 The Mall, Honolulu, HI 96822, U.S.A.

NAYLOR, Dr. S., Chemical Laboratories, University of Cambridge, Cambridge CB2 1EW, U.K.

STEYN, Dr. P. S., National Chemical Research Laboratory, CSIR, P.O. Box 395, Pretoria 0001, Republic of South Africa.

VLEGGAAR, Dr. R., National Chemical Research Laboratory, CSIR, P.O. Box 395, Pretoria 0001, Republic of South Africa.

Tremorgenic Mycotoxins

By P. S. Steyn and R. Vleggaar, National Chemical Research Laboratory, Council for Scientific and Industrial Research, Pretoria, Republic of South Africa

With 23 Figures

Contents

1. Introduction

The metabolic products of micro-organisms can be classified either as compounds of primary metabolic concern or else as secondary metabolites — substances which are apparently non-essential to the producing organism. Fungi have a remarkable capacity to produce such secondary metabolites, *e.g.* mycotoxins with a diverse array of structural and pharmacological properties (*1*).

The present resurgence of interest in all aspects of mycotoxin research (*2, 3*) can be related to the impact of the hepatotoxins (aflatoxin, sporidesmin, and phomopsin), nephrotoxins (ochratoxin and citrinin), and dermal toxins (trichothecenes) on human and animal health.

Some of the mycotoxins appear to act at the level of the central nervous system. Ergotism, the earliest known mycotoxicosis, that is a disease caused by mycotoxins, was attributed to the contamination of wheat by the parasitic neurotoxin-producing fungus, *Claviceps purpurea*. These neurotoxins elaborated by *C. purpurea* are collectively called the ergot toxins (*4*). The neurotoxin, citreoviridin (**1**) which causes paralysis in the extremities of laboratory animals, followed sometimes by convulsions and respiratory arrest, has been implicated in acute cardiac beri-beri in Japan (*5*). Tremoring has not been associated with citreoviridin or the structurally related aurovertins (**2**) (*6*) and asteltoxin (**3**) (*7*); these compounds are therefore excluded from this review. However, verrucosidin (**4**) (*8*), a tremorgenic compound which structurally resembles (**1**)—(**3**), will be described later.

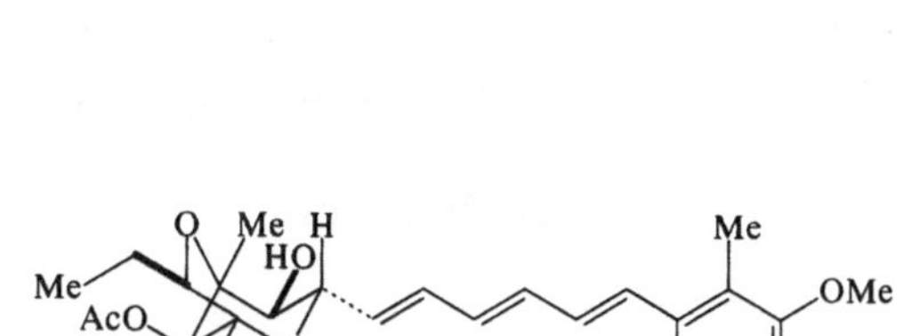

(**1**)

(**2**)

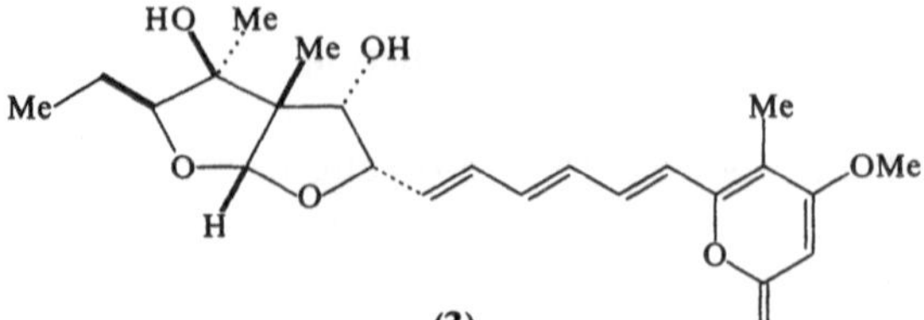

(3)

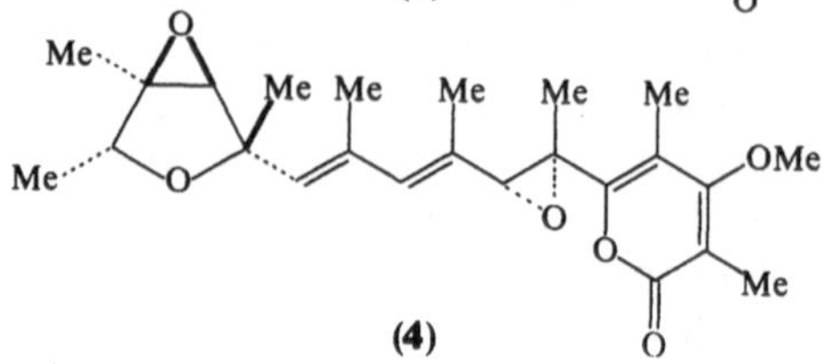

(4)

Some of these mycotoxins have been shown to induce neurological manifestations in vertebrate animals by eliciting a sustained or intermittent tremoring response. Natural products known to induce a tremoring in animals were rare until the recent discoveries of fungal tremorgens. This review is devoted solely to the chemistry and some biological properties of these tremorgens which are synthesized by certain species of *Aspergillus, Penicillium* and *Claviceps*.

It is of importance to note that all the tremorgens, with the exception of the territrems (*9*) and verrucosidin (*8*), contain an indole moiety, albeit in modified form, as a structural feature. The tremorgens have been formally classified in five groups based on their structural characteristics and biosynthetic relationships. The following details if available will be presented for each tremorgen or group of tremorgens *viz.* producing organism(s); chemical formulae; physical and chemical properties; mode of action; and biosynthesis.

The fungal tremorgens have been implicated in various naturally occurring syndromes and form the substance of reviews by COLE (*10, 11, 12*), CIEGLER *et al.* (*13*), MANTLE and PENNY (*14*), SHREEVE *et al.* (*15*), BETINA (*16*) and CYSEWSKI (*17*).

2. The Penitrems, Janthitrems, Lolitrems, Aflatrem, Paxilline, Paspaline, Paspalicine, Paspalinine, and Paspalitrems A and B

2.1 The Penitrems

2.1.1 Producing Organisms

The fungi which are capable of producing tremorgenic toxins can be found on several important agricultural commodities, including silage, maize and various forages. The taxonomic classification of fungi involved in the production of the penitrems however, appears to be still in a state of flux.

Early researchers (*18, 19*) referred to penitrem A as tremortin A but the trivial name penitrem A is used at present. WILSON *et al.* (*18*) first discovered penitrem A from three isolates of *P. cyclopium* shortly after their discovery of aflatrem by *Aspergillus flavus* (*20*) and commented on the nearly identical neurotoxic properties of the two substances. CIEGLER (*19*) and HOU *et al.* (*21*) described *P. palitans* as a producer of the penitrems. CIEGLER and PITT (*22*) surveyed the genus *Penicillium* and found that toxin production was confined to certain species of the subsection fasciculata, section Asymmetrica. The most intense synthesis was noted in strains of *P. crustosum, P. cyclopium, P. granulatum,* and *P. palitans.* PITT (*23*), however, concluded in 1979 that all isolates involved in the production of penitrem A belong to *P. crustosum.* FRISVAD (*24*) used physiological criteria and toxin production as aids in the identification of common asymmetric penicillia. The penitrem producers *P. crustosum* (NRRL 968, 1983, 516), *P. palitans* (NRRL 3465), *P. cyclopium* (NRRL 3476, 3477), and *P. verrucosum* var. *cyclopium* were therefore reclassified as *P. crustosum* p. A. PITT subsequently identified strain Sol-7, used in the structural and biosynthetic studies (*25—29*) of the penitrems, as *P. crustosum.*

PATTERSON *et al.* (*30*) studied the etiology of migram and ryegrass staggers and isolated *Penicillium* species from the soil; *P. canescens, P. clavigerum,* and *P. janthinellum* were reported as producers of penitrem A. SHREEVE *et al.* (*15*) described a condition resembling ryegrass staggers in adult cattle and sheep grazing pastures in eastern England. The penitrem A producers were identified according to a publication in 1983 by Dr. A. ONIONS of the Commonwealth Mycological Institute, Kew, as mainly *P. canescens* and to a lesser degree *P. clavigerum.* The coincident production of penitrems and the paralytic neurotoxin roquefortine was reported for *P. commune* (*31*), *P. cyclopium* (NRRL 6093) (*32*), and *P. crustosum* (*27, 33*). DORNER *et al.* (*34*) isolated a strain of *P. crustosum* from mouldered corn implicated in a natural intoxication in cattle. This strain, regarded as typical of *P. crustosum* Thom produced penitrem A.

The classification criteria of PITT (*23*) which link penitrem production to *P. crustosum* seems to be well established. However, it is rather bewildering to the non-mycologist.

The production of the penitrems on both liquid media and corn were adequately described in the abovementioned publications. The Czapek medium containing yeast extract was used by DE JESUS *et al.* (*27*) in fermentations, whereas WAGENER *et al.* (*31*) obtained higher toxin production with YES-1 medium, *viz. ca.* 8 mg of penitrem A per 100 ml medium.

2.1.2 Isolation and Chromatography of the Penitrems

The penitrems are unstable in chloroform when directly exposed to light, presumably as a result of acid formation in the solvent. Any contact of the penitrems with chloroform or mineral acid was avoided throughout the investigations (25—29).

P. crustosum (Sol-7) was grown in stationary culture at 25° C in conical flasks (380 × 500 ml) containing Czapek medium enriched with 2% yeast extract (100 ml) (27). After 10 days the cultures were filtered and the mycelium macerated with acetone in a Waring blender. The aqueous acetone solution was concentrated and partitioned between dichloromethane and water. The residue from the organic layer was partitioned between 90% methanol and hexane. The crude extract (137 g) from the 90% methanol layer was percolated through a silica gel column with benzene-acetone (4:1, v/v) to give a mixture (4.3 g) of the penitrems and roquefortine (27).

The above mixture of penitrems was purified by chromatography on silica gel using benzene-acetone (85:15, v/v) to yield in order of descending R_f values, six penitrems: Penitrem F, B, A, E, C and D.

Penitrem A (**5**), $C_{37}H_{44}ClNO_6$*, white amorphous solid** (1300 mg) [Wilson et al. (18) reported that the crystals decomposed at 180° to 200° C without melting.]; penitrem B (**6**), $C_{37}H_{45}NO_5$ (170 mg); penitrem C (**7**),

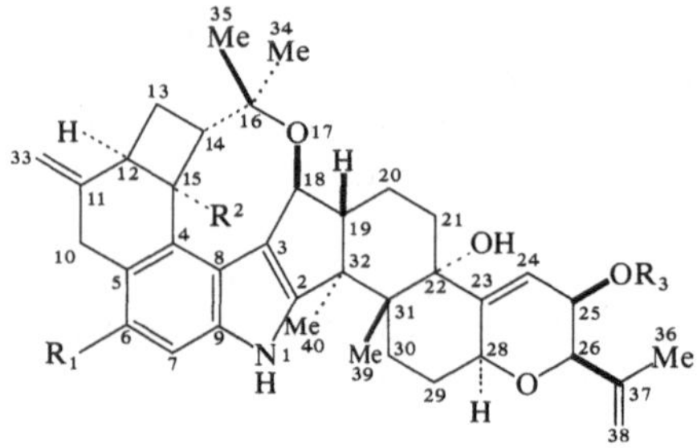

(**5**) R_1=Cl, R_2=OH, R_3=H; 23a,24a-epoxide
(**6**) R_1=R_2=R_3=H; 23a,24a-epoxide
(**7**) R_1=Cl, R_2=R_3=H
(**8**) R_1=R_2=R_3=H
(**9**) R_1=R_3=H, R_2=OH; 23a,24a-epoxide
(**10**) R_1=Cl, R_2=R_3=H; 23a,24a-epoxide
(**11**) R_1=Cl, R_2=OH, R_3=Ac; 23a,24a-epoxide
(**12**) R_1=Cl, R_2=OH, R_3=H; 23a,24a-epoxide; 11,33,37,38-tetrahydro

---

* The molecular formulae of the penitrems were established by high resolution mass spectrometry.

** Penitrems A, B, C, E and F were obtained as white amorphous solids.

$C_{37}H_{44}ClNO_4$ (60 mg); penitrem D (**8**), $C_{37}H_{45}NO_4$, crystallised from benzene as white needles, m.p. > 300°C (decomp.) (300 mg); penitrem E (**9**), $C_{37}H_{45}NO_6$ (100 mg); and penitrem F (**10**), $C_{37}H_{44}ClNO_5$ (150 mg).

The similarity in the molecular formulae of the penitrems indicates that these substances share the same basic structure and differ from each other only in the nature of substituents at certain carbon atoms.

GORST-ALLMAN and STEYN (*35*) and GIMENO (*36*) reported several solvent systems for the analysis of penitrem A. MAES *et al.* (*37*) developed three solvent systems for the separation of penitrems A—F (see Table 1). System c was the only system which permitted separation of all the penitrems. The best results were obtained by developing the chromatogram twice in this solvent system.

The penitrems are extremely sensitive towards spraying with 1% cerium(IV) sulphate in 3 M sulphuric acid and 10^{-7} g of penitrem A can be detected by this reagent on a chromatoplate. The spots turn blue on spraying and after heating at 120°C the colour changes to a stable dark purple. The R_f values of penitrems A—F in different solvent solutions are given in Table 1.

Table 1. R_f *Values of Penitrems A—F*

Penitrem	R_f		
	a	b	c
A	0.16	0.49	0.37
B	0.18	0.53	0.39
C	0.09	0.39	0.28
D	0.09	0.37	0.26
E	0.13	0.46	0.33
F	0.18	0.55	0.42

Solvents: a = hexane-ethyl acetate (70:30); b = dichloromethane-acetone (85:15); c = benzene-acetone (85:15).

The penitrems were also separated by reversed phase high-performance liquid chromatography, using a HP 79918 A RP-8 (10 μm) column and a u.v. detector (296 nm). Water-methanol (22:78) was used as eluant at a flow-rate of 1.5 ml/min. The following retention times (min) were obtained for the metabolites: penitrem A (4.66), penitrem B (11.33), penitrem C (14.68), penitrem D (9.56), penitrem E (3.75), and penitrem F (17.36).

The above approaches can be applied to the quantitative estimation of the penitrems in culture media.

2.1.3 Structure of the Penitrems

2.1.3.1 General Aspects, Ultraviolet and Infrared Spectroscopy

The elucidation of the molecular structure of penitrem A was a challenging problem since its discovery by Wilson *et al.* (*18*). Initial studies (*13*) were hampered by the extreme acid lability of the compounds, inadequate instrumentation, and the lack of sufficient material for structural studies. The advent of high-field nuclear magnetic resonance spectrometers provided new impetus to solving the problem of structure elucidation. The structures of penitrems A—F were eventually elucidated by integrating the data obtained from u.v., i.r, mass spectrometry, chemical reactions and mainly very high-field n.m.r. spectroscopy with the information obtained from specific biosynthetic labelling experiments. Structures (**5—10**) were proposed for penitrems A—F (*25—29*). The penitrems are relatively large molecules and contain some of the most complex features of natural products not solved by X-ray crystallography.

The u.v. spectral data of the penitrems were very similar, *e.g.* penitrem A had λ_{max} 233 (ε 37000) and 295 nm (ε 11 600). This spectral information indicated the presence of a substituted indole moiety in the molecule. Absorptions at 3580 and 3475 cm^{-1} in the i.r. spectrum of penitrem A were assigned to OH and NH groups, respectively. The presence of at least one hydroxy group in penitrem A was confirmed by acetylation to give 25-*O*-acetylpenitrem A (**11**).

2.1.3.2 Nuclear Magnetic Resonance Spectroscopy
^{1}H N.m.r. Spectra

Crystals of penitrems A—F suitable for single crystal X-ray crystallography were not obtained. N.m.r. spectroscopy and biosynthetic studies employing stable isotopes therefore played the major role in their structural elucidation. The utility of high-field resolution-enhanced proton spectra in the structural elucidation of complex molecules became well established in these studies.

The ^{1}H n.m.r. spectra of penitrem A were recorded at 500 MHz in [^{2}H$_6$]acetone. A signal at δ 10.03 was assigned to the NH proton of the indole moiety. The signals at δ 4.16 s, 3.40 d (J 7.5 Hz) and 3.32 s decreased upon the addition of deuterium oxide and were assigned to the hydroxy protons. A one-proton aromatic singlet appeared at δ 7.24. Five three-proton signals at δ 1.75, 1.71, 1.40, 1.22, and 1.07 were assigned to tertiary methyl groups. The remainder of the spectrum exhibited extensive fine structure. First-order analyses of these multiplets yielded the values of the proton chemical shifts and proton-proton coupling constants. The values of the coupling constants, as corroborated by extensive ^{1}H{^{1}H} homonuclear decoupling experiments, facilitated the identification of three fragments, called A, B, and C (Fig. 1).

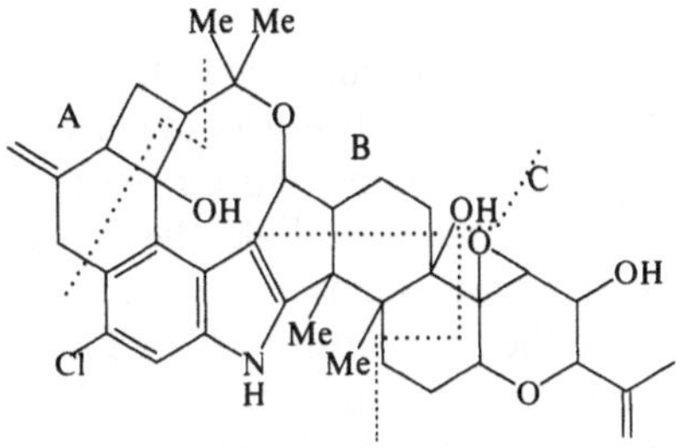

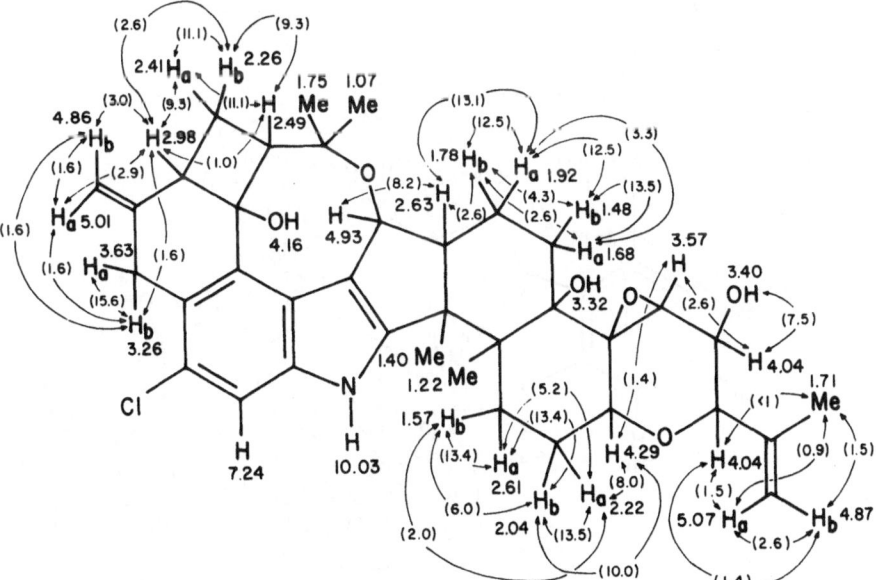

Fig. 1. Identification of three fragments (A, B and C) in penitrem A (**5**) by ¹H n.m.r. spectroscopy

Fragment C is characterized by the number of protons situated on oxygen-bearing carbon atoms as well as the presence of an exocylic methylene group. The geminal and vicinal coupling constants observed for the protons of C(28)-C(29)-C(30) indicate that they form part of a six-membered ring in a chair conformation. A feature of the fourteen-proton spin system is the chemical shift equivalence (δ 4.04) of 25-H and 26-H.

Irradiation at this frequency changed the patterns observed for 38-H$_a$, 38-H$_b$, 24-H and the protons of the C-36 methyl group. In 25-*O*-acetylpenitrem A (**11**) the chemical shifts of only 24-H, 25-H and 26-H were affected. A characteristic acetylation shift (Δδ 1.26 p.p.m.) was observed in the ¹H n.m.r. spectrum of (**11**) for 25-H which resonated at δ 5.30 (*J* 3.4, 1.6 and 0.6 Hz), whereas H-26 appeared as an unresolved multiplet at δ 4.14. The coupling constant of 1.6 Hz observed for 25-H must arise through spin-spin interaction with 26-H.

Fig. 2. The ¹H n.m.r. spectral data and (¹H, ¹H) coupling constants for penitrem A (**5**)

The ^{1}H n.m.r. data of penitrem A are depicted in Fig. 2. It is of importance to note that extensive use was made of nuclear Overhauser enhancement experiments, ^{13}C n.m.r. spectroscopy, biosynthetic information and mass spectral data to link the various fragments. The information obtained from penitrems B—F was also invaluable in the concerted effort to solve these structures (25—29).

The molecular formulae of penitrems A and E indicated that the chlorine atom is replaced in the latter by an hydrogen atom. This finding was evident from the ^{1}H n.m.r. spectrum of penitrem E which showed the presence of two *ortho* oriented aromatic protons at δ 6.70 and 7.09 (*J* 8.3 Hz). The chlorine atom apparently has a measurable influence on the C-10 protons. The chemical shift difference for these protons [δ(10-H$_a$) − δ(10-H$_b$)] is 0.57 ppm for penitrem A and −0.26 ppm in the case of penitrem E. Furthermore, the very small difference in chemical shift (0.01 ppm) for H-25 and H-26 in penitrem E was sufficient to determine the value of the coupling constant as 1.5 Hz (28).

Penitrem F (**10**) contains one oxygen atom less than penitrem A (**5**) (28). An analysis of the ^{1}H n.m.r. spectrum of penitrem F (**10**) showed a new proton signal at δ 3.87. The multiple coupling to H-15 gave rise to a very complex pattern (Fig. 3). A comparison of the proton data of penitrem A with those of penitrem F shows that the C-15 hydroxy group had a significant influence on both 18-H and the C-34 methyl protons, since these groups are located below the plane of the eight-membered ring.

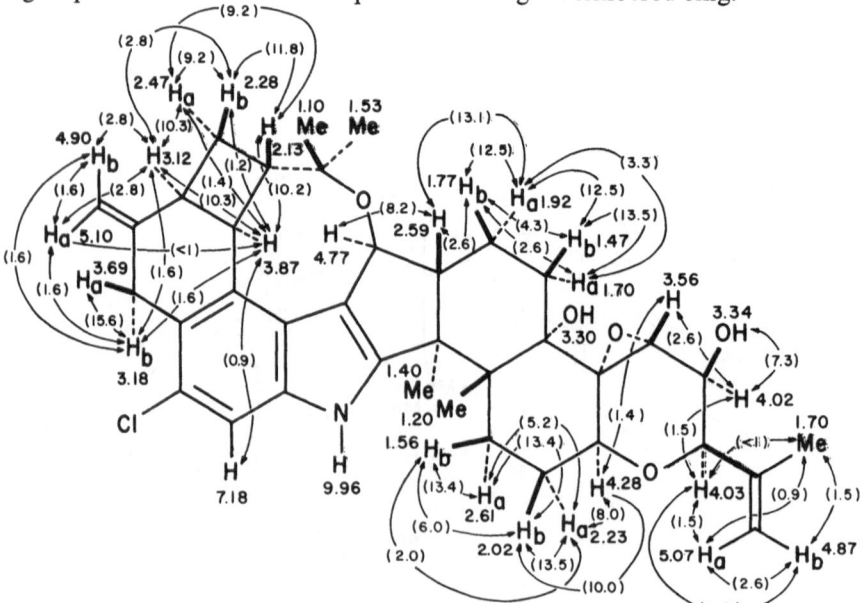

Fig. 3. The ^{1}H n.m.r. spectral data and (^{1}H, ^{1}H) coupling constants for penitrem F (**10**)

The molecular formula of penitrem D (**8**), $C_{37}H_{45}NO_4$, showed that the chlorine atom and the C-15 hydroxy group present in penitrem A have been replaced by hydrogen atoms and that a C(23)-C(24) double bond must be present in penitrem D. The ^{1}H n.m.r. data for penitrem C are not given since it differs from penitrem D only in the presence of the aromatic chlorine atom. The ^{1}H n.m.r. data for penitrem D are depicted in Fig. 4.

The ^{1}H n.m.r. spectra of 11,33,37,38-tetrahydro-penitrem A (**12**) showed changes consonant with the reduction of two exocyclic double bonds.

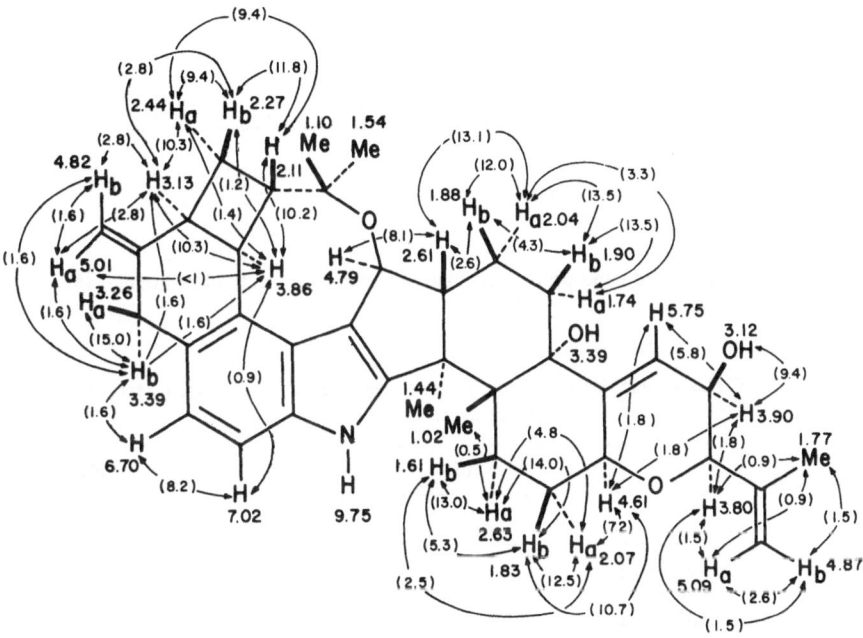

Fig. 4. The ^{1}H n.m.r. spectral data and (^{1}H,^{1}H) coupling constants for penitrem D (**8**)

^{13}C N.m.r. Spectra

The unambiguous assignment of the ^{13}C n.m.r. signals of the penitrems was essential in terms of the structural and biosynthetic studies.

The ^{13}C n.m.r. data of penitrem A (**5**) were obtained from proton-decoupled and single frequency nuclear Overhauser enhanced (n.O.e.) ^{13}C spectra. Single frequency off-resonance proton-decoupled and selective proton-decoupled ^{13}C spectra, selective population inversion (SPI) experiments, and the reported ^{13}C chemical shifts and (C,H) coupling constants of related substances were used in the assignment of the resonances of penitrem A and of the other penitrems (25—29). The deuterium shifts (38) refer to the separations observed between doubled signals in the proton-decoupled ^{13}C spectrum when the exchangeable protons were partially exchanged with deuterium upon the addition of $[^2H_4]$methanol.

The proton-decoupled spectrum of penitrem A is shown in Fig. 5. The assignments for penitrem A (**5**), 25-*O*-acetylpenitrem A (**11**) and tetrahydropenitrem A (**12**) are given in Table 2.

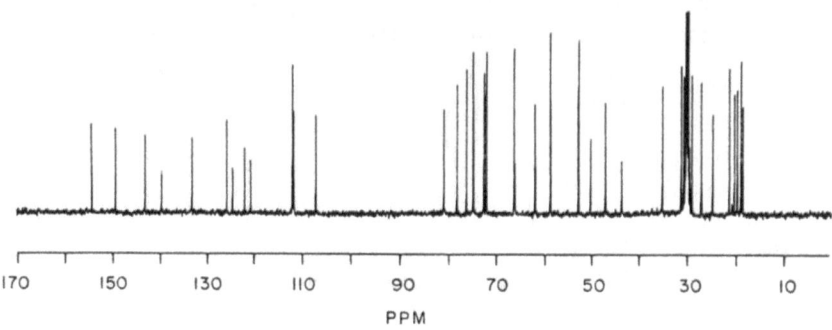

Fig. 5. The broad-band proton-decoupled ^{13}C n.m.r. spectrum of penitrem A (**5**)

Table 2. ^{13}C *N.m.r. (125.76 MHz) Data for Penitrem A* (**5**), *25-O-acetylpenitrem A* (**11**) *and Tetrahydropenitrem A* (**12**)

Carbon atom	(5)				(11)	(12)
	$\delta_C{}^a$	^1J	$^{>1}J$	$\Delta\delta^b$	$\delta_C{}^a$	$\delta_C{}^a$
2	154.36 Sq		4.2	−0.157	154.22	154.25
3	120.64 Sd		5.5	−0.034	120.67	120.51
4	133.29 Sm			−0.030	133.35	133.40
5	125.80 Sq		6.3		125.83	123.17
6	124.56 Std		4.9; 2.5		124.61	126.53
7	111.86 D	163.6		−0.047	111.84	111.60
8	121.99 St		5.8	−0.041	121.98	121.93
9	139.73 Sdd		3.9; <1.0^c	−0.148	139.78	139.96
10	35.06 DDdd	132.1; 126.5	11.5; 6.5		35.06	32.49
11	149.48 Sm				149.52	31.22
12	47.01 Dm	137.1		−0.063	47.02	45.17
13	24.67 Tm	135.9			24.67	23.20
14	52.71 Dm	129.5			52.75	53.64
15	81.01 St		7.7	−0.118	81.01	76.64
16	76.09 Sm				76.08	76.46
18	72.44 Dd	146.7	8.3		72.39	72.36
19	58.79 Dm	125.3			58.74	58.61
20	18.56 Tm	128.5			18.51	18.59
21	30.59 Tm	126.3			30.55	30.66

Table 2 *(continued)*

Carbon atom	(5)				(11)	(12)
	$\delta_C{}^a$	^1J	$^{>1}J$	$\Delta\delta^b$	$\delta_C{}^a$	$\delta_C{}^a$
22	78.24 Sm			−0.112	77.98	78.27
23	66.11 Sm				65.89	66.36
24	61.92 Ds, br	179.4		−0.024	58.79	62.25
25	66.31 Ds, br^d	144.0		−0.092	67.74	65.41
26	74.67 Dm	142.2			73.14	78.30
28	71.99 Ddt	147.4	8.2; 6.7		71.68	72.29
29	28.89 Tm	128.3			29.02	28.99
30	26.91 Tm	128.3			26.76	26.90
31	43.55 Sm			−0.030	43.71	43.52
32	50.08 Sm				50.04	49.98
33	107.10 Tq	155.6	4.8		107.11	18.33
34	20.32 Qqn	125.7	4.2		20.30	20.33
35	31.06 Qqd	125.8	4.7; 2.5		31.06	31.11
36	19.70 Qddd	126.2	11.1; 6.8; 1.7		19.61	14.09
37	143.27 Sm				142.58	28.88
38	111.64 Tm	156.6			111.98	20.33
39	18.98 Qdd	124.9	7.6; 1.5		18.51	19.10
40	21.35 Qd	127.5	5.8		21.30	21.12
25-OCOCH$_3$					170.20^e	
25-OCOCH$_3$					20.57	

^{a} Relative to internal Me$_4$Si; solvent (CD$_3$)$_2$CO. Measured from internal (CD$_3$)$_2$CO and corrected by using the expression δ(Me$_4$Si) = δ[(CD$_3$)$_2$CO] + 29.83. Capital letters refer to the pattern resulting from directly bonded (C, H) couplings [^1J(CH)] and small letters to that from (C, H) couplings over more than one bond [$^{>1}J$(CH)]. S or s = singlet, D or d = doublet, T or t = triplet, Q or q = quartet, qn = quintet, m = multiplet and br = broad (no fine structure but the line is noticeably broadened, indicating unresolved couplings).

^{b} Deuterium isotope shifts.

^{c} The small coupling was observed in a ^{13}C{^{1}H} SPI experiment when a 7-H transition was selectively irradiated.

^{d} After the exchangeable protons were replaced with deuterium this signal appears as Dd, ^1J(CH) = 144.0 Hz, $^{>1}J$(CH) = 3.3 Hz.

^{e} From the 25.2 MHz ^{13}C n.m.r. spectrum.

The assignment of the ^{13}C signals and their importance in the structural elucidation of penitrem A have been described in detail (25—29). The information to be gained from (C,H) coupling constants will be highlighted in a few cases. Although ^{1}H and ^{13}C chemical shift data indicated the presence of a proton-bearing oxirane carbon atom, confirmatory evidence was obtained from the directly bonded (C,H) coupling constant of 179.4 Hz observed for the resonance centred at δ 61.91. The resonance at δ 66.11 was assigned to the quaternary carbon atom (C-23) of the oxirane moiety on the basis of its chemical shift value and the change observed in this resonance upon irradiation of the 25-H/26-H protons. The exocyclic methylene carbon atoms δ 107.10 (C-33) and δ 111.64 (C-36) were identified by the triplet structure observed in off-resonance decoupled experiments and by the (C,H) coupling constants of 155.6 and 156.6 Hz, respectively.

The magnitude of directly bonded (C,H) coupling constants facilitated the assignment of the only proton-bearing aromatic carbon atom in penitrem A. This carbon atom resonates at δ 111.86 and exhibits a directly bonded (C,H) coupling of 163.6 Hz. On the basis of the chemical shift value this resonance could be assigned to C-3 or C-7. However, the reported ^1J(C,H) value for C-3 of indoles is 175 Hz (39); the signal at δ 111.86 was therefore assigned to C-7.

Selective $^{13}C\{^1H\}$ decoupling experiments gave invaluable information. Upon selective irradiation at $δ_H$ 2.63, the resonance position of 19-H and 30-H_a, (C,H) couplings of 5.8 and 7.6 Hz were removed from the methyl signals at δ 21.35 and 18.98, respectively. These couplings must be over three bonds since the corresponding proton resonances of these tertiary methyl groups are singlets. The results indicate that a two-carbon unit, consisting of a methyl group located on a quaternary carbon atom, must be located at both C-30 and C-19. The two quaternary carbon atoms concerned resonate at δ 43.55 and 50.08 and must be contiguous since a one-bond (C,C) coupling of 37.2 Hz is observed for these signals in the spectrum of [1-^{13}C]acetate-derived penitrem A (25, 29).

Penitrems A, B, E, and F contain the 23α,24α-epoxide as a common structural feature and their ^{13}C n.m.r. data are grouped together, whereas penitrems C and D contain the 23,24-double bond. The ^{13}C n.m.r. data of penitrems B, E and F are collated in Table 3.

Table 3. ^{13}C N.m.r. (125.76 MHz) Data for Penitrem B (6), Penitrem E (9), and Penitrem F (10)

Carbon atom	Penitrem B (6)				Penitrem E (9)	Penitrem F (10)	$ΔS_{Cl}{}^d$	$ΔS_{OH}{}^d$
	$δ_C{}^{a,b}$	^1J	$^{>1}J$	$Δδ^c$	$δ_C{}^a$	$δ_C{}^a$		
2	152.97 Sq		4.1	−0.154	153.48	153.86	0.89	0.51
3	119.40 Sd		5.8	−0.038	120.38	119.56	0.20	1.03
4	128.84 S br				131.52	130.76	1.88	2.61
5	128.13 Sq br		7.2		128.16	125.76	−2.36	
6	120.99 Dd	155.2	4.6		120.34	125.36	4.32	−0.73
7	110.26 D	158.3		−0.047	111.63	110.60	0.30	1.32
8	123.19 St		5.5	−0.040	122.75	122.33	−0.82	−0.39
9	139.34 Sdd		9.8; 3.9	−0.147	140.16	138.81	−0.49	0.87
10	38.75 DDdt	130; 125.3	11.6; 4.9			38.11	35.63	−3.10	0.61
11	150.22 S br				150.91	148.92	−1.34	0.64

Table 3 *(continued)*

Carbon atom	Penitrem B (6)				Penitrem E (9)	Penitrem F (10)	$\Delta S_{Cl}{}^d$	$\Delta S_{OH}{}^d$
	$\delta_C{}^{a,b}$	^1J	$^{>1}J$	$\Delta\delta^c$	$\delta_C{}^a$	$\delta_C{}^a$		
12	35.04 Dm	136			47.41	34.86	−0.25	12.26
13	26.69 Tm	133.2			24.72	26.61		−1.96
14	52.32 Dm	132			52.78	52.34		0.42
15	39.35 Ddt	134.6	8.2; 4.1		81.08	39.54		41.66
16	75.47 S br				76.09	75.51		0.59
18	72.10 Dd	144.8	8.4		72.52	71.97	−0.14	0.45
19	59.12 Dm	124			58.86	59.13		0.30
20	18.61 Tm	128.4			18.63	18.55		
21	30.58 Tm	124		−0.062	e	30.49		
22	78.26 S br			−0.113	78.28	78.23		
23	66.16 Sm				66.17	66.10		
24	61.95 Dm	178.8		−0.026	61.95	61.93		
25	66.33 Dt br	144.0	3.0	−0.096	66.34	66.31		
26	74.68 Dm	143			74.69	74.67		
28	72.03 Dm	147.6			72.05	71.97		
29	28.92 Tq br	128.7	3.2		28.94	28.88		
30	26.85 Tm	127			26.89	26.89		
31	43.60 Sm			−0.032	43.58	43.55		
32	49.68 Sbr				49.92	49.82	0.14	0.25
33	105.86 Tq	155.2	4.7		105.47	107.36	1.55	−0.33
34	18.61 Qqn	125.0	4.2		20.28	18.62		1.70
35	28.80 Qm	125.1			31.10	28.75		2.31
36	19.71 Qddd	126.2	11.4; 6.9; 1.8		19.70	19.70		
37	143.29 Sqn		6.1		143.31	143.26		
38	111.64 DDqd	159.8; 154.8	6.0; 3.9		111.62	111.65		
39	18.92 Qdd	125.2	6.8; 1.3		18.99	18.91		
40	21.24 Qd	127.1	5.6		21.53	21.06	−0.18	0.29

[a] Relative to internal Me_4Si; solvent $(CD_3)_2CO$. Measured from internal $(CD_3)_2CO$ and corrected by using the expression $\delta(Me_4Si) = \delta[(CD_3)_2CO] + 29.83$.

[b] Capital letters refer to the pattern resulting from directly bonded (C, H) coupling $[^1J(CH)]$ and lower case letters to that from (C, H) couplings over more than one bond $[^{>1}J(CH)]$. S = singlet, D or d = doublet, T or t = triplet, Q or q = quartet, qn = quintet, m = multiplet, and br = broad (no fine structure but the line is noticeably broadened).

[c] Deuterium isotope shifts in p.p.m. (see text).

[d] Average substituent shift in p.p.m. ΔS_{Cl} = average value of $\delta(5) - \delta(9)$, $\delta(10) - \delta(6)$, and $\delta(7) - \delta(8)$; ΔS_{OH} = average value of $\delta(5) - \delta(10)$ and $\delta(9) - \delta(6)$.

[e] Obscured.

The vitally important ^{13}C n.m.r. data of penitrems C (7) and D (8) are summarized in Table 4; the two compounds differ only in the presence of a chlorine atom in (7).

Table 4. ^{13}C N.m.r. (125.76 MHz) Data for Penitrem C (7) and Penitrem D (8)*

Carbon atom	Penitrem C (7)				Penitrem D (8)		
	$\delta_C{}^{a,b}$	^1J	^1J	$\Delta\delta^c$	$\delta_C{}^a$	$\Delta\delta^c$	$\Delta\delta_{oxirane}{}^d$
2	154.38 Sq		4.1	−0.157	153.49	−0.160	−0.52
3	119.33 Sd		5.6	−0.034	119.16		0.24
4	130.73 S br				128.80		
5	125.73 Sq br		5		128.09		
6	125.28 Sm				120.91		
7	110.56 D	163.9		−0.045	110.22	−0.044	
8	122.33 St		6	−0.040	123.18		
9	138.75 Sd		3.3	−0.147	139.27		
10	35.63 DDdd	132.0; 126.1	11.2; 6.7		38.75		
11	148.94 Sm				150.23		
12	34.86 Dm	134			35.04		
13	26.61 Tm	134.3			26.70		
14	52.37 Dm	134			52.36		
15	39.63 Ddt	134.8	8.4; 4.1		39.37		
16	75.49 Sm				75.44		
18	72.03 Dd	144.5	8.3		72.17		
19	58.89 Dm	124			58.87		0.25
20	19.06 Tm	126.0			19.11		−0.51
21	35.00 Tm	126		−0.060	35.08	−0.058	1.32
22	77.45 Sm			−0.116	77.48	−0.110	0.78
23	148.33 S br				148.44		−82.55
24	119.70 Dm	159.7			119.59		−57.71
25	64.27 Dm	142.7		−0.087	64.28	−0.088	2.04
26	74.36 Dm	143			74.40		0.30
28	80.38 Dm	137			80.39		−8.38
29	29.19 Tq	128.7	4.1		29.24		−6.14
30	27.71 Tm	129			27.68		−0.83
31	43.67 Sm			−0.025	43.71	−0.020	−0.12
32	49.91 Sm				49.77		
33	107.35 Tq	155.6	4.8		105.83		
34	18.64 Qqn	125.0	4.2		18.64		
35	28.77 Qm	125			* 28.81		
36	19.97 Qddd	125.7	11.3; 6.8; 1.8		19.98		−0.27
37	143.90 Sm				143.92		−0.64
38	110.78 Tqd	156.2	6; 4		110.75		0.88
39	20.10 Qd	125.9	8.0		20.11		−1.19
40	21.15 Qd	127.2	5.6		21.32		

* See Table 3 for footnotes.

The availability of the fully assigned ^{13}C n.m.r. data of penitrems A—F (5—10) enabled the study of substituent effects in this series of compounds (28).

Replacement of the chlorine atom and the C-15 hydroxy group by hydrogen atoms, as well as substitution of a double bond for an oxirane in the different penitrems, has a substantial influence on the ^{13}C chemical shifts of the neighbouring carbon atoms (see Tables 3 and 4). The effects are, however, largely restricted to that part of the molecule where the substitution takes place and very little interaction is observed between the different substituent changes. The influence of the chlorine atom is largely restricted to the carbon atoms of the indole ring as a result of electronic effects. The C-15 hydroxy group has a deshielding δ-substituent effect on both the C-34 and C-35 methyl groups which is, however, larger for the *anti* carbon atom, *i.e.* C-35. The effect of this moiety on the β- and γ-carbon atoms of the cyclobutane ring is only significant for C-12. The ^{13}C chemical shift changes induced by the oxirane can probably be associated with conformational changes on going from a double bond to an oxirane (*28*).

2.1.3.3 Mass Spectrometry

Mass spectrometry was employed to establish the molecular constitution of all the penitrems and their derivatives. The proposed structures are consistent with their mass spectral fragmentations.

FELLOWS *et al.* (*40*) reported the mass spectral fragmentation of penitrems A, B, D, E and F under electron impact. The compounds all give moderately intense molecular ions. The fragmentation patterns are derived by observation of first and second field free region metastable transitions. The former was derived by the linked scanning (B/E constant) technique.

The main fragmentations are shown as a-a and b-b on the proposed structures in Fig. 6.

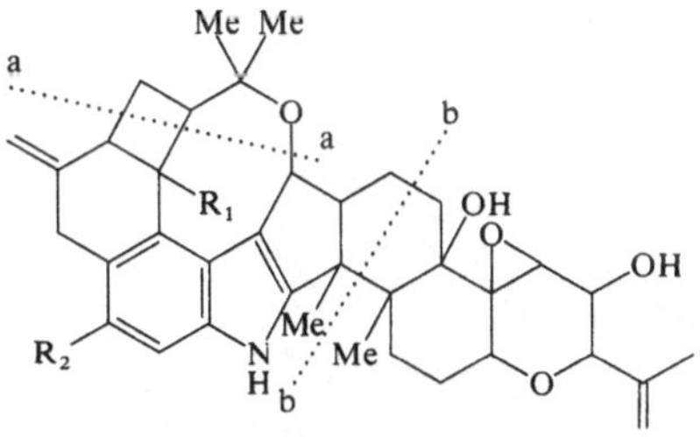

(**5**) R$_1$=OH, R$_2$=Cl
(**9**) R$_1$=OH, R$_2$=H
(**6**) R$_1$=R$_2$=H
(**10**) R$_1$=H, R$_2$=Cl

Fig. 6. Major mass spectral fragmentations of the penitrems

The loss of 86 u (C$_5$H$_{10}$O) from the molecular ions gives an abundant M^+-86 ion in all the spectra. Scission a-a takes place with the simultaneous transfer of a hydrogen atom to the C-5 unit. The loss of 251 u (C$_{14}$H$_{21}$O$_4$)

from M^+-18 and M^+-86 can be observed for penitrem A. This fragmentation can be represented by scission b-b with a transfer of hydrogen to the indole-containing residue. Fragments with the appropriate mass were lost in the other penitrems. FELLOWS et al. (40) tabulated the important ions in the mass spectra of the penitrems and some derivatives. The fragmentation of 25-O-acetylpenitrem A (11) was unusual in that the characteristic fragmentations of acetates, viz loss of CH_3CO_2H, CH_3CO or CH_2CO from the M^+ were not observed. FELLOWS et al. (40) explained this suppression of the normal acetate fragmentation as the result of the proximity of the acetoxy group to the epoxide ring.

2.1.3.4 Stereochemistry of the Penitrems: Conformation and Absolute Configuration

No X-ray structure was obtained for any of the penitrems. The conformation and relative configuration was deduced from the dihedral angular relations as indicated by proton-proton and carbon-proton coupling constants as well as proton-proton nuclear Overhauser effects (n.O.e.s). The n.O.e.s were measured in the difference mode. The observed n.O.e. connectivity pattern of penitrem A shown in Fig. 7 gave invaluable information on the conformation of (5), and furnished confirmatory evidence for the proposed structure.

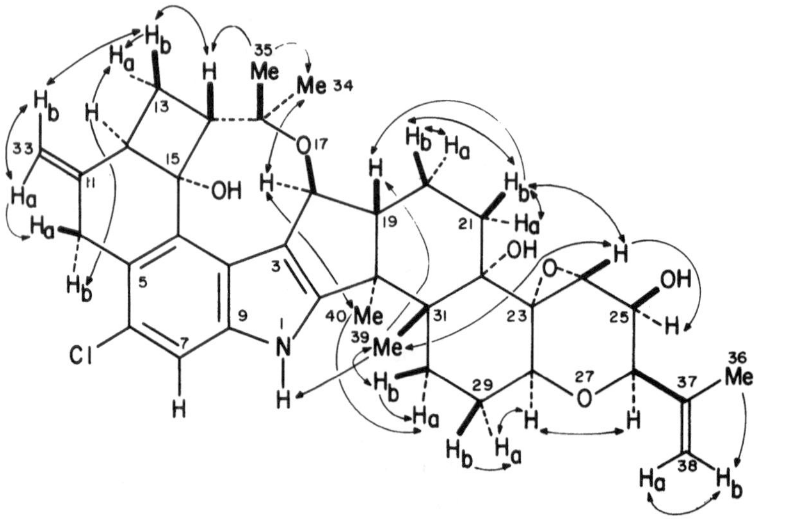

Fig. 7. The n.O.e. connectivity pattern observed for penitrem A (5)

The relative stereochemistry of rings F, G, H, and I is essentially the same as that observed in the crystal structures of the related metabolites paspalinine (13) (41) and paspalicine (14) (41). More emphasis will therefore be given to the stereochemistry of the novel ring systems C, D, and E.

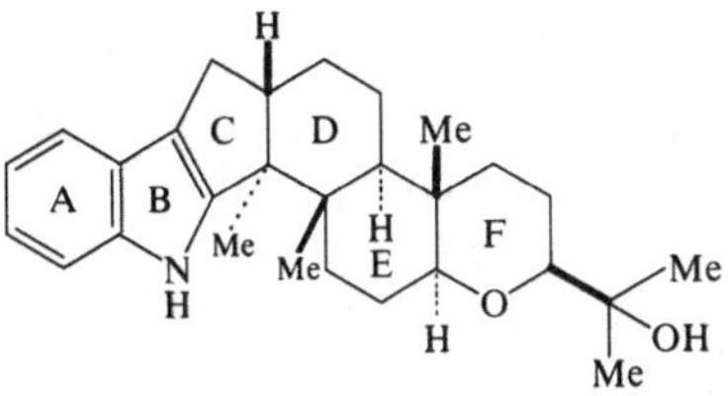

(13) 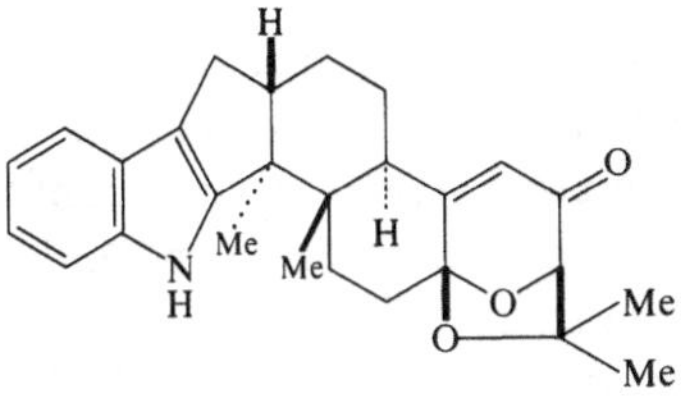   (14)

DE JESUS et al. (27) correctly assumed that C-15 of (5) has the R-configuration, i.e. the hydroxy group is located below the plane of the indole ring. The fusion between rings C and D is by necessity cis and the 12-H proton is therefore cis to the C-15 hydroxy group. The C-16 carbon atom can be either cis or trans to the C-15 hydroxy group of penitrem A. The trans relationship was established by the n.O.e. data and by the spin-spin coupling of 10.2 Hz between 14-H and 15-H observed in penitrems D and F; it is well known that enzymatic hydroxylation takes place with retention of absolute configuration. An n.O.e. is observed between 18-H and the methyl group protons which resonate at δ_H 1.76 (34-H) and between 14-H and δ_H 1.07 (35-H, Me). This information conforms with the indicated relative configuration and conformation of penitrem A.

An n.O.e. was observed between 18-H and the C-40 protons, but not between 18-H and 19-H. This finding shows that 18-H is cis to C-40 and that rings F and G are trans-fused. The relative configurations of rings F-I and the constituents are based on the observed n.O.e.s in conjunction with the proton-proton coupling constants. The construction of a Dreiding molecular model of the penitrem A molecule with the proposed relative configuration results in a fairly rigid structure.

The absolute configuration of the C-25 hydroxy group was determined by the "partial resolution" method of HOREAU (42) and this result established the absolute configuration of penitrem A (5). Penitrem A was esterified with α-phenylbutyric acid anhydride and 4-dimethylamino-pyridine to give the 25-O-phenylbutyryl penitrem A. The recovered α-phenylbutyric acid had $[\alpha]_D^{23} - 7.5°$. Penitrem A therefore has the 25S-configuration and the configuration shown in Fig. 8, viz 12R, 14S, 15R,

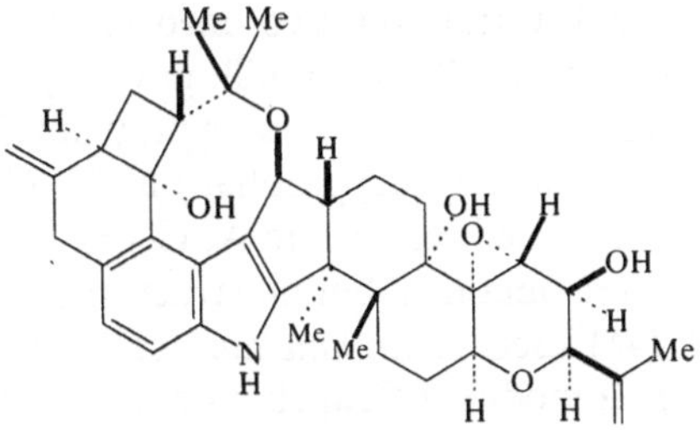

Fig. 8. Absolute configuration of penitrem A (5)

$18S, 19R, 22S, 23S, 24R, 25S, 26R, 28S, 31R, 32S$. Penitrem E differs from penitrem A only in the lack of a chlorine atom and the two compounds therefore have the same conformation and absolute configuration.

Both penitrem B (6) and penitrem F (10), the 6-chloro derivative of penitrem B (6), lack the C-15 hydroxy group present in penitrem A (5). The relative configuration of the 15-H proton in (6) and (10) must be the same as that of the C-15 hydroxy group of (5). This supposition was verified by the proton-proton coupling constants and the requirement of a cis-fusion between rings C and D.

Penitrems C (7) and D (8) both lack the C-15 hydroxy group and the oxirane moiety. The relative configurations of rings C-E in penitrem D are based on proton-proton coupling constants and are the same as for penitrems B (6) and F (10). The chirality of the C-25 hydroxy group was determined as $25R$ by the partial resolution method of Horeau (42). Penitrem C and D therefore have the following absolute configuration: $12S, 14S, 15S, 18S, 19R, 22S, 25R, 26R, 28S, 31R, 32S$.

2.2 The Janthitrems

2.2.1 Producing Organisms

In studies on the role of tremorgenic mycotoxins in ryegrass staggers, Lanigan et al. (43) isolated Penicillium janthinellum Biourge as the most frequently occurring tremorgenic species. Two tremorgens, identified as verruculogen and fumitremorgin (see later), were isolated from the mycelial mats.

In similar studies, Gallagher et al. (44) studied numerous isolates of Penicillium species isolated from ryegrass pastures. Twenty-one isolates of P. janthinellum Biourge were screened and more than half of these were found to produce new fluorescent tremorgenic toxins, called janthitrems.

2.2.2 Isolation and Chromatography of the Janthitrems

Gallagher et al. (44) examined the P. janthinellum isolates for tremorgen production by cultivation on two media: (1) potato broth (3% dehydrated mashed potatoes, 2% skim milk powder, 2% sucrose) at 25°C for 16 to 18 days; (2) potato-dextrose agar tryptophan (20% fresh potatoes, 2% dextrose, 12% agar, 0.025% tryptophan) at 25°C for 12 to 15 days. Three tremorgens, designated janthitrem A, $C_{37}H_{47}NO_6$; B, $C_{37}H_{47}NO_5$; and C, $C_{37}H_{47}NO_4$ were obtained, but no structural details were reported.

De Jesus et al. (45) used P. janthinellum TDD$_4$ which was grown in stationary culture on a modified Czapek medium at 25°C for 5 days and then for 15 days at 18°C. Three tremorgens, janthitrems E, F, and G, were isolated as colourless amorphous solids by chromatography on silica gel using benzene-acetone (70 : 30, v/v) as eluant. The molecular composition of janthitrems E, F, and G was determined by mass spectrometry as $C_{37}H_{49}NO_6$, $C_{39}H_{51}NO_7$, and $C_{39}H_{51}NO_6$, respectively.

LAUREN and GALLAGHER (46) developed a method of analysis for the janthitrems using reversed-phase high-performance liquid chromatography on a Zorbax C_8 column with water-methanol (20 : 80 v/v) as eluant at a flow-rate of 1 ml/min and u.v. detection at 265 nm. The use of a fluorescence detector allowed smaller amounts of janthitrems to be detected. A 50-fold increase in sensitivity could be obtained by using a 254 nm line excitation and a 370 nm cut-off emission filter. LAUREN and GALLAGHER (46) also reported on janthitrem D, but no molecular formula was reported.

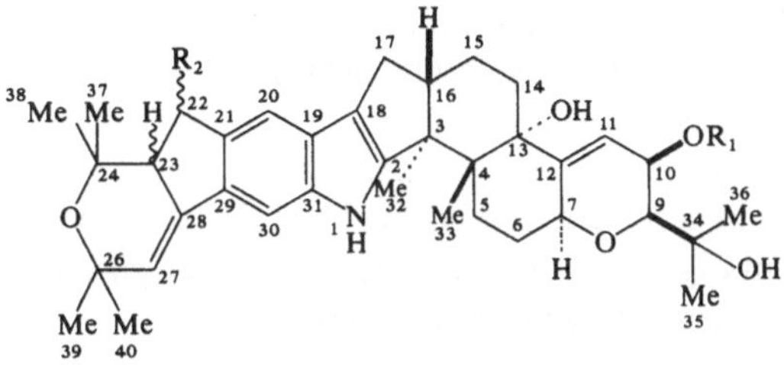

(15) R₁=H, R₂=OH
(16) R₁=Ac, R₂=OH
(17) R₁=Ac, R₂=H

2.2.3 Structure of the Janthitrems

2.2.3.1 General Aspects, Ultraviolet and Infrared Spectroscopy

The similarity in the molecular formulae and the biological action of the janthitrems and the penitrems strongly indicated closely related structural and biosynthetic features. Absorptions at λ_{max} 228 (ε 17 700), 258 sh (ε 27 300), 265 (ε 30 000), and 330 nm (ε 17 000) in the u.v. spectrum of the janthitrems, e.g. janthitrem E (15) (45) suggested the presence of a 2,3-disubstituted indole nucleus with a double bond in conjugation with the indole as in paspalitrem B. The janthitrems are highly fluorescent when irradiated with long wavelength u.v. light; the fluorescence emission maximum of janthitrem B occurs at 385 nm (in methanol) (44). The infrared spectra lacked structural information. The acetate carbonyl moiety absorbed at 1720 cm⁻¹ in janthitrems F and G.

Only the structures of janthitrems E (15), F (16) and G (17) were reported by DE JESUS et al. (45) but janthitrems A—C evidently have closely related structures.

2.2.3.2 Nuclear Magnetic Resonance Spectroscopy

¹H N.m.r. Spectroscopy

The ¹H n.m.r. spectrum of janthitrem E exhibited eight three-proton signals at δ 1.406, 1.318, 1.292, 1.281, 1.247, 1.247, 1.095 and 0.868 assigned to eight tertiary methyl groups. First-order analysis of the remaining resonances in the ¹H spectrum gave the values of the proton chemical shifts and proton-proton coupling constants. DE JESUS *et al.* (*45*) used these values to identify three proton-linked fragments.

Fragment A, comprising the protons of rings A—C, was quite different from that present in the penitrems (*27, 28*) and will be discussed in some detail. An important characteristic of this spin system is the small coupling (0.7 Hz) for the two *para* oriented aromatic protons at δ 7.373 and δ 7.379. The additional small coupling (1.0 Hz) observed in the latter resonance and that at δ 4.902 (*J* 7.8, 6.3 and 1.0 Hz) provided a link between the indole moiety and the carbocyclic ring B. The protons of rings E and F constituted fragment B, whereas rings G and H comprise fragment C. The observed coupling patterns for these two fragments are similar to those observed for the related rings in penitrem D (**8**) (*28*). The ¹H n.m.r. data of janthitrem E (**15**) are shown in Fig. 9. ¹³C N.m.r. spectroscopy was used to determine the connections between the various fragments.

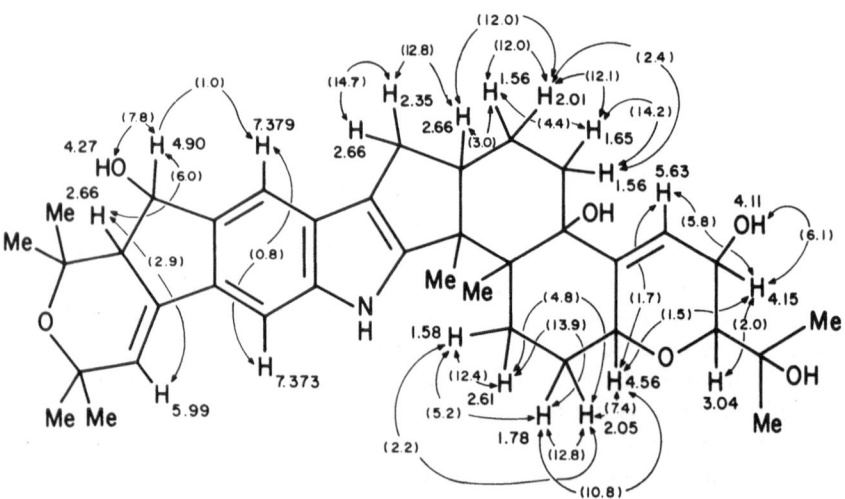

Fig. 9. The ¹H n.m.r. spectral data and (¹H, ¹H) coupling constants for janthitrem E (**15**)

Janthitrem F (**16**) which contains an additional C_2H_2O moiety was identified as 10-*O*-acetyljanthitrem E. The presence of an acetate group was evident from the additional three-proton singlet at δ 2.010 and from the effects on the chemical shifts at 7-H, 9-H and 10-H.

Janthitrem G (**17**) lacks the C-22 hydroxy group which is present in janthitrem F. This observation was corroborated by its (^{1}H,^{1}H) connectivity pattern and by the ^{13}C n.m.r. data.

^{13}C N.m.r. Spectroscopy

The analyses of the ^{13}C n.m.r. data of the janthitrems (*45*) are based on the same techniques as described for the penitrems (*27, 28*). Extensive use was made of heteronuclear ^{13}C{^{1}H} selective population inversion (SPI) experiments (*47*), particularly in the linkage of the various fragments.

Selective irradiation of the C-36 proton transitions (δ_H 1.198) in a SPI experiment on janthitrem G (**17**) affected the resonances at δ 27.83 (C-35), 71.62 (C-34) and 82.47 (C-9) (see Fig. 10). These affected carbon resonances must be two and/or three bonds removed from 36-H. The chemical shift values indicated the presence of the two oxygen-bearing carbon atoms (C-34 and C-9); in addition a deuterium isotope shift was observed for C-34. The data evidence for the location of a hydroxylated isopropyl moiety at C-9 of janthitrem G (**17**) (*45*).

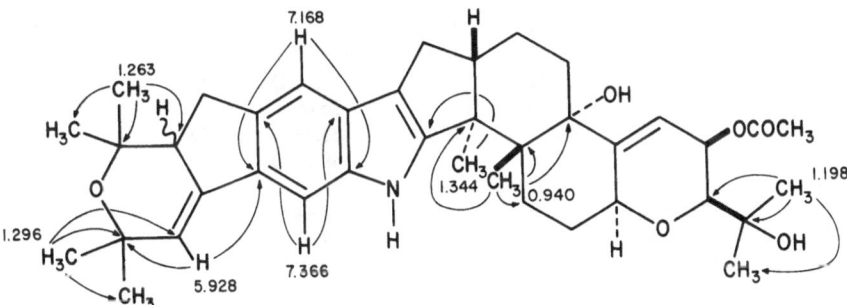

Fig. 10. The (^{1}H, ^{13}C) connectivity pattern for janthitrem G (**17**) as determined by SPI experiments

The plethora of structural information obtained from a number of SPI experiments on (**17**) is depicted in Fig. 10. A case in point is provided by the effects observed on irradiation of the 33-proton transitions (δ_H 0.940), which provided that these protons couple to C-5 (δ 27.74), C-4 (43.45), C-3 (51.70) and C-13 (77.47). The location of the C-32 methyl group relative to C-2 (δ 155.38) was also established by a SPI experiment.

SPI Experiments were invaluable in the construction of rings A and B and their linkage to ring C (*45*) (see Fig. 10). The deuterium isotope shifts observed for janthitrems F and G were useful in the assignment of the aromatic carbon atoms and of those carbon atoms in the proximity of the hydroxy groups. The fully assigned ^{13}C n.m.r. data of janthitrems E—G (**15—17**) are given in Table 5.

Table 5. ^{13}C N.m.r. (125.76 MHz) Data for Janthitrem E (**15**), Janthitrem F (**16**) and Janthitrem G (**17**)

Carbon atom	(15) $\delta_C{}^a$	(16) $\delta_C{}^a$	(16) $\Delta\delta^b$	(17) $\delta_C{}^a$	(17) $^1J(CH)/Hz$	(17) $\Delta\delta^b$
2	155.92 S	155.64	−0.160	155.38 S		−0.164
3	51.75 S	51.72		51.70 S		
4	43.39 S	43.37		43.45 S		
5	27.85 T	27.70		27.74 T	129	
6	29.18 T	29.14		29.15 T	127	
7	75.03 D	75.03		75.06 D	143	
9	81.82 D	82.44		82.47 D	135	
10	64.48 D	66.00		66.08 D	148	
11	119.15 D	115.14		115.16 D	163	
12	148.56 S	151.26		151.28 S		
13	77.49 S	77.42	−0.114	77.47 S		−0.113
14	34.51 T	34.50	−0.059	34.56 T	125	−0.053
15	22.02 T	21.88		21.92 T	126	
16	50.33 D	50.27		50.39 D	127	
17	28.04 T	27.85		27.87 T	128	
18	116.85 S	116.84	−0.048	116.46 S		−0.049
19	127.77 S	127.67	−0.036	127.60 S		−0.038
20	113.99 D	114.08		114.11 D	156	
21	139.92 S	139.74		136.69 S		
22	76.45 D	76.35	−0.107	33.53 T	128	
23	60.32 D	60.10		49.79 D	127	
24	74.26 S	74.22		74.74 S		
26	72.78 S	72.70		72.94 S		
27	120.13 D	120.13		119.53 D	158	
28	137.11 S	136.85		140.79 S		
29	131.51 S	131.57		133.31 S		
30	103.73 D	103.36	−0.048	103.96 D	157	−0.047
31	142.11 S	142.02	−0.145	141.17 S		−0.144
32	16.58 Q	16.60		16.60 Q	125	
33	20.20 Q	19.87		19.86 Q	126	
34	72.65 S	71.61	−0.109	71.62 S		−0.104
35	27.28 Q	27.81		27.83 Q	127	
36	27.21 Q	25.95		25.94 Q	125	
37	30.43 Q	30.25		29.82 Q		
38	23.76 Q	23.68		22.53 Q	125	
39	30.63 Q	30.61		30.49 Q		
40	32.15 Q	32.40		32.30 Q	126	
COCH$_3$	—	170.86		170.69 S		
COCH$_3$	—	21.37		21.33 Q	130	

^{a} Relative to Me$_4$Si; solvent [^{2}H$_6$]acetone. Capital letters refer to the pattern resulting from one-bond (C, H) couplings; S = singlet, D = doublet, T = triplet and Q = quartet.

^{b} Deuterium isotope shifts (p.p.m.) observed when the exchangeable protons were partially exchanged with deuterium.

2.2.3.3 Stereochemistry of the Janthitrems:
The Relative Configuration

The relative configuration of rings E—H in the janthitrems was deduced by examination of the proton-proton coupling constants and by comparison with the penitrems. The configurations of C-22 and C-23 were not determined (45). The secondary hydroxy group at C-22 is located at a sterically congested part of the janthitrem E molecule as esterification with acetic anhydride and therefore also by the Horeau method failed. It was unfortunately not possible to determine the steric relationship of rings A—B to rings E—H.

2.3 The Lolitrems

2.3.1 Producing Organisms

Perennial ryegrass staggers is a neurotoxic syndrome affecting sheep, cattle, and horses grazing pastures in which perennial ryegrass (*Lolium perenne* L.) is dominant (48). It was a disease of unknown etiology until the recent findings of LATCH (49) that an *Acremonium* species, an endophytic fungus which infects ryegrass, could be associated with production of neurotoxins.

2.3.2 The Structure of Lolitrems B and C

The structure elucidation of lolitrem B (18) was reported in a recent communication by GALLAGHER et al. (50). The molecular formula of lolitrem B indicated the presence of an additional isoprene C_5 unit compared with the penitrems or janthitrems.

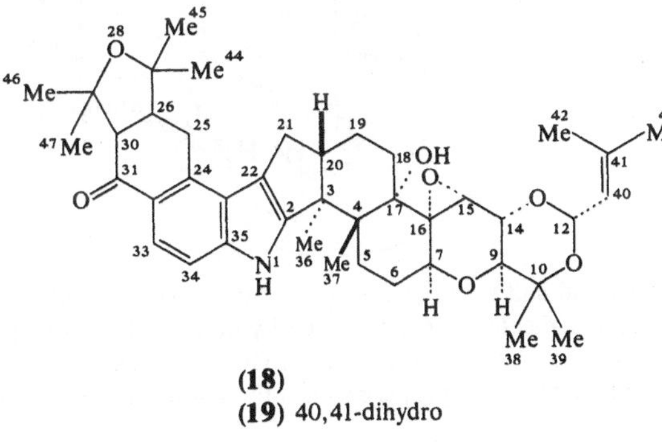

(18)
(19) 40,41-dihydro

Lolitrem B (18), $C_{42}H_{55}NO_7$, has a melting point of 303—304°C. A major mass spectral fragmentation arises through cleavage of the C-3—C-4 and C-18—C-19 bonds with hydrogen transfer to the indole containing fragment, leading to an abundant ion at m/z 348. Similar fragmentations were observed in the structurally related aflatrem (51) and penitrems (27, 40).

The i.r. spectrum showed the presence of OH and NH groups and carbonyl absorption at 1664 cm^{-1}. The u.v. absorptions at 267 (ε 26 800) and 290 nm (ε 6 700) indicated a 2,3-disubstituted indole nucleus with a carbonyl group in conjugation with the aromatic ring. Sodium borohydride reduction of (18) afforded a diastereoisomeric mixture (M^+, 687 and m/z 350), each with typical indole u.v. absorption.

The 500.13 MHz ^{1}H n.m.r. spectrum of lolitrem B showed the presence of ten tertiary methyl groups of which eight appeared as singlets (δ 1.515, 1.370, 1.301, 1.276, 1.276, 1.266, 1.240 and 1.134) whereas two of the methyl groups showed three-bond coupling δ 1.717d, (J 1.3 Hz) and 1.711 d, (J 1.3 Hz). The remaining resonances exhibited extensive fine structure. Full analysis of the coupling pattern as corroborated by extensive ^{1}H\{^{1}H\} homonuclear decoupling experiments showed the presence of four coupling patterns. GALLAGHER et al. (50) established the (^{1}H,^{1}H) connectivity pattern of lolitrem B shown in Fig. 11.

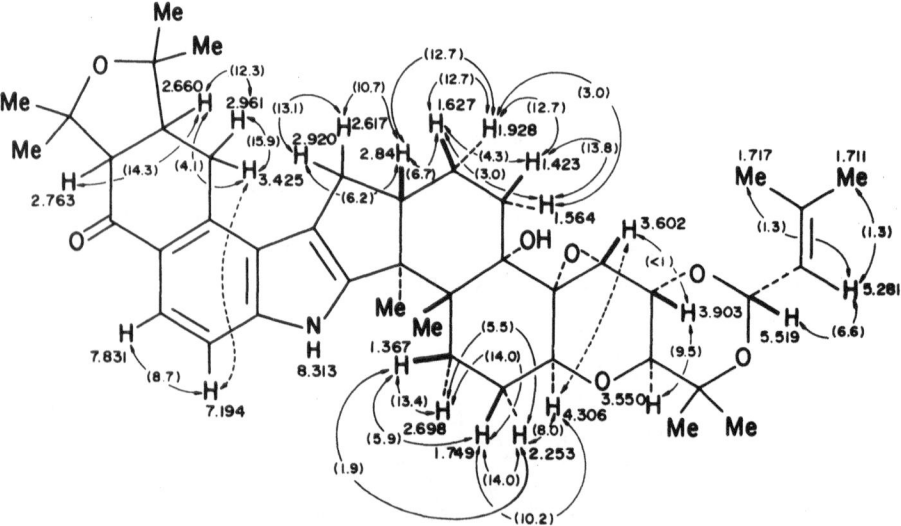

Fig. 11. The ^{1}H chemical shifts and coupling constants (Hz) for lolitrem B. The (^{1}H, ^{1}H) connectivity pattern as indicated was determined by homonuclear decoupling experiments and 2D (^{1}H, ^{1}H) correlation spectroscopy (COSY). The broken lines show cases where effects were observed during decoupling experiments although no splittings were measurable

The ^{13}C n.m.r. assignments of lolitrem B (18) are based on the same techniques as those of the penitrems (27, 28). In addition, the multiplicities of the different ^{13}C resonances were determined by generating the proton-decoupled CH, CH_2, and CH_3 subspectra using the DEPT sequence (52). The signals of all the proton-bearing carbon atoms were correlated in turn with specific proton resonances in a two-dimensional ($^{13}C, ^1H$) shift correlation experiment (53). The ^{13}C n.m.r. data are collated in Table 6.

Table 6. ^{13}C N.m.r. (125.76 MHz) Data for Lolitrem B (18)[a]

Carbon atom	δ_C[b]	$^1J(CH)$/Hz	$\Delta\delta$[c]	Carbon atom	δ_C[b]	$^1J(CH)$/Hz	$\Delta\delta$[c]
2	152.77 S		−0.162	26	49.87 D	128	
3	50.69 S			27	79.26 S		
4	42.36 S			29	79.93 S		
5	27.38 T			30	59.93 D	126.7	
6	27.98 T			31	196.51 S		
7	71.45 D	149		32	136.97 S		
9	71.21 D	147		33	120.23 D	162.3	
10	74.70 S			34	110.42 D	161.0	
12	92.66 D	160.6		35	142.00 D		−0.146
14	71.11 D	139.6		36	15.91 Q[d]		
15	61.13 D	181.2		37	16.59 Q[d]		
16	67.73 S			38	18.87 Q[d]		
17	78.04 S		−0.113	39	25.01 Q[d]		
18	30.26 T			40	121.96 D	158.3	
19	20.49 T			41	139.54 S		
20	50.08 D	122		42	25.65 Q		
21	29.16 T			43	18.63 Q		
22	118.58 S			44	25.10 Q[d]		
23	126.96 S			45	28.28 Q[d]		
24	123.89 S			46	29.13 Q[d]		
25	28.28 T			47	30.63 Q[d]		

[a] Recorded on a Bruker WM-500 spectrometer.
[b] Relative to Me_4Si, solvent $CDCl_3$.
[c] Deuterium isotope shifts (in p.p.m.) observed on addition of H_2O-D_2O (1:1).
[d] May be interchanged.

Location of the carbonyl group at C-31 rather than C-25 is based on the chemical shift of the 33-H proton (δ_H 7.831) in lolitrem B (18). In addition, irradiation of the C-33 proton transition in a SPI experiment affected the signal at δ_C 196.51 assigned to C-31. The location of the additional C_5 unit was evident from the n.m.r. data.

The ^1H n.m.r. and mass spectrum of a minor metabolite lolitrem C (19) ($C_{42}H_{57}NO_7$, M^+ 687) indicated that the C-40−C-41 double bond present in lolitrem B has been reduced.

2.3.2.1 The Stereochemistry of Lolitrem B

GALLAGHER et al. (50) determined the relative configuration of lolitrem B (18) from proton-proton coupling constants as well as proton-proton n.O.e.s. Comparison of these data with those of the penitrems (27, 28) indicates that the relative configuration at C-3, C-4, C-7, C-15, C-16, C-17 and C-20 is the same as for the penitrems. The n.O.e. observed between 9-H and 7-H, but not 14-H, shows that rings H and I are *trans*-fused with 9-H *cis* to 7-H. The *trans* configuration of 9-H and 14-H is based on the fact that no n.O.e. is observed between these protons and the vicinal (H,H) coupling constant of 9.5 Hz. The C-14 chiral centre in lolitrem B (18) must therefore have the *R*-configuration. This supposition was corroborated by the vicinal (H,H) coupling constant of <1 Hz for 14-H and 15-H, which is indicative of a dihedral angle of *ca.* 90°. The relative configuration at C-12 was established by the n.O.e. observed between 14-H and 12-H (50).

The *trans*-fusion of rings A and B was deduced from the vicinal (H,H) coupling (J 14.3 Hz) for 26-H and 30-H. The steric relationship of these two centres to the rest of the molecule is not known at present.

2.4 Aflatrem

2.4.1 Producing Organism

In 1964, WILSON and WILSON (20) reported the important discovery that a number of strains of *A. flavus* produced a tremorgen when grown on foodstuffs such as potatoes, corn, millet, and rice. The isolated toxin called aflatrem has the ability to cause a hypertensive state in dosed animals (54). Mice given the crude toxin at first became inactive, but then responded to auditory and tactile stimuli and exhibited marked tremors of the entire body when movement was attempted (54).

It is of importance to note that many strains of *A. flavus* also have the ability to produce the hepatotoxic aflatoxins and α-cyclopiazonic acid (55). The coproduction of these toxins can lead to a multiple toxigenic threat.

2.4.2 The Structure of Aflatrem

The presence of a 2,3-disubstituted indole moiety in aflatrem (20), $C_{32}H_{39}NO_4$, was indicated by the u.v. spectrum, λ_{max} (EtOH) 231 (ε 27 700), 282 (ε 9 000) and 292sh nm (ε 7 800), whereas absorption at 250 nm (ε ∼ 10 000) (56) was attributed to an α,β-unsaturated carbonyl moiety. The i.r. spectrum showed NH and OH stretching bands at 3400 and 3390 cm^{-1}, respectively, and intense absorption at 1680 cm^{-1} assigned to the α,β-unsaturated carbonyl group (56).

(20) R=—$CMe_2CH=CH_2$
(21) R=H

GALLAGHER *et al.* (*56*) made efficacious use of mass spectrometry in the structural studies of aflatrem when its close similarity to paspalinine (21), $C_{27}H_{31}NO_4$, became self-evident. The mass spectrum of aflatrem showed the molecular ion at m/e 501 and a major fragment at m/e 250, suggesting that a C_5H_8 unit was attached to the aromatic portion. An analogous fragmentation pattern with associated proton transfer was observed in the other structurally related tremorgens (*40*).

The presence of an α,α-dimethylallyl substituent in aflatrem was evident from the ^1H n.m.r. data: δ 6.21 dd (1H, *J* 11 and 17 Hz), δ 4.44 dd (1H, *J* 11 and 1.5 Hz), δ 4.80 dd (1H, *J* 17 and 1.5 Hz) and δ 1.41 s (6H), and ^{13}C n.m.r. data: δ_C 29.3, 33.3, 40.8, 110.6, and 149.0. The downfield shift (20 p.p.m.) observed for one aromatic carbon signal indicated attachment of the C_5 unit to that part of the molecule. The location of the C_5 unit at C-4 is based on ^1H n.m.r. data and the presence of the diagnostic proton-bearing carbon signal (C-7) at δ_C 111.2 p.p.m. GALLAGHER *et al.* (*56*) reported the following ^{13}C chemical shift values (in [$^2H_6$]-DMSO) for aflatrem: δ 196.7, 169.4, 151.8, 149.0, 140.5, 138.5, 122.8, 118.6, 116.6, 115.3, 114.0, 111.2, 110.6, 103.9, 86.7, 77.7, 75.8, 50.0, 47.8, 40.8, 38.9, 33.3, 31.5, 29.3, 29.1, 28.3, 27.8, 25.7, 22.7, 22.3, 20.6 and 16.0. A detailed description of the structure of aflatrem (20) including the complete ^{13}C n.m.r. assignments has not been published.

The co-occurrence of aflatrem (20) and paspalinine (21) in cultures of *A. flavus* lends supplementary evidence for the characterization of aflatrem. Previously, paspalinine has only been isolated from the sclerotia of *C. paspali*.

COLE *et al.* (*57*) and GALLAGHER *et al.* (*58*) isolated two indole-mevalonate metabolites from *A. flavus*, called aflavinine (22), $C_{28}H_{39}NO$, and dihydroxyaflavinine (23), $C_{28}H_{39}NO_3$. The metabolites are structurally unrelated to the tremorgens and showed no tremorgenic effects. The structures were established by *X*-ray crystallography (*57, 58*). The *X*-ray structures showed that the extra-annular double bond was essentially orthogonal (78°) to the indole ring, thereby explaining the lack of conjugation apparent from the u.v. spectra.

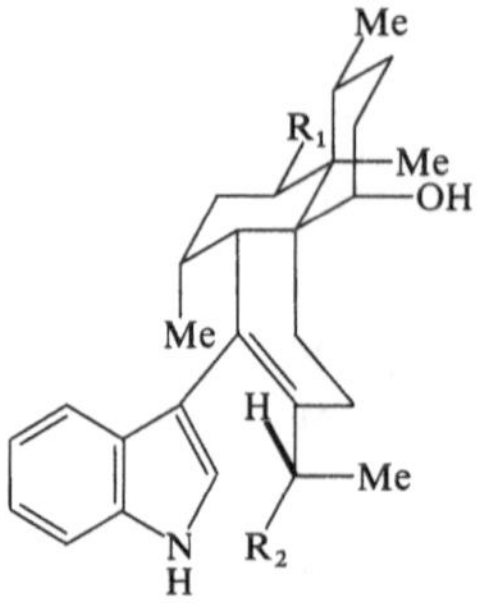

(22) R_1=H, R_2=Me
(23) R_1=OH, R_2=CH$_2$OH

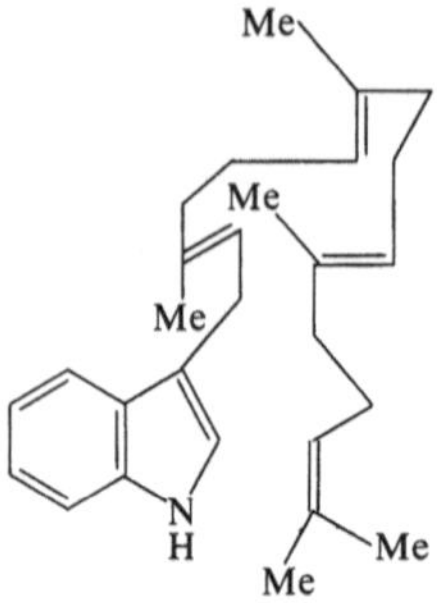

(24)

Aflavinine (22) and aflatrem (20) might have a common bioprecursor (24), but a different mode of cyclization and migration of the methyl group is required.

2.5 Paxilline

2.5.1 Producing Organism

COLE et al. (59) reported the discovery of a new tremorgen-producing mould, *Penicillium paxilli* (ATCC 26601), which was isolated from insect-damaged pecans. The tremorgen was subsequently called paxilline (60).

2.5.2 Production and Isolation of Paxilline

P. paxilli was grown in Fernbach flasks, each containing 100 g shredded wheat, 200 ml Difco mycological broth (pH 4.8), plus YES medium. The fermentation was carried out for 14 days at 25° C. Paxilline was isolated by homogenizing the fungal culture with chloroform and subsequent separation by chromatography on silica gel. The tremorgenic metabolite was recrystallized from acetone and had a melting point of 252° C. A yield of 1.0 g of paxilline was obtained from each 1.5 kg of culture medium.

Paxilline can be separated on silica thin layer chromatoplates, using toluene-ethyl acetate-formic acid (5 : 4 : 1 v/v/v) as developing solvent. In this system paxilline appears at R_f 0.75.

COLE et al. (59) reported on the clinical signs observed in cockerels and mice upon administration of paxilline. Severe tremors were induced in mice at 25 mg/kg.

2.5.3 Structure of Paxilline

The mass spectrum of paxilline (25) confirmed its molecular formula C$_{27}$H$_{33}$NO$_4$ (59). Its u.v. spectrum, λ_{max} (MeOH) 230 (ε 41 500) and 281 nm (ε 8 000), indicated the presence of a 2,3-disubstituted indole moiety, whereas strong i.r. absorption at $1650\,\text{cm}^{-1}$ was evidence in favour of an α,β-unsaturated ketone group.

The structure of paxilline (**25**) was based on X-ray crystallography (*60*). Paxilline belongs to the orthorhombic space group $P2_12_12_1$ with $a = 31.009(3)$, $b = 11.522(1)$, $c = 7.707(1)$ Å. The structure was solved by application of a multiple solution tangent formula approach (*61*).

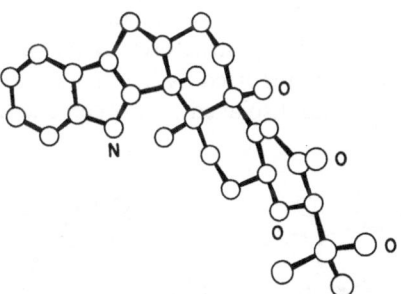

(**25**)

A computer generated perspective drawing of paxilline is depicted in Fig. 12. It shows that ring C is fused in a *trans* fashion to the cyclohexane ring D. An additional *trans* equatorial fused ring at C-4 and C-13 leads to a bond in the molecule which is completed by a β-pyrone ring. The cyclohexane rings D and E are in the chair conformation. The same conformation, based on n.m.r. spectroscopy, was deduced for the diterpene part of penitrem A (*27*).

Fig. 12. Perspective drawing of paxilline (**25**)

Some details of the 100 MHz ^{1}H n.m.r. spectrum of paxilline were reported. However, limited structural information was obtained. The ^{13}C n.m.r. assignment of paxilline is reported in Table 7 (see p. 36).

2.6 Paspaline, Paspalicine, Paspalinine and Paspalitrems A and B

2.6.1 Producing Organism

The neurological disorder, Dallis grass poisoning, or paspalum staggers, occurs in cattle that graze *Paspalum dilatatum* infected with the ergot
fungus *C. paspali*. This syndrome occurs sporadically in the southern parts
of the United States of America. The metabolites paspaline, paspalicine,
paspalinine and paspalitrems A and B have been isolated from cultures of
the fungus.

2.6.2 Structural Studies

Structural studies on paspaline and paspalicine by ARIGONI's group
were carried out in the era preceding very high field n.m.r. spectroscopy and
the ready availability of single crystal *X*-ray crystallography (*62*). Their
studies are based on elegant and detailed chemical degradations which serve
as a model for these types of compounds: some of the salient reactions will
therefore be reported.

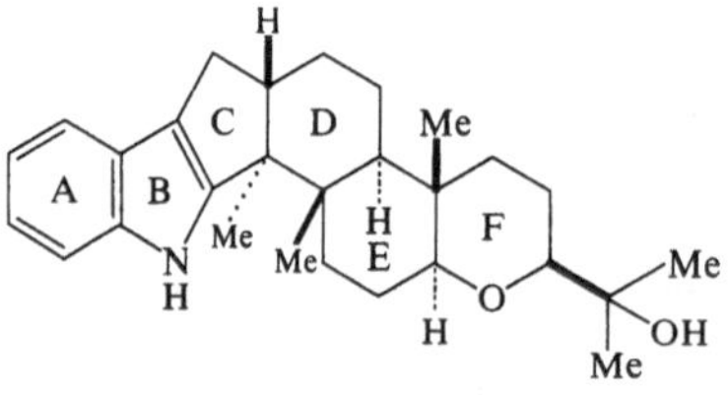

(13) 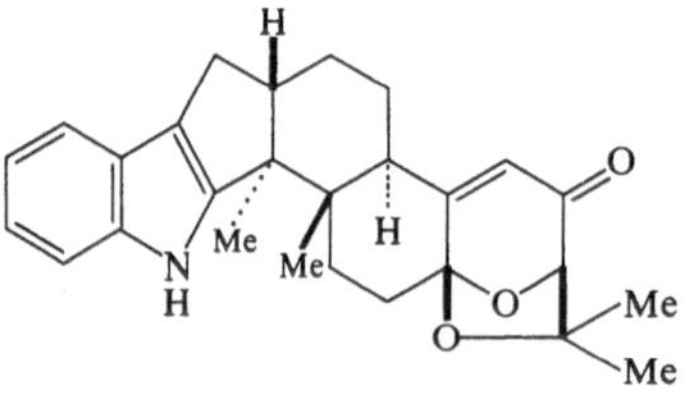   (14)

Paspaline (**13**), $C_{28}H_{39}NO_2$, and paspalicine (**14**), $C_{27}H_{31}NO_3$, were
obtained from cultures of *C. paspali* and had melting points of 254°C and
230°C, respectively. The *X*-ray structures of paspaline, paspalicine and
paspalinine were recently reported (*41, 66*) and are given in Fig. 13.
However, only paspalinine displayed tremorgenic activity.

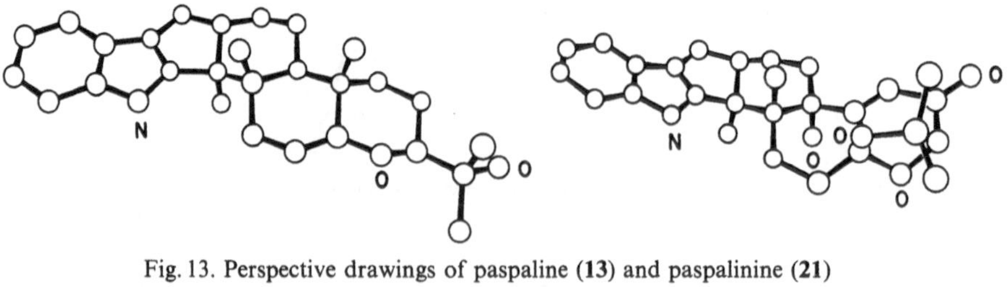

Fig. 13. Perspective drawings of paspaline (**13**) and paspalinine (**21**)

GYSI (*63*) described the degradation reactions of ring F of paspaline (**13**)
shown in Scheme 1.

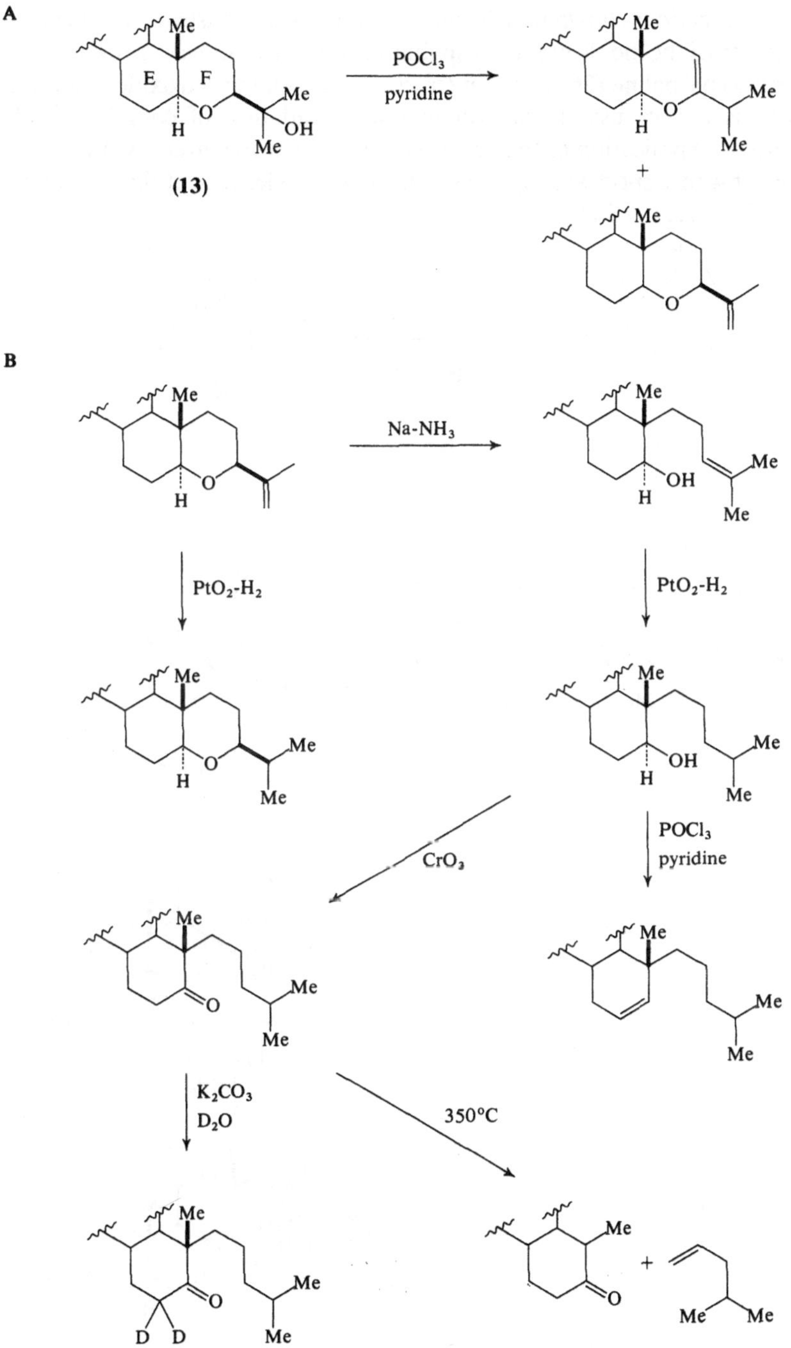

A

(13)

B

Scheme 1. Chemical reactions on ring F of paspaline (13)

The reactions shown in Scheme 2 adequately demonstrate the chemical reactivity of the indole part of paspaline (*63*). The reactions were carried out on deoxypaspaline (**26**) using, *inter alia,* a ruthenium tetroxide oxidation which is a very useful reaction in the degradation of complex indole alkaloids. Application of this reaction to tetrahydropenitrem A (**12**) gave a product whose spectral properties indicated that cleavage of the 2,3-double bond had occurred (*67*).

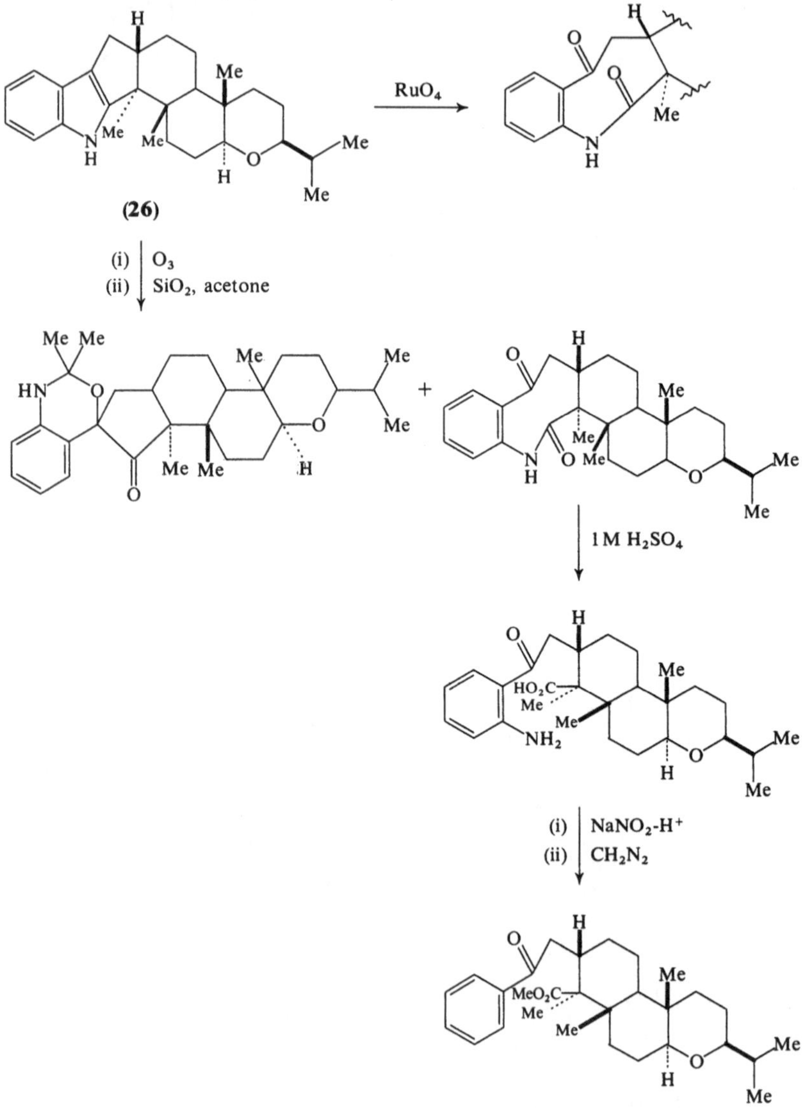

Scheme 2. Chemical reactions of the indole moiety of deoxypaspaline (**26**)

In a separate series of reactions (Scheme 3) the nature of the D ring was investigated (63).

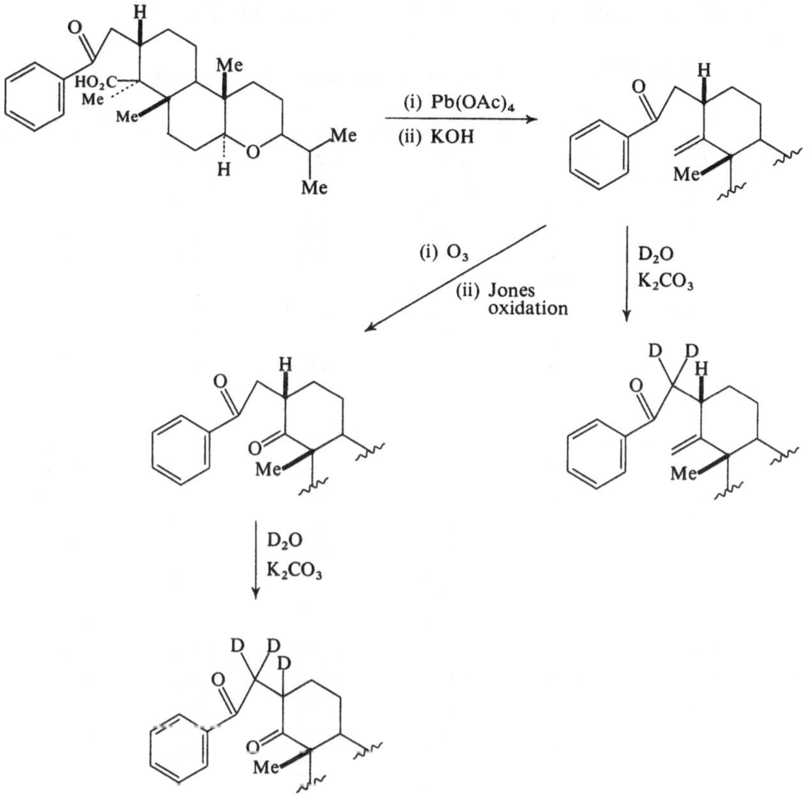

Scheme 3. Chemical reactions involving the D-ring of deoxypaspaline (26)

GYSI (63) proposed the correct structure for paspaline (13) based on the chemical degradations and the available spectroscopic data.

The complete stereochemistry of paspaline (13) and paspalicine (14) is based on *X*-ray crystallography (41). The pertinent crystal data are as follows: for paspaline: $a = 49.388(5)$, $b = 6.527(1)$, $c = 7.891(1)$ Å, space group $P2_12_12_1$, asymmetric unit $C_{28}H_{39}NO_2$ $(CH_3OH)^1/_2$; for paspalicine: $a = 9.706(1)$, $b = 10.670(1)$, $c = 21.775(2)$ Å, space group $P2_12_12_1$, asymmetric unit $C_{27}H_{31}NO_3$.

The absolute configuration of these compounds (13) and (14) is based on the circular dichroism exhibited by these compounds, *viz.* a positive Cotton effect at 348 nm with $[\theta] = 3.9 \times 10^4$ for paspalicine (14) and correlation with paxilline (25).

Gallagher *et al.* (*66*) isolated a new tremorgen, called paspalinine (**21**) (formula see p. 29) from the dried mycelia of *C. paspali*. Its molecular formula $C_{27}H_{31}NO_4$ as established by mass spectrometry showed that it contained one oxygen atom more than paspalicine (**14**) (*66*).

Table 7. ^{13}C *Chemical Shifts of Paspaline and Related Compounds*

Carbon	Paspaline (13)	Paspalicine (14)	Paspalinine (21)	Paxilline (25)	Paspalitrem A (27)	Paspalitrem B (28)
1	150.7	149.4	152.1	152.1	151.2	152.3
2	52.9	51.5	51.5	50.0	51.2	51.2
3	42.2	39.9	39.9	39.9	39.9	39.9
4	21.9[a]	21.6[a]	21.1[a]	20.6[a]	21.1[a]	21.1[a]
5	24.6[a]	27.7[a]	26.3	25.8[a]	27.0[a]	27.0[a]
6	84.7	104.4	104.5	72.2	104.4	104.4
7	85.7	88.4	87.9	83.1	88.0	87.9
8	37.7	197.6	197.2	196.9	197.3	197.3
9	25.2[a]	118.4	117.5	117.9	117.6	117.6
10	36.4	171.8	169.9	169.9	169.8	169.7
11	46.4	37.5	77.4	75.4	77.6	77.5
12	21.9[a]	28.6[a]	27.4[a]	26.7[a]	28.3[a]	28.3[a]
13	27.5[a]	29.4[a]	28.2[a]	28.1[a]	29.4[a]	29.9[a]
14	48.7	48.9	48.6	49.1	48.7	48.6
15	33.8	32.1	32.8	32.4	33.8	33.8
16	118.3	118.4	117.1	114.5	116.7	116.4
17	125.1	125.3	125.2	124.1	124.6	124.2
18	118.3	118.6	118.4	117.2	117.6	116.0
19	119.4	119.9	119.5	118.4	127.9	128.6
20	120.4	120.8	120.3	118.8	120.9	120.9
21	111.3	111.6	111.5	111.4	109.4	110.5
22	139.9	140.2	139.9	139.3	139.8	140.3
23	14.5	14.9	16.2	16.2	16.3	16.3
24	19.9	23.2	23.0	18.6	23.1	23.0
25	71.9	78.2	78.6	70.8	78.7	78.7
26	23.7[b]	23.8	23.5	25.7[b]	23.6	23.6
27	26.1[b]	29.0	28.8	25.8[b]	28.9	28.8
28	12.7				32.0	124.7
29					123.7	137.5
30					133.0	71.2
31					18.0	29.9
32					25.8	29.9

[a] Assignments of these carbons uncertain at this time.
[b] These assignments may be reversed.

The presence of a 2,3-disubstituted indole chromophore and of an α,β-unsaturated carbonyl group was deduced from the u.v. spectrum of (**21**) which exhibited maxima at 232 (ε 25 000), 250 (ε 14 800) and 274 nm (ε 8 000). Strong peaks in the i.r. spectrum at 1669 and 1620 cm^{-1} were assigned to the α,β-unsaturated carbonyl moiety. The ^{13}C n.m.r. data of paspalinine were reported (66) and assigned in the publication of COLE et al. (65) (see Table 7). The most striking features of the ^{13}C n.m.r. spectrum were the absence of the doublet at δ 39.2, assigned to the C-13 carbon atom of paspalicine (14), and the appearance of a singlet at δ 76.0 assigned to an oxygen-bearing carbon atom in paspalinine. The ^{1}H n.m.r. data of paspalinine also corroborated the point of attachment (C-13) of the hydroxy group.

The proposed structure of paspalinine was verified by X-ray crystallography (67). The crystals were orthorhombic with $a = 9.801(4)$, $b = 10.555(3)$ and $c = 21.605(7)$ Å, space group $P2_12_12_1$. Attempts to solve the structure by conventional direct procedures failed. This was ascribed to the poor scattering ability of the crystals at higher 2θ values. The structure was solved by GALLAGHER et al. (66) by using structural information from paspalicine. The structure refined to an R-factor of 0.084.

The absolute configuration of paspalinine as indicated in (21) was based on that of paspaline and the observed biosynthetic transformation of paspaline to paspalicine (68).

COLE et al. (65) isolated three tremorgens from the sclerotia of C. paspali collected from P. dilitatum in toxic pastures. One substance was identified as the known tremorgen, paspalinine, whereas the two unknown substances called paspalitrems A and B each contained an additional C_5 unit located at C-19 of paspalinine. Paspalitrem A (27), $C_{39}H_{39}NO_4$, and paspalinine (21) each possessed u.v. and i.r. data in accordance with their structures. The structure of paspalitrem A was based on ^{1}H and ^{13}C n.m.r. data (see Table 7).

(27) R = Me₂C=CHCH₂—
(28) R = Me₂C(OH)CH=CH—

The u.v. spectrum of paspalitrem B (28), $C_{32}H_{39}NO_5$, λ_{max} (MeOH) 227, 248, 305 and 335 nm (no extinction coefficient values given) indicated the presence of an extended conjugated chromophore. The ^{1}H and ^{13}C n.m.r. data of paspalitrem B identified the additional C_5 unit as a 3-methyl-3-hydroxybutenyl group and its location at C-19 (see Table 7).

It is noteworthy that paspalinine, paspalitrem A and B all contain a tertiary hydroxy group at C-11, the functionality which seems to be associated with their tremorgenic activity. The ED_{50} values of the substances were ≤ 14 mg/kg in mice dosed intraperitoneally (65).

2.7 Biosynthesis of the Penitrems, Janthitrems, Lolitrems, Aflatrem, Paxilline, Paspaline, Paspalicine, Paspalinine, and Paspalitrems A and B

A brief survey of the structural properties of this interesting group of fungal tremorgens points to their close biogenetic relationship (see Fig. 14). All these substances contain an indole nucleus linked to a diterpenoid unit. However, in some cases [aflatrem (20) and paspalitrems A (27) and B (28)] an additional C_5 unit is attached to C-4 or C-5 of the indole moiety. Some of the more complex substances, e.g. the penitrems and janthitrems, contain two C_5 units linked to the indole part of the molecule. The lolitrems contain even more elaborate features by the presence of a further C_5 unit linked to the diterpenoid moiety through ether bridges.

The biosynthesis of paspaline (13) and related metabolites was initially studied by ARIGONI's group (69). ACKLIN et al. (69) recognized that these substances originated from combination of indole and a diterpene unit. The hypothesis was examined by supplementing cultures of C. paspali with $[1-^{13}C]$-, $[2-^{13}C]$-, and $[1,2-^{13}C_2]$acetate. The enriched paspaline (13) showed that acetate was readily incorporated into the diterpenoid part (C_{20}) of the molecule: $[1-^{13}C]$acetate enriched 8 carbon atoms, whereas $[2-^{13}C]$acetate enriched 12 carbon atoms.

The incorporation of $[1,2-^{13}C_2]$acetate led to the presence of eight intact acetate units in paspaline, and to four enriched carbon atoms, C-4, C-9, C-13 and C-26, which must have been derived from C-2 of mevalonate (69).

A postulated 1,2-migration during the biosynthesis leads to the observed coupling of C-13 to C-14 and C-2 to C-3 upon administration of $[2-^{13}C]$- and $[1-^{13}C]$acetate, respectively. The identity of the two methyl groups constituting the hydroxyisopropyl group was retained during the biosynthesis of paspaline (69).

ACKLIN et al. (69) briefly reported on the structure of (29) but no details were given. This metabolite of C. paspali is a likely precursor in the biosynthesis of paspaline. WEIBEL (68) showed in biosynthetic experiments that paspaline (13) is a precursor in the formation of both paspalicine (14) and paspalinine (21). The results are summarized in Scheme 4.

TANABE (70) proposed a biosynthetic scheme for paxilline (25) based on the results obtained from labelling experiments with $[1,2-^{13}C_2]$acetate and analysis of the product by ^{13}C n.m.r. spectroscopy. The labelling pattern is shown in Figure 15.

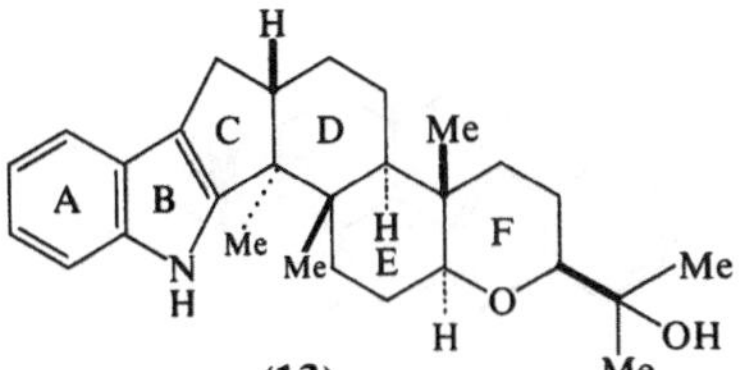

(13)

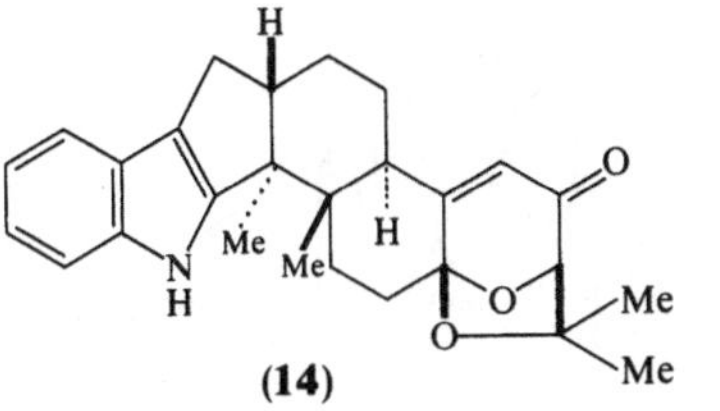

(14)

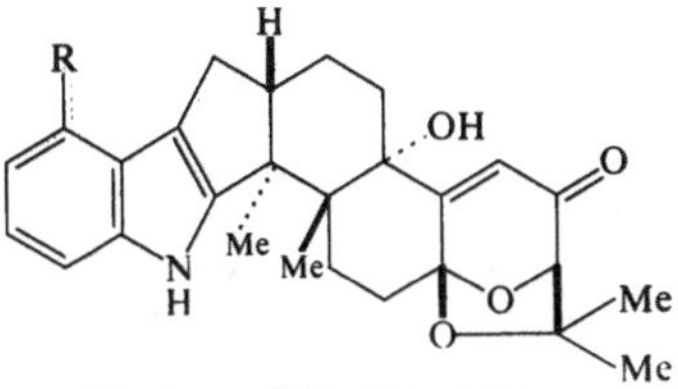

(20) R=—CMe₂CH=CH₂
(21) R=H

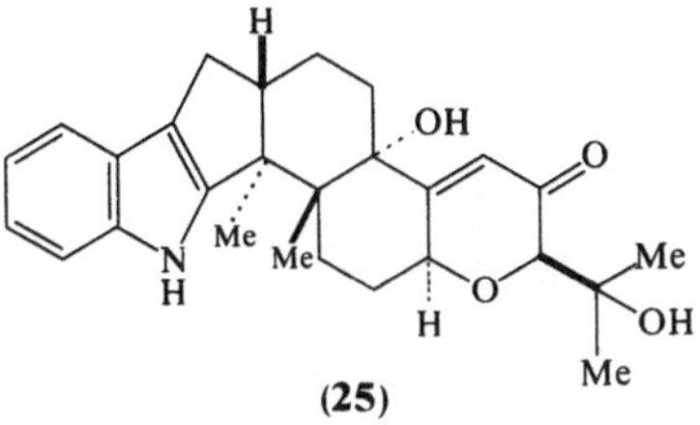

(25)

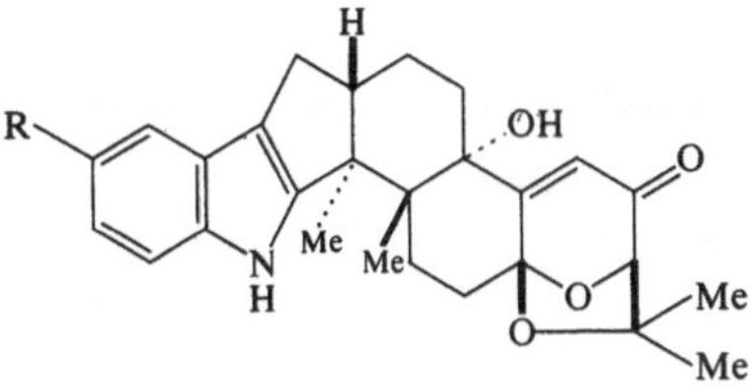

(28) R=Me₂C(OH)CH=CH—

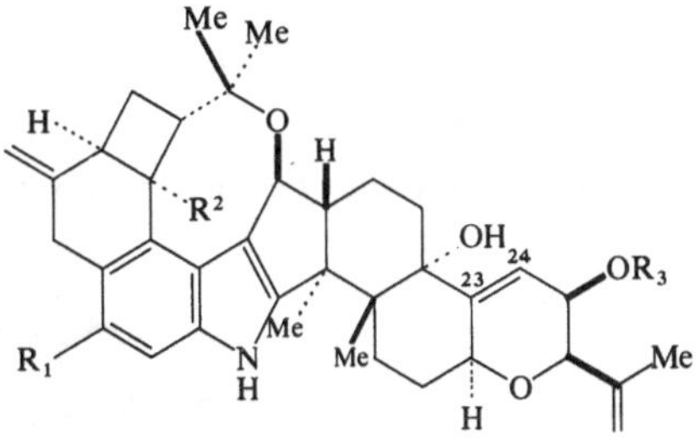

(5) R₁=Cl, R₂=OH, R₃=H; 23a,24a-epoxide
(8) R₁=R₂=R₃=H

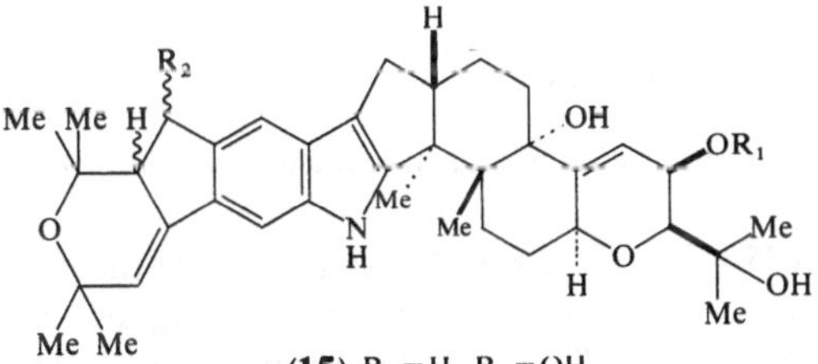

(15) R₁=H, R₂=OH

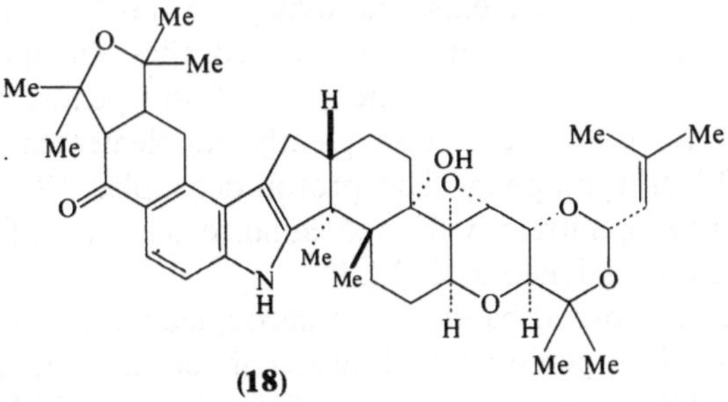

(18)

Fig. 14. Representative structures of group 1 tremorgens

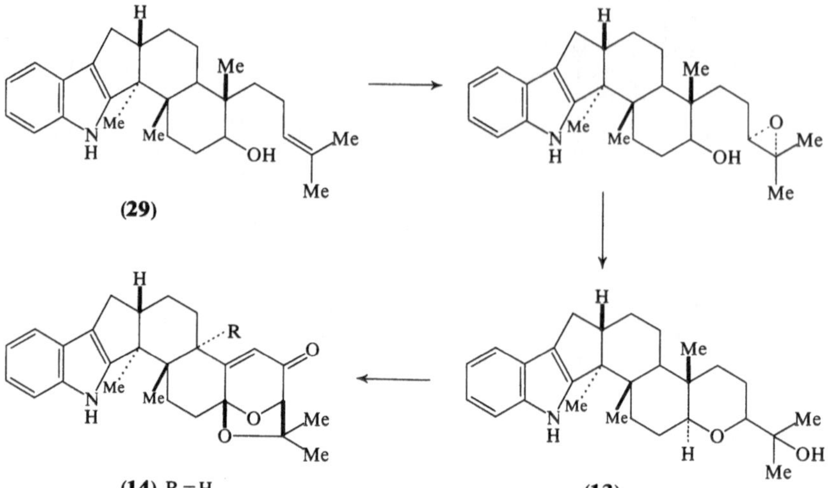

(29)

(14) R=H
(21) R=OH

(13)

Scheme 4. Postulated steps in the biosynthesis of paspalicine (14) and paspalinine (21)

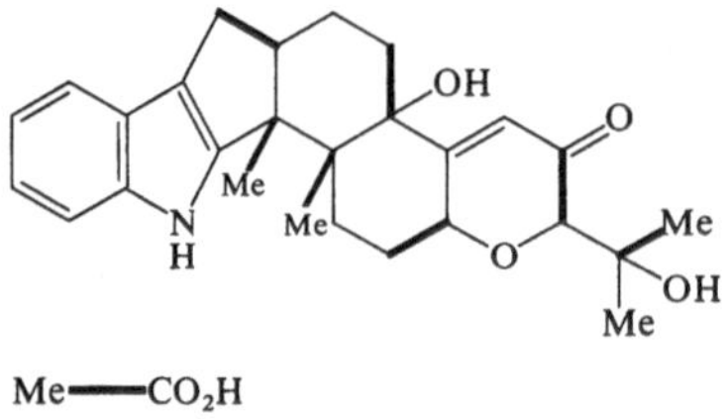

Me——CO₂H

Fig. 15. The distribution of intact acetate units in paxilline (25)

DE JESUS et al. (25—29) elucidated the structures of the penitrems by the simultaneous application of advanced physical techniques and biosynthetic studies. The biosynthetic studies contributed importantly to this effort.

Separate feeding experiments employing $(2S)$-[3-^{14}C]tryptophan and $(2RS)$-[benzene-ring-U-^{14}C]tryptophan showed that the indole part of tryptophan contributes the aromatic part of the penitrems (29). The cultures of P. crustosum were subsequently supplemented with $(2RS)$-[indole-2-^{13}C,2-^{15}N]tryptophan. The proton-decoupled ^{13}C n.m.r. spectrum of the derived penitrem A showed enhancement only of the signal at δ 154.36 which was assigned to C-2 (29).

The 100.62 MHz broad-band proton-decoupled ^{13}C n.m.r. spectrum of [1-^{13}C]acetate-derived penitrem A showed enhancement of the signals of twelve carbon atoms, viz. C-11, C-13, C-15, C-16, C-18, C-21, C-23, C-25, C-29, C-31, C-32 and C-37: six mevalonate units are therefore involved in the formation of the non-indole part of this compound. A one-bond (C,C) coupling (37.1 Hz) was observed for the carbons resonating at δ 43.55 (C-31) and δ 50.08 (C-32) (see Fig. 16). This information indicates that a 1,2-shift must occur in the course of biogenesis—the same conclusion was previously arrived by ACKLIN *et al.* (*69*). Complementary results were obtained from the 125.76 MHz proton-decoupled spectrum of [2-^{13}C]acetate-derived penitrem A which showed enhancement of seventeen carbon atoms, *viz.* C-10, C-12, C-14, C-19, C-20, C-22, C-24, C-26, C-28, C-30, C-33, C-34, C-35, C-36, C-38, C-39 and C-40. A single carbon atom derived from C-2 of acetate and originally present as the methyl group of a mevalonate unit, is lost during the bioformation. This spectrum also showed strong interacetate (C,C) coupling between C-19 and C-20 [^1J(C,C) 35.2 Hz], due to bond migration. The same phenomenon was observed for paspaline by ACKLIN *et al.* (*69*). The formation of [1,2-^{13}C$_2$]acetate from [2-^{13}C]acetate was observed by DE JESUS *et al.* (*26, 29*) and is the result of frequent cycling of [2-^{13}C]acetate in the Krebs' citric acid cycle.

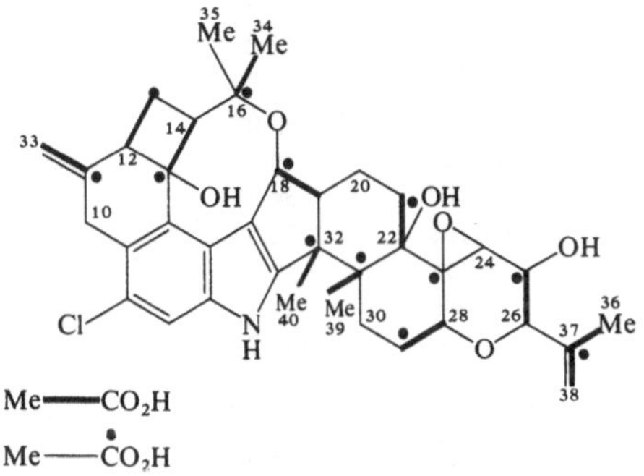

Fig. 16. The labelling pattern of penitrem A derived from [1-^{13}C]- and [1,2-^{13}C$_2$]acetate labelling experiments

The (C,C) coupling constants of [1,2-^{13}C$_2$]acetate-derived penitrem A showed the presence of eleven intact acetate units (see Scheme 5 and Table 8). The carbon atoms which constitute the isopropylidene moiety showed a complex coupling phenomenon: C-37 exhibited two pairs of satellite peaks due to coupling with C-38 [^1J(CC) 73.2 Hz] and C-36 [^1J(CC) 42.7 Hz]. This conundrum was further explored by supplementing cultures of *P. crustosum* with (3*RS*)-[2-^{13}C]mevalonolactone and with (3*RS*)-[2,3-^{13}C$_2$]mevalonolactone: approximately equal enrichment of C-36 and C-38 was observed. In the studies of the biosynthesis of paspaline and paxilline (*69, 70*), the stereochemical identity of the two diastereotopic methyl groups was retained. The isomerization of the Δ^{37,38} bond in penitrem A (**5**) apparently occurs during the course of the isolation or chromatography on silica gel (*29*).

Table 8. [13]C N.m.r. (125.76 MHz) Data for Penitrem A Derived
from Biosynthetic Labelling Experiments

Carbon atom	δ_C[a]	[1]J(CC)[b]	[1]J(CC)[c]
10	35.06§	(40.3)	40.7
11	149.48*	73.9, (40.3)	40.7
12	47.01§	29.9	
13	24.67*	29.9	
14	52.71§	33.6	
15	81.01*	33.6	
16	76.09*	39.1	41.0
18	72.44*	39.7	
19	58.79§	39.7	
20	18.56§	(34.8)	
21	30.59*	37.8	
22	78.24§	39.1	
23	66.11*	(29.9)	29.9
24	61.92§	(29.9)	29.8
25	66.31*	39.3	
26	74.67§	39.6	
28	71.99§	37.2	
29	28.89*	37.8	
30	26.91§	(33.6)	34.0
31	43.55*	36.6	34.3
32	50.08*	34.8	
33	107.10§	73.9	
34	20.32§	39.7	
35	31.06§	(40.3)	41.0
36	19.70§	(42.7)	43.0
37	143.27*	73.2, (42.7)	73.2, 42.8
38	111.64§	73.2	73.0
39	18.98§	36.0	
40	21.35§	34.8	

[a] Relative to internal Me_4Si; solvent $(CD_3)_2CO$. * = enriched by [1-^{13}C]acetate; § = enriched by [2-^{13}C]acetate.

[b] Values obtained from the broad-band proton-decoupled spectrum of penitrem A derived from [1,2-[13]C_2]acetate. Values in parentheses are due to multiple labelling.

[c] Values obtained from the broad-band proton-decoupled spectrum of penitrem A derived from [2,3-[13]C_2]mevalonolactone.

The initial stages of the biosynthetic pathway leading to the penitrems must be very similar to those proposed for paspaline (69) and paxilline (70). A biosynthetic scheme (29) for penitrem A (5) is shown in Scheme 5. The results on the conversion of paspaline into paspalinine and paspalicine are very relevant, particularly since paspaline still contains the methyl group at C-10 which is lost during the formation of the other tremorgens, e.g. penitrem A.

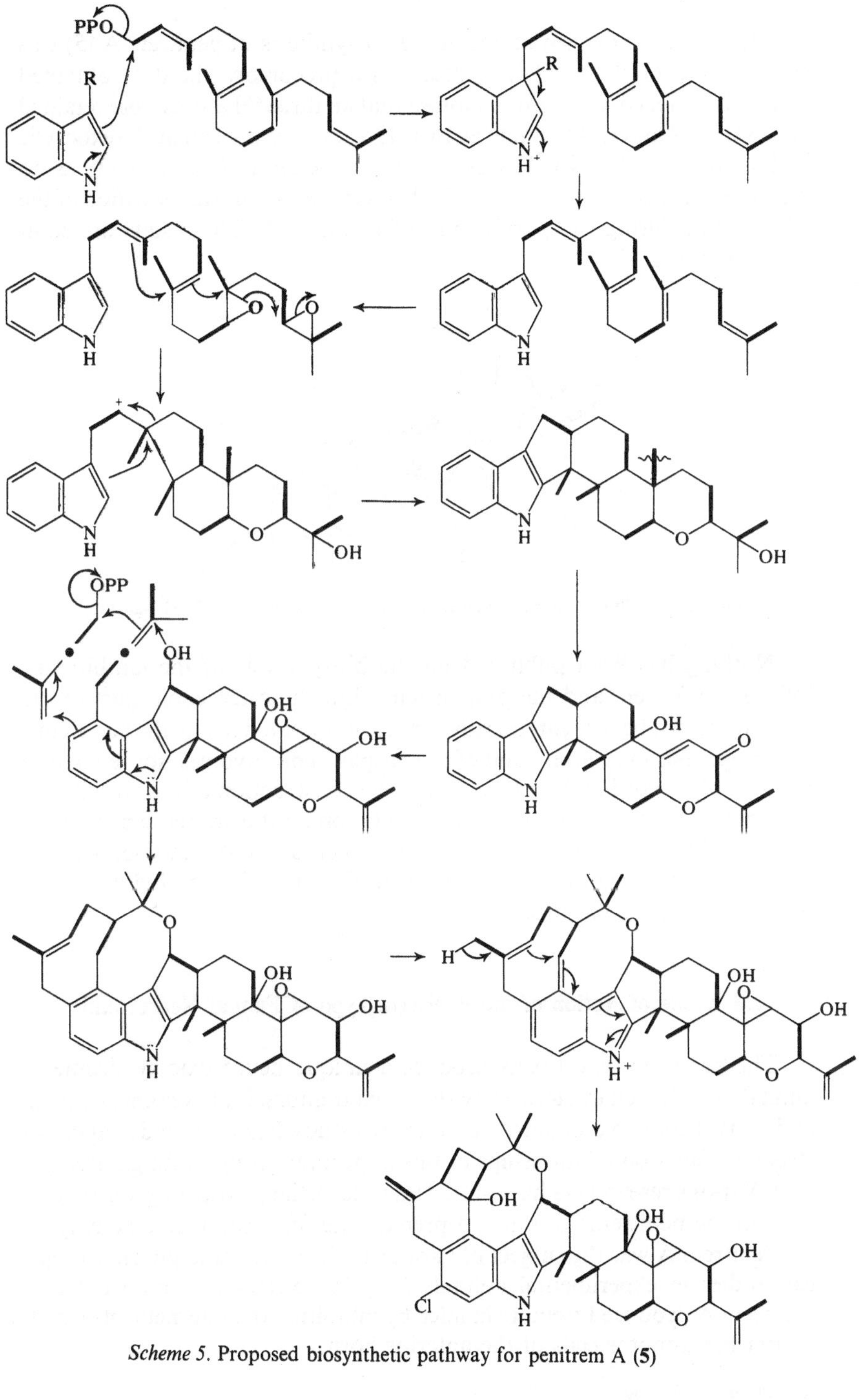

Scheme 5. Proposed biosynthetic pathway for penitrem A (**5**)

The fate of hydrogen atoms in the biosynthesis of penitrem A (**5**) was studied by using $[1\text{-}^{13}C,2\text{-}^2H_3]$acetate as a precursor. The data obtained from this enriched penitrem A showed that all three ^2H atoms were retained at C-34, C-39 and C-40. Similarly two ^2H atoms are present at the exocyclic carbon atom, C-33. No ^2H was present at the either C-36 or C-38 as no β-isotope shift is observed for C-37. This is evidence for isomerization of the $\Delta^{37,38}$ bond during the purification of penitrem A. The results are summarized in Fig. 17.

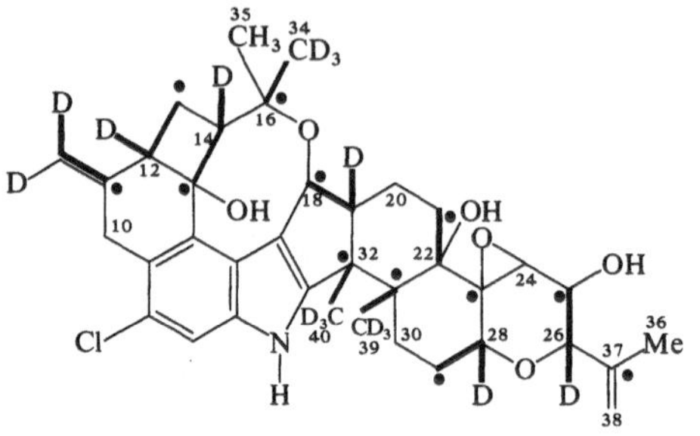

Fig. 17. Labelling pattern of penitrem A enriched with $[1\text{-}^{13}C,2\text{-}^2H_3]$-acetate

Nothing has been published on the biosynthesis of the janthitrems, lolitrems, aflatrem and the paspalitrems, but the diterpenoid part of the molecules must be biosynthetically very closely related to *e.g.* the penitrems. The biogenesis of aflatrem and of the paspalitrems involves isoprenylation at C-4 and C-5 of indole, respectively. The construction of the janthitrems and lolitrems involves a double isoprenylation on the aromatic part of the indole. In the janthitrems, isoprenylation occurs at C-5 and C-6, whereas closely related reactions takes place at C-4 and C-5 of indole in the formation of the lolitrems.

2.8 Mode of Action of the Penitrem Type of Fungal Neurotoxins

The penitrem-type toxins produce a unique neurotoxic syndrome in animals which is characterized by sustained tremors, limb weakness, ataxia and convulsions. No detailed comparative study has been undertaken on structure-function relationships of this important group of fungal tremorgens. Various researchers have stated that the tertiary hydroxy group at *e.g.* C-22 in the penitrems is in fact a prerequisite for tremorgenic activity.

Penitrem A was the subject of various toxicological and pharmacological studies in experimental animals (71—76). STERN (77) reported that penitrem A produced tremors in mice by inhibiting the interneurons which inhibit the α-motor cells of the anterior horn.

The neurochemical effects of penitrem A were studied by using sheep and rat synaptosomes (78). It was observed that penitrem A increased the spontaneous release of endogenous glutamate, γ-aminobutyric acid and aspartate by 213%, 455% and 277%, respectively. These results suggested that penitrem A acts by interfering with the amino acid neurotransmitter release mechanisms.

These neurotoxins seem to play an important role in neurotoxicoses, e.g. ryegrass staggers under field conditions. Ryegrass staggers is a nervous disorder which sporadically affects sheep, cattle and horses in New Zealand when these animals are grazing pastures in which there is a predominance of perennial ryegrass (L. perenne)—this condition occurs in summer and autumn. The background to this problem was adequately reviewed by GALLAGHER (79). Only some pertinent literature references to ryegrass staggers are given in this review (14, 48, 80—84).

Mycotoxins have also been implicated in the etiology of Dallis grass poisoning, also called "paspalum staggers" (65), and in "Bermuda grass tremors" (85).

3. The Territrems

3.1 Producing Organisms

In a mycological survey during 1968 of aflatoxin-producing fungi present on stored, unhulled rice in Taiwan, TUNG et al. (86) and CHUNG et al. (87) found that among 206 specimens of aspergilli, 11 isolates of Aspergillus terreus were capable of producing tremorgenic metabolites. These metabolites, collectively called territrems, induce tremors and convulsions in mice upon intraperitoneal administration (88, 89) but differ from the other tremorgenic mycotoxins, e.g. the penitrems (5)—(10), in that they contain no nitrogen.

3.2 Isolation and Chromatography of the Territrems

Cultures of A. terreus 32-1 were maintained on rice (90). An aqueous spore suspension, prepared by inoculation of a modified Czapek-Dox medium with A. terreus and incubation in stationary culture at 28—30°C for 1 month, served as the inoculum in the large scale production of the territrems on polished rice (91). The rice cultures were incubated at 28—30°C for 14 days. The resultant mouldy rice was extracted with warm chloroform and the chloroform extracts applied to a column consisting of silica gel (100 g) and on top anhydrous sodium sulphate (150 g). Territrems A and B were eluted with a mixture of chloroform-acetone

(4 : 1 v/v). Fractions of the eluate were monitored by thin layer chromatography on silica gel using toluene-ethyl acetate-formic acid (5 : 4 : 1 v/v/v).
The crude mixture of territrems A and B could be separated by column
chromatography on silica gel using benzene-ethyl acetate (65 : 35 v/v) as
eluant (91) or by preparative thin layer chromatography on silica gel with
benzene-ethyl acetate (7 : 3 v/v) (90). Territrem B crystallizes from chloroform as colourless needles (90).

After elution of territrems A and B, the column was developed with
chloroform-acetone (4 : 1 v/v) (92). The fractions containing territrem C
were combined and purified by column chromatography first on silica gel
using benzene-ethyl acetate (1 : 1 v/v) and then on Sephadex LH-20 with
absolute ethanol. Territrem C crystallised from absolute ethanol (92).

Ling et al. (93) developed several solvent systems for separating
territrem A and B by thin layer chromatography and also for differentiating
between the territrems and aflatoxin B_1 and B_2. Territrems A and B can also
be purified by high-performance liquid chromatography on a μ-Porasil
silica gel (10 μm) column using water-saturated chloroform-cyclohexane-
acetonitrile (25 : 75 : 1 v/v/v) with 0.4% ethanol as eluant at a flow rate of
2 ml/min (93).

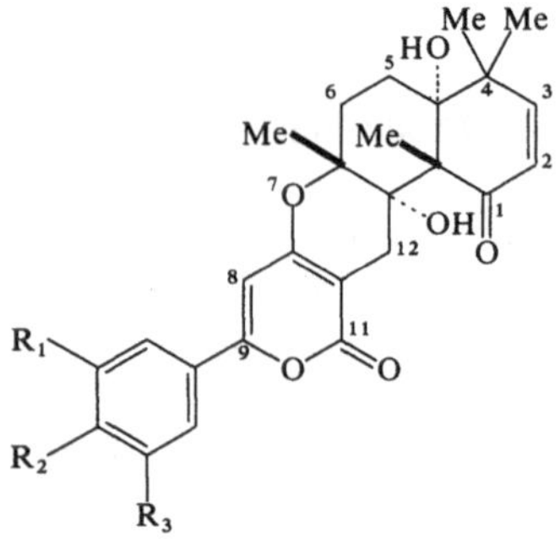

(30) R_1 = OMe; R_2, R_3 = —OCH_2O—
(31) R_1 = R_2 = R_3 = OMe
(32) R_1 = R_3 = OMe; R_2 = OH

3.3 Structure of the Territrems

The molecular formulae, as determined by high resolution mass
spectrometry for territrem A (30) ($C_{28}H_{30}O_9$), territrem B (31) ($C_{29}H_{34}O_9$)
and territrem C (32) ($C_{28}H_{32}O_9$) (90, 92), indicate that these compounds
share the same basic structure and differ from each other only in the nature
of the substituents at certain carbon atoms. The u.v. absorption and
fluorescence data for the territrems, collated with some other physicochemical data in Table 9, indicate that these compounds have similar chromo-

Table 9. *Some Physicochemical Data for the Territrems*

	Territrem A	Territrem B	Territrem C
Molecular formula	$C_{28}H_{30}O_9$	$C_{29}H_{34}O_9$	$C_{28}H_{32}O_9$
Molecular weight	510	526	512
Melting point (°C)	288 – 290 (dec)	200 – 203	172.5 – 173.5
$[\alpha]_D$ (CHCl₃)	+ 102° (c 0.10)	+ 131° (c 0.60)	+ 120° (c 0.10)
U.v. spectrum (MeOH)	219 nm (ε 43 000)	219 nm (ε 39 000)	219 nm (ε 36 000)
	338 nm (ε 19 600)	331 nm (ε 18 400)	344 nm (ε 18 800)
Fluorescence (MeOH)			
λ_{exc}	375 nm	375 nm	375 nm
λ_{em}	420 nm	420 nm	430 nm

phores. The absorption band at 344 nm (ε 18 800) in the u.v. spectrum of territrem C was shifted to 398 nm (ε 21 300) in 0.1M methanolic sodium hydroxide which suggested the presence of a phenolic hydroxy group. The i.r. spectrum of territrem B showed major absorptions at 3470 and 3340 (OH); 1705 and 1685 (C=O); 1640, 1585 and 1505 (C=C); and 1130 cm⁻¹ (C−O−C) (90). These absorptions also appeared in the i.r. spectra of the other two territrems (90).

The structure of the territrems is based on *X*-ray crystallography of territrem B (**31**) (9) and ¹H n.m.r. spectroscopy (90, 92).

Territrem B crystallized from methanol as orthorhombic crystals, space group $P2_12_12_1$, $a = 8.681(2)$, $b = 12.477(2)$, and $c = 25.314(2)$ Å, $Z = 4$, and $D_x = 1.27\,g\,cm^{-3}$. The relative stereochemistry of territrem B is illustrated in Fig. 18. The pyrone ring B is essentially planar. Rings C and E are both in a half-chair conformation and the junction with ring D is *trans* in each case. This results in ring D adopting a chair conformation. There are no intermolecular hydrogen bonds. The oxygen atom of the C-13 hydroxy group is within hydrogen-bonding distance (~ 2.7Å) of both the C-17 hydroxy group and the C-19 carbonyl group oxygen atom. An intramolecular hydrogen bond, however, was found only between the proton of the C-13 hydroxy group and the C-19 carbonyl oxygen atom (9).

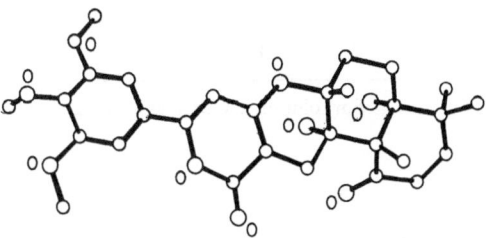

Fig. 18. Perspective drawing of territrem B (**31**)

The ^{1}H n.m.r. data for the territrems have been reported (90, 92) but no assignments were made by the authors. The chemical shifts and coupling constants for the protons of territrem A (30) are summarized in Table 10. The two *meta* oriented aromatic protons appear as doublets (J 2 Hz) at δ 6.88 and 7.02. The ^{1}H n.m.r. spectrum of territrem B (31) exhibited a singlet resonance at δ 3.90 (9H), which is assigned to the protons of the three aromatic methoxy groups, as well as a singlet at δ 6.95 (2H) for the aromatic protons. The resonance at δ 5.97, assigned to the protons of the methylenedioxy moiety in territrem A (30), is absent in the spectra of both territrem B (31) and C (32). The symmetrical 3,5-dimethoxy-4-hydroxy substitution pattern of the aryl group in territrem C followed from the singlet resonances in the ^{1}H n.m.r. spectrum at δ 3.92 (6H, two methoxy groups), 6.95 (2H, *meta* aromatic protons), and 5.82 (1H, disappears on addition of deuterium oxide).

Methylation of territrem C (32) using dimethyl sulphate gave territrem B (31) (92).

Table 10. ^{1}H N.m.r. Data for Territrem A
(30)^a

Proton	$δ_H$	J/Hz
2	5.74 d	10
3	6.26 d	10
5	1.70 – 2.07 m	
6	2.27 m	
8	6.23 s	
12	3.41 d	17
	2.78 d	17
OMe	3.93 s	
OCH$_2$O	5.97 s	
Ar-H	7.02 d	2
	6.88 d	2
12a-OH	5.83 s	
4a-OH	3.83 s	
C-Me	1.21 s	
	1.30 s	
	1.47 s	
	1.53 s	

^{a} For solution in CDCl$_3$ at 60 MHz.

4. Verrucosidin

4.1 Producing Organism

WILSON *et al.* (*94*) reported the isolation of a tremorgenic mycotoxin named verrucosidin from the fungus *Penicillium verrucosum* var. *cyclopium* implicated in a neurologic disease in cattle.

(4)

4.2 Structure of Verrucosidin

Verrucosidin (4) crystallizes from ether as colourless plates, m.p. 90—91° C, $[\alpha]_D^{26} +92.4°$ (*c* 0.25, methanol) (*8*). The empirical formula $C_{24}H_{32}O_6$ was determined by mass spectrometry and elemental analysis. The pyrone carbonyl group absorbed at 1700 cm^{-1} in the i.r. spectrum. The ^{1}H and ^{13}C n.m.r. spectra* revealed the presence of 9 methyl groups in the metabolite. The presence of a 6-substituted 4-methoxy-3,5-dimethylpyrone moiety in verrucosidin (4) was inferred from a comparison of the ^{13}C chemical shifts with those of citreoviridin (1) and suggested a possible structural relationship between verrucosidin (4) and citreoviridin (1), aurovertin (2), and asteltoxin (3) (see p. 3 for formulae 1—3). A major difference, however, is that the u.v. spectrum of verrucosidin (4), λ_{max} 241 (ε 21 000) and 294 nm (ε 13 000) is consistent with an isolated pyrone chromophore, whereas extended conjugation in citreoviridin (1) leads to a u.v. absorption maximum at 388 nm (ε 48 000).

The $C_{16}H_{23}O_3$ fragment attached to C-6 of the pyrone moiety must contain the two trisubstituted double bonds, six of the methyl groups, and three oxygen atoms all of which must be present as epoxides or cyclic ethers. Degradation of (4) with 0.5 M potassium hydroxide in water-methanol (1 : 1 v/v) gave the aldehyde (33). The u.v., i.r., ^{1}H and ^{13}C n.m.r. spectra of

---

* N.m.r. data: ^{1}H (CDCl$_3$): δ 1.21 d (3H, *J* 7 Hz), 1.42 s (3H), 1.44 s (3H), 1.48 s (3H), 1.93 s (3H), 1.98 s (3H), 2.05 s (6H), 3.45 s (1H), 3.50 s (1H), 3.85 s (3H), 4.12 q (1H, *J* 7 Hz), 5.50 s (1H), and 5.88 s (1H); ^{13}C (CDCl$_3$): δ 8.9 q, 10.0 q, 13.5 q, 15.0 q, 15.2 q, 18.2 q, 18.5 q, 21.6 q, 60.0 q, 60.3 s, 64.4 d, 67.1 s, 67.1 d, 76.4 d, 79.7 s, 110.0 s, 110.6 s, 127.7 s, 131.1 d, 132.7 d, 134.2 s, 155.6 s, 164.5 s, and 167.5 s.

(33)* established that the aldehyde was present as a 5-substituted-2,4-dimethylpenta-2,4-dienal unit. The linkage between the pyrone and side-chain moieties was assigned as an epoxidized alkene in order to isolate the pyrone and diene chromophores (8).

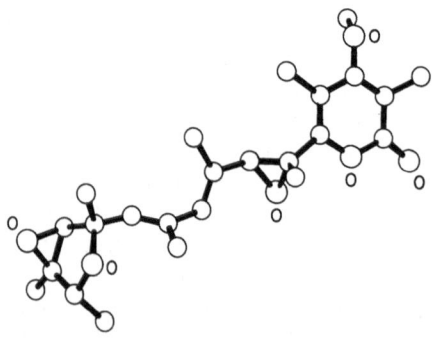

(33)

Confirmation of the structural deductions as well as the assignment of the relative stereochemistry of the six chiral centres present in verrucosidin (4), was obtained by X-ray crystallography. The compound crystallizes in the orthorhombic space group $P2_12_12_1$ with $a = 5.7971(8)$, $b = 11.912(2)$ and $c = 33.634(5)$ Å, $Z = 4$, and $D_c = 1.19 \, \text{g cm}^{-3}$. The structure was solved by the direct methods routine and is illustrated in Fig. 19.

Fig. 19. Perspective drawing of verrucosidin (4)

---

* λ_{max} (MeOH) 277 nm (ε 11 800) (α,β-unsaturated aldehyde); ν_{max} (CHCl$_3$) 2975, 2930, 1670 (O=CH−C=C−C=C), and 1610 cm^{-1}; ^{1}H n.m.r. (CDCl$_3$): δ 1.19 d (3H, J 7 Hz), 1.45 s (3H), 1.48 s (3H), 1.95 s (3H), 2.16 s (3H), 3.44 s (1H), 4.15 q (1H, J 7 Hz), 5.91 s (1H), 6.68 s (1H), and 9.41 s (1H, −CHO); ^{13}C n.m.r. (CDCl$_3$): δ 10.80 q, 13.78 q, 17.73 q, 18.87 q, 21.53 q, 67.19 d, 67.47 s, 76.80 d, 80.03 s, 135.24 s, 137.13 s, 140.22 d, 154.14 d, and 195.80 d.

References, pp. 74—80

5. Verruculotoxin

5.1 Producing Organism

The fungus *Penicillium verruculosum* Peyronel, isolated from green peanuts, produces a toxic metabolite with an LD_{50} of 20 mg/kg (*95*).

5.2 Isolation and Chromatography

P. verruculosum was grown on shredded wheat (100 g), supplemented with Difco mycological broth (pH 4.8) plus YES medium (200 ml), contained in Fernbach flasks (2.8 l) at 27° C for 2—3 weeks (*95*). The toxin was extracted by homogenization of the fungal cultures with chloroform in a Waring blender, followed by vacuum filtration through anhydrous sodium sulphate and concentration under vacuum at 50° C. The crude toxic extract was chromatographed on a silica gel column eluted sequentially with 1 l each of hexane, chloroform, ethyl acetate, acetone and methanol. The toxin eluted in the ethyl acetate fraction. The ethyl acetate fraction was concentrated and chromatographed on a second silica gel column using a linear gradient from chloroform to ethyl acetate. Column fractions were monitored for toxicity with 1-day-old cockerels. The toxin-containing fractions were combined and evaporated to dryness. The residue was purified by chromatography on Sephadex LH-20 using methanol-chloroform (1 : 4 v/v) as the eluting solvent. Verruculotoxin was crystallized from benzene.

The toxin was analyzed by thin layer chromatography on silica gel using chloroform-acetone (93 : 7 v/v) and visualized by spraying the chromatoplates with ninhydrin solution and heating for 5 min at 100° C (R_f 0.26).

5.3 Structure of Verruculotoxin

Verruculotoxin (**34**) had m.p. 152° C (from benzene) and analyzed for $C_{15}H_{20}N_2O$ (*96*). High-resolution mass spectral analysis of the toxin showed the molecular ion at m/e 244.1575 with a molecular formula of $C_{15}H_{20}N_2O$ and prominent fragment ions at m/e 215, 200, 187 and 153.

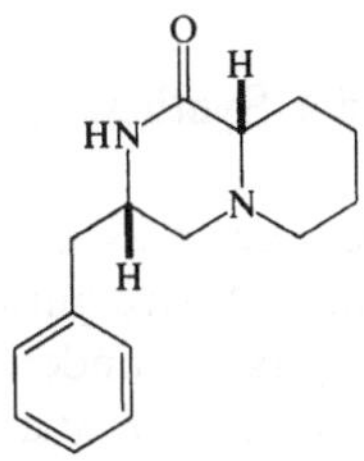

(**34**)

The u.v. spectrum of verruculotoxin showed strong end absorption and a series of absorptions of relatively low intensity in the 240—260 nm region, similar to the $\pi - \pi^*$ transitions observed in benzene. The i.r. spectrum indicated that the only oxygen atom was part of a carbonyl function. The prominent band at 3320 cm^{-1} (NH stretching) together with the carbonyl absorption at 1688 and 1640 cm^{-1} (amide I and II bands) was indicative of a secondary amide moiety.

The presence of fragment ions at m/e 153 ($M^+ -91$) and m/e 125 (base peak, $M^+ -119$) in conjunction with the aforementioned data, was indicative of phenylalkyl cleavage $i.e.$ $M^+ - (C_6H_5\text{-}CH_2\text{-})$ and $M^+ - (C_6H_5CH_2CH_2CH_2\text{-})$, respectively.

The 60 MHz ^{1}H n.m.r. spectrum of verruculotoxin (34) (96) showed a singlet resonance at δ 7.20 (5H, aromatic protons) and a series of unstructured multiplets in the δ 1—3 region.

Verruculotoxin (34) crystallizes from dichloromethane-heptane as large needles in the space group $P2_12_12_1$ with $a = 14.340(1)$, $b = 12.211(1)$ and $c = 7.896(1)$ Å, and one molecule of $C_{15}H_{20}N_2O$ per asymmetric unit. The X-ray crystal structure is illustrated in Fig. 20. The piperazine ring exists in a flattened chair conformation with the benzyl group in the axial position. The piperidine ring is also in a chair conformation and the bridgehead nitrogen is puckered to give a $trans$ ring junction. The phenyl group is planar and no hydrogen bonds were detected.

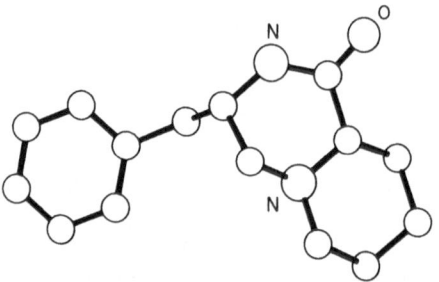

Fig. 20. Perspective drawing of verruculotoxin (34)

5.4 Synthesis

Verruculotoxin (34) can be viewed as a modified cyclic dipeptide of pipecolic acid and phenylalanine. A straightforward synthetic approach based on this retrosynthetic analysis was undertaken (96). Phenylalinol (35) was prepared from L-phenylalanine by quantitative esterification with thionyl chloride in methanol and reduction of the resulting methyl ester hydrochloride with sodium borohydride in aqueous ethanol. A neat solution of equimolar amounts of phenylalinol and methyl picolinate was heated at 150° C for 1 h to give the amide (**36**). This amide was transformed into the chloride (**37**) by treatment with thionyl chloride in methylene chloride. The chloride salt (**38**) was produced by refluxing a dimethyl-formamide solution of (**37**) for 9 h. Hydrogenation at 40 psi of this salt (**38**) in methanol using PtO_2 as catalyst yielded verruculotoxin (**34**) and its C-10 epimer. Thin layer chromatography on silica gel with chloroform-acetone (1 : 1 v/v) and visualization with ninhydrin showed two spots of which the major (~80%, $R_f 0.64$) was identical with natural verruculotoxin. Recrystallization of the crude reaction mixture from benzene gave pure (**34**) which possessed physical and biological properties identical with natural (**34**). In contrast the C-10 epimer had no observable biological effect at 25 mg/kg (oral, 1-day-old cockerel) dose levels.

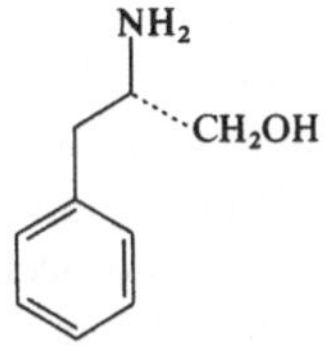

(**35**)

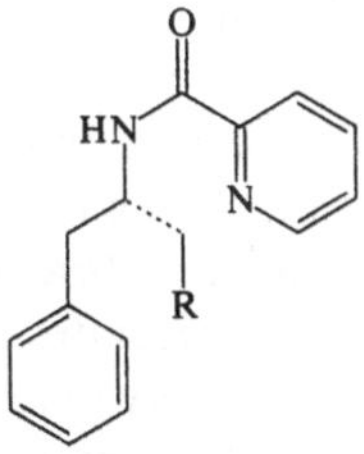

(**36**) R = OH
(**37**) R = Cl

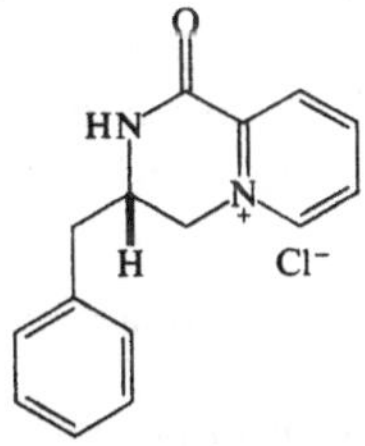

(**38**)

The absolute configuration was established (*96*) by observing that both natural and synthetic verruculotoxin (**34**) has a positive Cotton effect for the 220 nm band ($\theta = +3300$). Since L-phenylalanine was used as a starting material and since epimerization was considered unlikely in the synthesis, the absolute configuration is as shown in (**34**). This is reasonable as verruculotoxin is most probably derived biogenetically from the two L-amino acids, phenylalanine and pipecolinic acid.

6. Tremorgens Containing a 6-Methoxyindole Moiety: The Verruculogens and Fumitremorgins

6.1 Producing Organisms

In 1972 COLE *et al.* (*97*) isolated a new mycotoxin, verruculogen, which caused severe tremors when administered orally to mice or 1-day old cockerels, from a strain of *P. verruculosum* Peyronel isolated from peanuts. Two cultures of *Aspergillus caespitosus,* one isolated from cotton seed and another, NRRL 1929 were found by SCHROEDER *et al.* (*98*) to produce both verruculogen and fumitremorgin B. A fungal isolate obtained from the air in Kochi, Japan, and subsequently identified as *P. paraherquei* Abe *ex* G. Smith also produced verruculogen (*99*). In studies on the role of tremorgenic mycotoxins in ryegrass staggers, LANIGAN *et al.* (*43*) isolated *P. janthinellum* Biourge as the most frequently occurring species. Two tremorgens identified as verruculogen and fumitremorgin were isolated from the mycelial mats of *P. janthinellum.* These two metabolites were also isolated from cultures of *P. piscarium* Westling isolated from ryegrass pastures by GALLAGHER and LATCH (*100*). In a similar study in Australia, *P. paxilli* was isolated from the faeces of cattle in Victoria and from soil from pastures on which an outbreak of ryegrass staggers had occurred in South Australia. Both these isolates of *P. paxilli* produced substantial quantities of verruculogen as the only detected tremorgen (*101*). Strains of *A. fumigatus* isolated from moulded maize silage by COLE *et al.* (*102*) were found to produce both verruculogen and TR-2.

In the course of investigations on toxigenic food-borne fungi, YAMAZAKI *et al.* (*103*) obtained fumitremorgin A and B from a strain of *A. fumigatus* Fres. isolated from rice. Lanosulin, produced by *P. lanosum* Westling (*104*) was found to be identical with fumitremorgin B (*105*).

6.2 Isolation and Chromatography

COLE *et al.* (*97*) examined the production of verruculogen by *P. verruculosum* Peyronel by cultivation on shredded wheat (100 g) and Difco mycological broth (pH 4.8) supplemented with 1.6% yeast extract at 28°C. Cultures were extracted with chloroform and verruculogen was isolated and purified by column chromatography first on silica gel using a linear gradient from toluene to ethyl acetate and then on Florosil with benzene-ethyl acetate (95 : 5 v/v) as eluant. Approximately 2.1 g of verruculogen was recovered from 6.75 kg of culture medium.

Verruculogen was also produced by cultures of *P. paraherquei* at a level of 1 mg/g of dried mycelium when grown on a Czapek-Dox medium (pH 6.8) enriched with 0.5% peptone (*99*).

Cultures of *P. piscarium* Westling when grown on oats in Erlenmeyer flasks incubated at 25°C for 12 to 14 days produce a mixture of verruculogen and fumitremorgin B (*100*). The cultures were extracted with chloroform-methanol (2 : 1 v/v) and the residue obtained after evaporation was purified by multiple column chromatography on silicic acid using first ethyl ether-cyclohexane (3 : 1 v/v) as eluant and subsequently with dichloromethane-acetone (95 : 5 v/v). Both these metabolites were also obtained from cultures of *A. caespitosus* grown on cracked maize at 25°C for 12 to 14 days (*98*).

When *A. fumigatus* Fres. (IFM 4482) was cultured on a liquid medium containing 25 g glucose, 1.6 g ammonium succinate, 0.5 g KH_2PO_4, 0.5 g $MgSO_4$ and 0.1 g yeast extract in 1 000 ml of water and supplemented with 1 ml of a solution of 100 mg $FeSO_4 \cdot 7H_2O$, 15 mg $CuSO_4 \cdot 5H_2O$, 100 mg $ZnSO_4 \cdot 7H_2O$, 10 mg $MgSO_4 \cdot 7H_2O$ and 10 mg $(NH_4)_6Mo_7O_{24} \cdot H_2O$ in 100 ml of water, the presence of fumitremorgins was almost non-detectable. However, 49 mg/l of the fumitremorgins was produced on addition of 250 mg of L-tryptophan to the culture medium (*103*). The fumitremorgins were isolated from the ethyl acetate extract of the cultures by column chromatography on silica gel with benzene-acetone (4 : 1 v/v) (*103*).

6.3 Structure of Verruculogen

Verruculogen (**39**) crystallizes from benzene-ethanol as colourless plates, m.p. 233—235°C (dec.), $[\alpha]_D -27.7°$ (*97*). Elemental analysis (*101*) and high-resolution mass spectral analysis of the molecular ion, m/e 511.236 (*106, 107*) indicated the molecular formula as $C_{27}H_{33}N_3O_7$. The u.v. spectrum is suggestive of a 6-*O*-methylindole with λ_{max} (EtOH) 226 (ϵ 47 500), 277 (ϵ 11 000), and 295 nm (ϵ 9 750) (*97, 98, 99, 106, 107*). The c.d. spectrum showed two Cotton effects at 265 nm ($\Delta\epsilon +0.56$) and 290 nm ($\Delta\epsilon +0.16$), corresponding to two u.v. absorption bands; the third Cotton effect was not observed (*107*). The i.r. spectrum showed absorption bands at 3520 and 3460 (OH and NH), and 1655 cm^{-1} (amide carbonyl) (*97*).

(**39**) R=H
(**40**) R=OAc

Chemical shifts and coupling constants for the resonances in the ^{1}H n.m.r. spectrum (60 MHz, solvent CDCl$_3$) of verruculogen (**39**) have been reported by FAYOS et al. (107) and YOSHIZAWA et al. (99) and assigned as follows (107, 108): δ 1.01 s (3H, 23-H), 1.72 s (6H, 24-H and 28-H), 1.99 s (3H, 29-H), 1.8—2.6 m (6H, 15-H, 16-H, 21-H), 3.61 t (2H, 14-H), 3.82 s (3H, 30-H), 4.13 s (1H, 11-OH), 4.48 m (1H, 17-H), 4.79 d (1H, J 3 Hz, 10-OH), 5.05 d (1H, J 8 Hz, 26-H), 5.64 d (1H, J 3 Hz, 10-H), 6.05 d (1H, J 10 Hz, 20-H), 6.58 d (1H, J 2 Hz, 7-H), 6.63 d (1H, J 8 Hz, 25-H), 6.81 dd (1H, J 9 and 2 Hz, 5-H), 7.89 d (1H, J 9 Hz, 4-H). The spectrum recorded in DMSO-d$_6$ shows four sharp singlets at δ 0.95, 1.58, 1.70, and 1.99. The appearance of the C-20 methine proton as a doublet is consistent with the geometry of the verruculogen molecule as determined by X-ray crystallography (see below and Fig. 21).

The resonances in the ^{13}C n.m.r. spectrum of verruculogen have been assigned by COLE and COX (108) and COLE et al. (109)*.

The structure and relative stereochemistry of verruculogen (**39**) was deduced from a single crystal X-ray diffraction study. Verruculogen crystallizes from a benzene-ethanol solution as the benzene solvate C$_{27}$H$_{33}$N$_3$O$_7$ · C$_6$H$_6$ in the space group $P2_12_12_1$ with $a = 9.88(1)$, $b = 10.86(2)$, and $c = 28.52(3)$ Å. A computer generated perspective drawing of verruculogen is depicted in Fig. 21. The indole portion of the molecule is planar and the diketopiperazine is folded into a boat conformation. The C-10 and C-11 hydroxy groups are cis oriented.

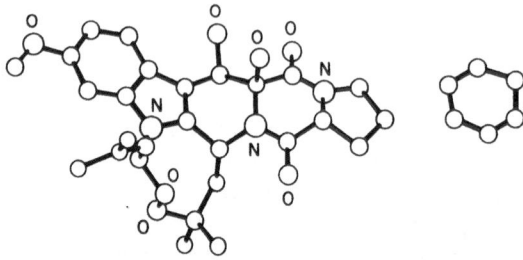

Fig. 21. Perspective drawing of verruculogen (**39**)

A closely related metabolite was recently isolated from the culture broth of P. verruculosum (108). This compound, 15-acetoxyverruculogen (**40**), C$_{29}$H$_{35}$N$_3$O$_9$, m.p. 217—218° C, had spectra similar to those of verruculogen except for data which indicated the presence of an acetoxy group

---

* δ$_C$(CDCl$_3$ at 25 MHz): C-2, 131.6 s; C-3, 105.7 s; C-4, 121.7 d; C-5, 109.3 d; C-6, 156.4 s; C-7, 93.9 d; C-8, 136.3 s; C-9, 121.1 s; C-10, 68.7 d; C-11, 82.6 s; C-12, 170.7 s; C-14, 51.3 t; C-15, 22.7 t; C-16, 29.1 t; C-17, 58.7 d; C-18, 166.2 s; C-20, 48.9 d; C-21, 45.3 t; C-22, 82.1 s; C-23, 27.1 q; C-24, 25.6 q; C-25, 85.8 d; C-26, 118.7 d; C-27, 142.9 s; C-28, 18.8 q; C-29, 24.2 q; C-30, 55.7 q.

in this compound: v_{max} 1735 and 1235 cm^{-1}; m/e 527 $(M^+ - 42)$; δ_H (CDCl$_3$) 2.10 s (3H) and δ_c (CDCl$_3$) 20.9 and 170.0. Hydrolysis of (**40**) afforded hydroxyproline suggesting the position of the acetoxy group in the proline moiety of this compound. The structure and stereochemistry were finally determined by X-ray crystallography (*110*).

Hydrogenation of verruculogen (**39**) in ethanol over palladium on charcoal (5%) afforded two products TR-2 (**41**) and isovaleraldehyde (*99, 106*), the latter being isolated and identified as its 2,4-dinitrophenyl-hydrazone derivative. TR-2 (**41**) crystallises from benzene-ethyl acetate, m.p. 150—152°C and analyses for $C_{22}H_{27}N_3O_6$ (*106, 109*). The major argument for the assignment of structure (**41**) to TR-2 came from comparisons of the ^{1}H and ^{13}C n.m.r. spectra of verruculogen and TR-1 (*109*). TR-2 was subsequently isolated from the culture broth of *A. fumigatus* (*102*).

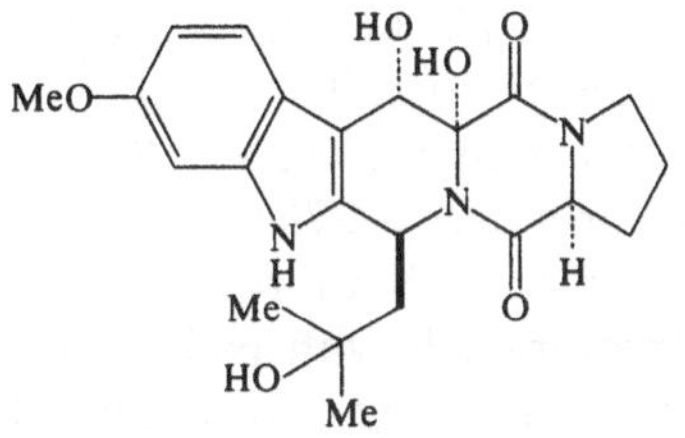

(**41**)

6.4 Structure of the Fumitremorgins

Two tremorgenic mycotoxins fumitremorgin A and B were isolated by YAMAZAKI *et al.* (*103*) from a strain of *A. fumigatus* Fres. found growing on rice. A comparison between lanosulin, produced by *P. lanosum* Westling (*104*), and fumitremorgin B showed that they were identical although an alternative structure had been proposed for the former (*105*).

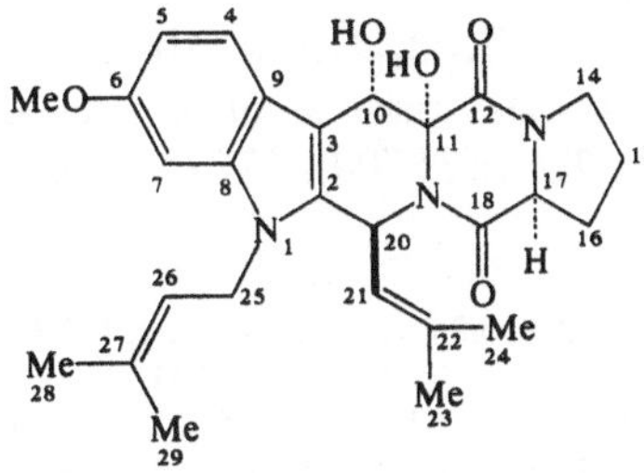

(**42**)

Fumitremorgin B (**42**), $C_{27}H_{33}N_3O_5$, $[\alpha]_D +24°$ (c 0.91, $CHCl_3$) (*104*) has m.p. 211—212°C (from methanol) (*105*). The i.r. spectrum possesses two intense bands at 1685 and 1645 cm^{-1} which suggested the presence of a dioxopiperazine ring in the molecule (*105*). Its u.v. spectrum $[\lambda_{max} 228$ (ε 39 960), 278 (ε 11 000) and 295 nm (ε 12 290)] indicated the presence of an indole nucleus and the ^1H n.m.r. data [δ 7.83 (4-H), 6.74 (5-H) and 6.65 (7-H); $J_{4,7}$ 0.5 Hz, $J_{5,7}$ 2.2 Hz, and $J_{4,5}$ 8.6 Hz] showed that a 5- or 6-methoxyindole was present (*104*). Comparison with the n.m.r. data for 2,3-dimethyl-5- and -6-methoxyindole suggested that the methoxy group was at C-6 and this was confirmed by a nuclear Overhauser effect (n.O.e.) observed for 7-H upon irradiation of the C-25 protons. An n.O.e. was also observed for 5-H and 7-H upon irradiation of the protons of the C-6 methoxy group. The presence of the 3,3-dimethylallyl group at N-1 (*105*) was confirmed by comparing the relevant ^1H n.m.r. data for fumitremorgin B (**42**) [δ 5.04 (26-H), 4.51 (25-H), 1.84 (29-H) and 1.70 (28-H); $J_{25,26}$ 5.1 Hz, $J_{26,28}$ 1.3 Hz, and $J_{26,29}$ 1.0 Hz] with those for N-(3,3-dimethylallyl)-3-methylindole (*104*). The position of the other isopentyl group was established by a signal at δ 5.97 (20-H) which showed vicinal coupling with an olefinic proton at δ 4.76 (*105*). The assignement of the ^{13}C n.m.r. spectrum of fumitremorgin B was reported by COLE and COX (*108*)*.

Reduction of fumitremorgin B with palladium on charcoal gave the 26,27-dihydroderivative, $C_{27}H_{35}N_3O_5$, m.p. 179—180.5°C whereas hydrogenation with PtO_2 led to the formation of a tetrahydro derivative $C_{27}H_{37}N_3O_5$, m.p. 97—100°C (*105*).

Crystals of fumitremorgin B, crystallized from methanol, analyzed for $C_{27}H_{33}N_3O_5 \cdot H_2O$ and are orthorhombic, space group $P2_12_12_1$ with $a = 14.771$, $b = 24.925$, and $c = 7.321$ Å, $Z = 4$. The relative stereochemistry as determined by X-ray crystallography is depicted in Fig. 22 (*111*).

Hydrolysis of fumitremorgin B (**42**) with 6M hydrochloric acid gave proline (*104, 105, 111*) which was purified by paper chromatography using a n-butanol-acetic acid-water (4:1:1 v/v/v) solvent system. The optical rotary dispersion spectrum showed a negative plain curve identical to that obtained for (2S)-proline thereby establishing the absolute configuration shown in Fig. 22 (*111*).

Fumitremorgin A (**43**), $C_{32}H_{41}N_3O_7$, $[\alpha]_D^{10} +61°$ (acetone) crystallizes from methanol as colourless prisms, m.p. 206—209°C (*112*). Its u.v. spectrum, λ_{max} (EtOH) 226 (ε 31 700), 277 (ε 5 300), and 296 nm (ε 4 900),

---

* δ_C(CDCl$_3$, 25 MHz): C-2, 130.8s; C-3, 104.2s; C-4, 121.1d; C-5, 104.0d; C-6, 155.8s; C-7, 93.7d; C-8, 137.6s; C-9, 120.3s; C-10, 68.8d; C-11, 82.8d; C-12, 170.0s; C-14, 48.9t; C-15, 22.5t; C-16, 28.9t; C-17, 58.6d; C-18, 165.8s; C-20, 58.2d; C-21, 122.7d; C-22, 134.8s; C-23, 25.5q; C-24, 18.3q; C-25, 45.1t; C-26, 120.0d; C-27, 134.4s; C-28, 25.5q; C-29, 18.3q; C-30, 55.6q.

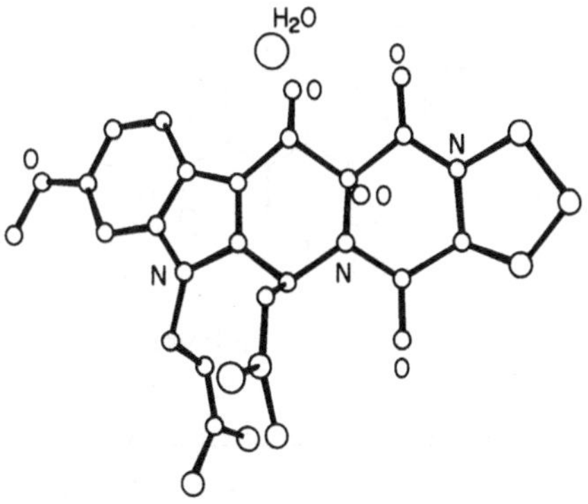

Fig. 22. Perspective drawing of fumitremorgin B (42)

was similar to that of fumitremorgin B (42) and indicated the presence of a 6-methoxyindole moiety (108). This supposition was confirmed by the ^{1}H n.m.r. spectrum of the compound (112). The presence of a 3,3-dimethylallyl ether moiety in fumitremorgin A (43) was deduced from the resonances at δ 1.71 s and 1.81 s (Me$_2$C=C), 5.60 t (C=CH), and 4.71 dd (−OCH$_2$) in the ^{1}H n.m.r. spectrum. The signal at δ 4.48 s, which disappeared on addition of D$_2$O, was assigned to the proton of the C-11 hydroxy group (112).

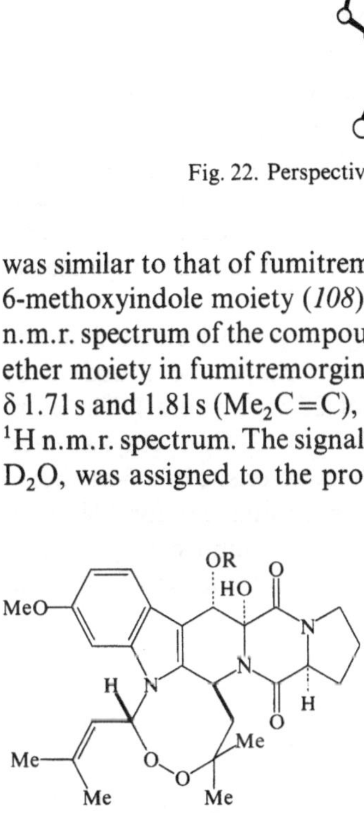

(43) R = —CH$_2$CH=CMe$_2$ 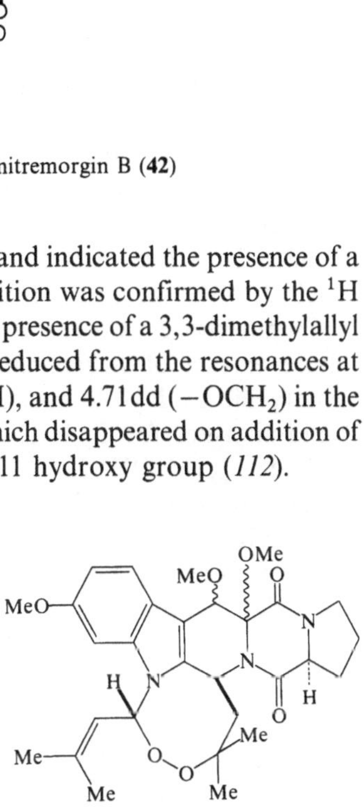   (44)

Treatment of fumitremorgin A (43) with 0.1% sulphuric acid in methanol gave a degradation product (44), C$_{29}$H$_{37}$N$_3$O$_7$. The ^{1}H n.m.r. spectrum of (44) indicated that cleavage of the 3,3-dimethylallyl ether occurs under the reaction conditions as the signals of the protons of this moiety are absent. The two signals at δ 3.16 and 3.38 are assigned to the protons of the two methoxy groups (112).

Hydrogenation of fumitremorgin A (43) over PtO$_2$ in ethyl acetate results in the loss of the substituted 3,3-dimethylallyl moiety present at N-1 by reductive cleavage to give the tertiary alcohol (45). The same cleavage is observed for verruculogen (39) (99, 106).

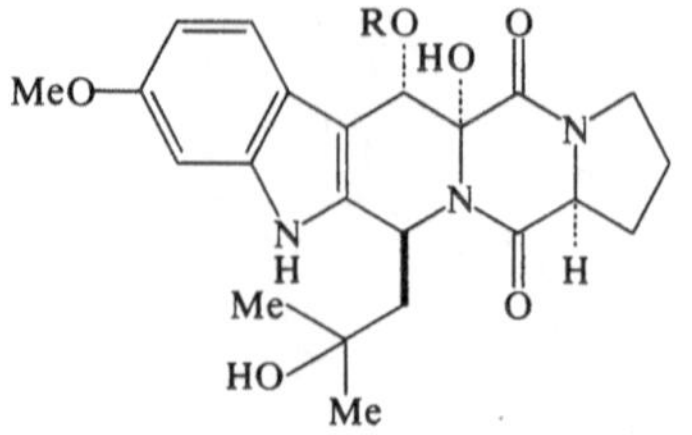

(45) R = CH₂CH₂CHMe₂

The structure and relative stereochemistry of fumitremorgin A as depicted in structure (43) is based on X-ray crystallography (113). Crystals of fumitremorgin A belong to the space group $P2_12_12_1$ with $a = 26.862(2)$, $b = 15.718(2)$, and $c = 7.412(1)$ Å, $Z = 4$. The structure was solved routinely by direct methods and is best described as the verruculogen structure (39) extended by a five carbon mevalonate derived 3,3-dimethylallyl moiety attached to the secondary hydroxy group at C-10.

Fumitremorgin C (46), also called SM-Q (102), was isolated from the culture extract of A. fumigatus obtained from silage and found to be tremorgenic to day-old cockerels at levels of 25 mg/kg (oral). The compound analyzed for $C_{22}H_{25}N_3O_3$, m.p. 125—130° C (from ethyl acetate) and had λ_{max} (MeOH) 224, 272 and 294 nm, characteristic of a 6-methoxy-indole chromophore. The ¹H and ¹³C n.m.r. resonances for (46) were assigned by Cole and Cox (108)*. The structure of fumitremorgin C (46) was determined by X-ray analysis (114). Epoxyfumitremorgin C (47) was also isolated from A. fumigatus (102, 114).

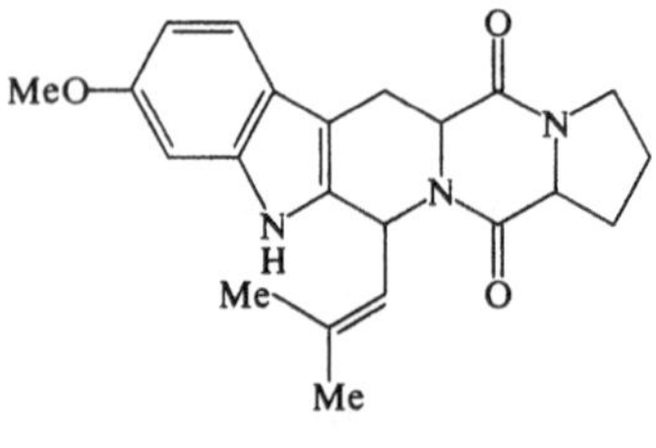

(46)

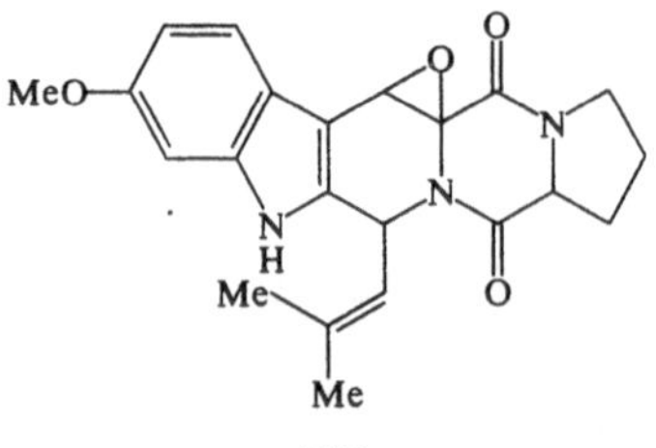

(47)

---

* δ_H: 1.72 s (3H, 24-H), 2.07 s (3H, 23-H), 3.91 s (3H, 25-H), 4.16 m (1H, 17-H), 4.98 d (1H, J 9.0 Hz, 20-H), 6.06 d (1 H, J 9.0 Hz, 21-H), 6.86 dd (1 H, J 9.0 and 2.0 Hz, 5-H), 6.92 d (1 H, J 2.0 Hz, 7-H), , 7.48 d (1 H, J 9.0 Hz, 4-H), and 8.32 s (1H, 1-H).

δ_C: C-2, 132.1 s; C-3, 106.2 s; C-4, 118.8 d; C-5, 109.4 d; C-6, 156.4 s; C-7, 95.2 d; C-8, 136.9 s; C-9, 120.7 s; C-10, 45.4 t; C-11, 51.0 d; C-12, 169.4 s; C-14, 45.4 t; C-15, 23.0 t; C-16, 28.6 t; C-17, 59.2 d; C-18, 165.6 s; C-20, 56.8 d; C-21, 124.1 d; C-22, 133.9 s; C-23, 18.1 q; C-24, 25.7 q; C-25, 55.7 q.

6.5 Biosynthesis of the Fumitremorgins and Verruculogen

When *A. fumigatus* was grown in a synthetic basal medium in stationary culture, the production of fumitremorgins was extremely poor. However, addition of L-tryptophan to the culture medium caused an abundant production of fumitremorgins A (**43**) and B (**42**) (*103*). This result strongly suggested that tryptophan is an efficient precursor of the fumitremorgins and forms the dioxopiperazine ring with proline. In fact both DL-[3-^{14}C]-tryptophan and L-[U-^{14}C]proline as well as (3RS)-[2-^{14}C]mevalonic acid were efficiently incorporated into fumitremorgin A (**43**) and B (**42**) (*115*).

Sodium [1,2-^{13}C$_2$]acetate added to the culture medium of *P. verruculosum* was incorporated into verruculogen (**39**). The proton-decoupled ^{13}C n.m.r. spectrum of the enriched verruculogen indicated the presence of intact acetate units at C-14−C-15, C-17−C-18, C-20−C-21, C-22−C-23, C-25−C-26 and C-27−C-29.

The results indicate that two acetate units were incorporated into the proline moiety by *de novo* synthesis of proline : acetate → tricarboxylic acid cycle → glutamic acid → proline, and that the other four acetate units were incorporated into the two mevalonate-derived prenyl groups in verruculogen (*116*).

7. The Tryptoquivalines

7.1 Producing Organisms

Tryptoquivaline (synonyms: tryptoquivaline A and C) and nortryptoquivalone (synonyms: tryptoquivalone and tryptoquivaline B) are two tremorgenic metabolites produced by cultures of *Aspergillus clavatus* NRRL 5890 collected from a sample of mould damaged rice obtained from a Thai household in which a child died of an unindentified toxicosis (*117, 118*). A number of tryptoquivalines, in addition to fumitremorgin A (**43**) and B (**42**), were isolated by YAMAZAKI *et al.* (*119—122*) from a culture of *A. fumigatus* isolated from rice.

7.2 Isolation and Chromatography

A. fumigatus, strain 0011, was incubated in stationary culture on rice at 25° C for 14 days (*120*). The mouldy rice was extracted with ethyl acetate. The solvent was evaporated and the residue extracted with *n*-hexane. The hexane-insoluble material containing the tryptoquivalines was purified by column chromatography on silica gel using mixtures of benzene and acetone as eluants (120).

An improved procedure for the production of tryptoquivaline and nortryptoquivalone by cultures of A. clavatus growing on pearled barley for 12 days at 30° C in shake culture was developed by DEMAIN et al. (123) and is based on the solid substrate fermentation technique of HESSELTINE (124). The growth medium was extracted with dichloromethane and filtered and the filtrate evaporated in vacuo. After the residue was suspended in petroleum ether the pale yellow precipitate was collected and chromatographed on silica gel with mixtures of hexane-chloroform, chloroform, chloroform-ethyl acetate, and ethyl acetate to give six fractions, fraction I being eluted first. Fraction IV was rechromatographed on silica gel (Woelm) using benzene-acetone (9 : 1 v/v) as eluant to give five fractions (IVA—IVE). Fraction IVA was rechromatographed by high-performance liquid chromatography on Porasil A using hexane − 2-propanol (97 : 3 v/v) as solvent to give tryptoquivaline and nortryptoquivalone (125).

7.3 Structure of Tryptoquivaline A

Tryptoquivaline A (48), crystallizes from dichloromethane-hexane as colourless prisms, m.p. 155—157° C, $[\alpha]_D^{25}$ +130° (c 0.22, CHCl$_3$) (118, 125). High-resolution mass spectral analysis of the molecular ion, m/e 546.2155, gave the molecular formula as $C_{29}H_{30}N_4O_7$. The i.r. spectrum exhibited absorption bands at 3490 (OH), 1786 (lactone CO), 1728 (acetate CO), and 1672 (amide CO) cm^{-1} (125). The u.v. spectrum had λ_{max} (EtOH) 228 (ε 41 200), 232 sh (ε 40 200), 252 sh (ε 19 500), 268 sh (ε 10 900), 279 (ε 9 500), 307 (ε 3 700) and 319 nm (ε 3 000) (125).

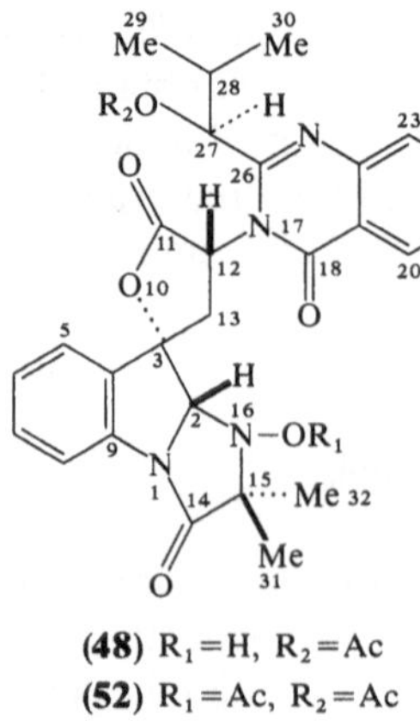

(48) R$_1$=H, R$_2$=Ac
(52) R$_1$=Ac, R$_2$=Ac
(54) R$_1$=R$_2$=H

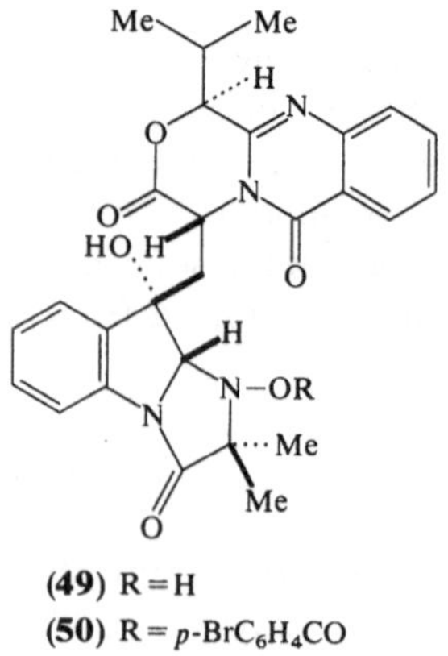

(49) R=H
(50) R= p-BrC$_6$H$_4$CO
(51) R= p-BrC$_6$H$_4$NHCO
(53) R=Ac

Tryptoquivaline A (**48**) gave a positive triphenyltetrazolium chloride (TTC) test suggesting the presence of a hydroxylamine moiety (*118, 125*). Methanolysis of (**48**) with 0.5% HCl at 5°C for 8 h gave a deacetyl product (**49**) which was characterized as both the *p*-bromobenzoate (**50**) and the *p*-bromophenylurethane derivative (**51**). The *p*-bromophenylurethane derivative (**51**) crystallized from a dichloromethane-heptane mixture in the space group $P2_12_12_1$ with cell constants $a = 28.493(3)$, $b = 20.386(2)$, and $c = 6.6377(6)$ Å. The data refined to a final R-factor of 10.5%. A difference Fourier map indicated that a badly disordered heptane molecule existed in the asymmetric unit (*118*).

Comparison of the carbonyl absorptions in the i.r. spectra of tryptoquivaline A (**48**) and its methanolysis product (**49**) (ν_{max} 1765, 1720, 1670 and 1605 cm^{-1}) excluded the possibility that the former was simply the acetate of the latter (*118*). Indeed, acetylation of (**48**) with acetic anhydride in pyridine yielded a diacetate (**52**) (m.p. 171—172°C; ν_{max} 1795, 1735 and 1685 cm^{-1}) while acetylation of (**49**) afforded a monoacetate (**53**) (m.p. 187—189°C; ν_{max} 1765, 1735 and 1685 cm^{-1}). The intermediate hydroxy γ-lactone (**54**) must have undergone acyl transfer to the more stable *trans* disubstituted hydroxy δ-lactone (**49**). Tryptoquivaline A (**48**) thus becomes a spiro-γ-lactone, and related γ-lactones with electron withdrawing α-substituents do indeed exhibit carbonyl absorptions between 1790 and 1800 cm^{-1} (*118*).

Base hydrolysis of tryptoquivaline A (**48**) with 0.2 M sodium carbonate in aqueous dioxane (20°C for 45 min) produced an uncharacterized carboxylate which on acidification was transformed to the hydroxy γ-lactone (**55**) [m.p. 175—178°C, $[\alpha]_D^{25}$ −142° (*c* 1.02, CHCl$_3$)]. Acetylation afforded a diacetate (**56**) (m.p. 144—146°C). The new γ-lactone (**55**) appeared to be a C-12 epitryptoquivaline A derivative and this assumption was verified by formation of the deuteriolactone (**57**) when the hydrolysis was performed in D$_2$O (*118*).

(**55**) R$_1$=R$_2$=R$_3$=H
(**56**) R$_1$=R$_2$=Ac; R$_3$=H
(**57**) R$_1$=R$_2$=H; R$_3$=D

The resonances in the ^{1}H n.m.r. spectrum of tryptoquivaline A (**48**) in CDCl$_3$ have been assigned (*108, 125*): δ 1.01 d (3H, *J* 7 Hz, 30-H), 1.13 d (3H, *J* 7 Hz, 29-H), 1.47 s and 1.49 s (each 3H, 31-H and 32-H), 2.14 s (3H, C-27 OAc), 2.57 m (1H, 28-H), 2.96 dd (1H, *J* 13 and 10 Hz, 13-H), 3.16 dd (1H, *J* 13 and 10 Hz, 13-H), 4.92 s (1H, 2-H), 5.52 d (1H, *J* 9 Hz, 27-H), 5.63 dd (1H, *J* 10 and 10 Hz, 12-H), 7.00—7.78 m (7H, aromatic protons), and 8.67 m (1H, 20-H).

7.4 Structure of Nortryptoquivalone (Tryptoquivaline B or Tryptoquivalone)

Nortryptoquivalone (**58**), isolated as a cometabolite of trypto-quivaline A (**48**) from cultures of *A. clavatus* by Clardy *et al.* (*118*) and Büchi *et al.* (*125*), had m. p. 208—209° C (from dichloromethane-hexane), [α]$_D$ +255° (*c* 0.30, CHCl$_3$) and analyzed for C$_{26}$H$_{24}$N$_4$O$_6$. The metabolite gave positive TTC and 2,4-dinitrophenylhydrazone tests indicative of the presence of a hydroxylamine and carbonyl group. The i.r. spectrum had carbonyl absorption bands at 1783 (spirolactone CO), 1725 (acetate CO), 1705 (ketone CO), and 1672 (amide CO) cm^{-1}. A comparison of the u.v. data for nortryptoquivalone (**58**) and tryptoquivaline A (**48**) suggested that these two tremorgens have the same chromophore (*118, 125*). On the basis of these data and the ^{1}H n.m.r. spectrum of nortryptoquivalone* the structure (**58**) was assigned to this tremorgin (*118, 125*).

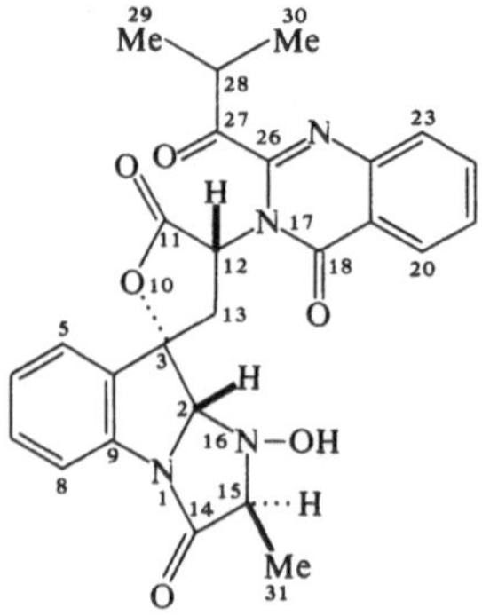

(**58**)

---

* δ$_H$ (CDCl$_3$): 1.27 d (3H, *J* 7 Hz, 29-H), 1.32 d (3H, *J* 7 Hz, 30-H), 1.60 d (3H, *J* 7 Hz, 31-H), 3.10 dd (1H, *J* 13 and 10 Hz, 13-H), 3.48 dd (1H, *J* 13 and 10 Hz, 13-H), 4.13 m (1H, *J* 7 Hz, 28-H), 4.36 q (1H, *J* 7 Hz, 15-H), 5.24 s (1H, 2-H), 5.51 dd (1H, *J* 10 and 10 Hz, 12-H), 7.12—7.95 m (7H, aromatic protons), and 8.52 m (1H, 20-H).

7.5 Structures of Toxic Metabolites Related to Tryptoquivaline A and Nortryptoquivalone

BÜCHI *et al.* (*125*) also isolated, in addition to tryptoquivaline A (**48**) and nortryptoquivalone (**58**), four new tryptoquivaline-related metabolites from cultures of *A. clavatus* which were called nortryptoquivaline (**59**), deoxytryptoquivaline (**60**), deoxynortryptoquivaline (**61**) and deoxy-nortryptoquivalone (**62**). These four metabolites were reported to be toxic but details of their toxicity have not been described. YAMAZAKI *et al.* (*120, 122*) have shown that tryptoquivaline D and N are identical with nortrypto-quivaline (**59**) and deoxynortryptoquivalone (**62**), respectively.

(**59**) R₁=OH, R₂=H
(**60**) R₁=H, R₂=Me
(**61**) R₁=R₂=H

(**62**)

Nortryptoquivaline (**59**), m.p. 256—258°C, analyzes for $C_{28}H_{28}N_4O_7$ and forms an acetate derivative on treatment with acetic anhydride and pyridine at 0°C. The u.v. spectrum of (**59**) was indistinguishable from that of tryptoquivaline A (**48**) and all i.r. absorptions associated with functional groups were identical.

The ^{1}H n.m.r. spectrum of nortryptoquivaline (**59**) was similar to that of tryptoquivaline A (**48**) except that the resonances associated with the geminal dimethyl group in the latter were replaced by those of a secondary methyl function (δ_H: 1.58 d, J 7 Hz, 31-H; 4.28 q, J 7 Hz, 15-H). The presence of a spiro-γ-lactone moiety and the relative stereochemistry of nortryptoquivaline (**59**) was established by X-ray crystallography (*126*). Crystals of nortryptoquivaline formed as large prisms from dichloro-methane-hexane mixtures. Preliminary X-ray experiments indicated that the space group of the crystals was either $P4_12_12$ or $P4_32_12$. Subsequent calculations revealed that the space group $P4_32_12$ was correct with $a = 10.343(2)$, and $c = 49.366(6)$ Å, $Z = 8$ for a calculated density of 1.34 g/ml. Fig. 23 shows a perspective drawing of nortryptoquivaline (**59**)

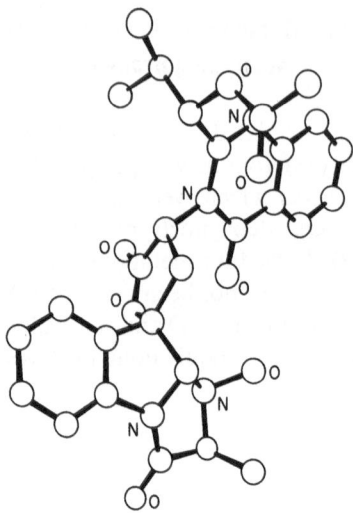

Fig. 23. Perspective drawing of nortryptoquivaline (**59**)

with the relative and correct absolute stereochemistry. The absolute stereochemistry followed from the work of YAMAZAKI *et al.* (*119—121*) on tryptoquivaline D, identical to nortryptoquivaline (**59**), which established the chirality at C-15 as *S* as L-alanine was obtained from (**59**) on reduction with zinc and acetic acid followed by 6 M hydrochloric acid hydrolysis.

Deoxytryptoquivaline (**60**), $C_{29}H_{30}N_4O_6$, differs from tryptoquivaline A (**48**) by the absence of an oxygen atom. The metabolite gives a negative triphenyltetrazolium chloride test for hydroxylamines and is only weakly basic as it was not extracted from organic solvents with 1 M hydrochloric acid. Structure (**60**) was confirmed by oxidation to tryptoquivaline A (**48**) with *m*-chloroperbenzoic acid.

A negative colour test for hydroxylamines suggested that the third metabolite, m.p. 158—160° C, $C_{28}H_{28}N_4O_6$, isolated by BÜCHI *et al.* (*125*) is also a secondary amine. Its oxidation to nortryptoquivaline (**59**) with *m*-chloroperbenzoic acid established the structure as deoxynortryptoquivaline (**61**).

The last of the toxins analyzed for $C_{26}H_{24}N_4O_5$ and was identified as deoxynortryptoquivalone (**62**) because oxidation with *m*-chloroperbenzoic acid formed nortryptoquivalone (**58**).

7.6 Structures of Nontoxic Metabolites Related to Tryptoquivaline A and Nortryptoquivalone

The isolation and structure elucidation of eleven tryptoquivaline-related metabolites from *A. fumigatus*, tryptoquivalines C—J and L—N, have been reported by YAMAZAKI *et al.* (*120—122*).

Tryptoquivaline C analyzed for $C_{29}H_{30}N_4O_7$ and had m.p. 215—217°C, $[\alpha]_D + 168°$ (*120*). Acetylation of tryptoquivaline C afforded an acetate derivative m.p. 194—195°C, $[\alpha]_D + 125°$. The physicochemical properties of tryptoquivaline C are very similar to those of tryptoquivaline A (**48**) although they are not identical (*122*). However, the acetate derivatives of tryptoquivaline C and A (**48**) were found to be identical by direct comparison. Tryptoquivaline C was therefore assumed to be an isomer of tryptoquivaline A (**48**) insofar as the location of the *O*-acetyl group is concerned. However, this assumption is not tenable as the chemical shift and coupling constant for 27-H in tryptoquivaline C (δ 5.61 d, J 9 Hz) is indicative of the presence of a C-27 acetoxy group (*122*). Direct comparison of tryptoquivaline C and A, however, did not clearly confirm the identity of these two metabolites (*122*).

A direct comparison between tryptoquivaline D and nortryptoquivaline (**59**) showed these compounds to be identical (*122*).

Tryptoquivaline E (**63**), $C_{22}H_{18}N_4O_5$, m.p. ~257°C, $[\alpha]_D + 257°$, gave a positive triphenyltetrazolium chloride test for a hydroxylamine moiety (*121*). In the ^{1}H n.m.r. spectrum (solvent pyridine-d_5) the proton of the hydroxylamine function appears as a broad singlet at δ 10.42 which disappears on addition of D_2O to the sample. The assignment of the other resonances is as follows (*121*): δ 1.63 d (3H, J 7 Hz, 31-H), 3.56 dd and 3.38 dd (each 1H, J 13 and 10 Hz, 13-H), 4.26 q (1H, J 7 Hz, 15-H), 5.42 s (1H, 2-H), 6.48 dd (1H, J 10 and 10 Hz, 12-H), 6.83—7.80 m (7H, aromatic protons), 8.17 d (1H, J 8 Hz, 20-H), and 8.66 s (1H, 26-H).

(**63**) $R_1 = OH$, $R_2 = H$
(**64**) $R_1 = OH$, $R_2 = Me$
(**65**) $R_1 = H$, $R_2 = H$

Tryptoquivaline G (**64**) crystallized from acetone as colourless prisms, m.p. 240—241.5°C, $[\alpha]_D + 215°$ and analyzed for $C_{23}H_{20}N_4O_5$ (*121*). The n.m.r. spectrum of tryptoquivaline G closely resembled that of tryptoquivaline E (**63**) but lacked the resonances of the C-15 secondary methyl group. Instead two three-proton singlet resonances were observed at δ 1.50 and 1.62 indicative of a geminal methyl group (*121*).

Tryptoquivaline J (65), m.p. 254—258°C, $[\alpha]_D^{14}$ +135° analyzed for $C_{22}H_{18}N_4O_4$ and gave a negative triphenyltetrazolium chloride test. The signal at δ 3.76 in the ^{1}H n.m.r. spectrum which disappeared on addition of D_2O to the sample, was assigned to the proton of a secondary amine group (121).

Yamazaki et al. (121) found that treatment of tryptoquivalines E (63), G (64), and J (65) with 0.1% potassium hydroxide in methanol led to the formation of tryptoquivalines H, L, and F, respectively, metabolites isolated also from cultures of A. fumigatus. When the reaction was carried out in deuteriated solvents, a single deuterium atom was incorporated in each case at C-12, indicating that epimerization occurs at C-12 with base, converting tryptoquivalines E, G, and J into H, L, and F, respectively (120, 121). The equivalent reaction could occur in vivo in the fungus but the possibility remains that tryptoquivalines H, L, and F are artifacts. It is of interest to note that the optical rotations of tryptoquivalines H, L, and F are opposite (negative) to those of E, G, and J (positive), respectively although their chemical structures, except for the stereochemistry at C-12, are identical.

Tryptoquivaline I (66) analyzed for $C_{27}H_{26}N_4O_6$ and had m.p. 232—235.5°C, $[\alpha]_D^{14}$ +239°. The presence of a hydroxylamine group was inferred from the positive triphenyltetrazolium chloride test. The i.r. spectrum showed carbonyl absorption bands at 1780, 1732, 1710, and 1675 cm^{-1}. The u.v. spectrum closely resembled that of nortryptoquivalone (58). The presence of an isobutyl sidechain in tryptoquivaline I (66) was deduced from the doublet signals (J 7 Hz) at δ 1.22 and 1.28, assigned to two secondary methyl groups, and the quintet (J 7 Hz) at δ 4.07, assigned to the C-28 methine proton. The presence of a C-15 geminal dimethyl group followed from the six-proton singlet at δ 1.49 (121).

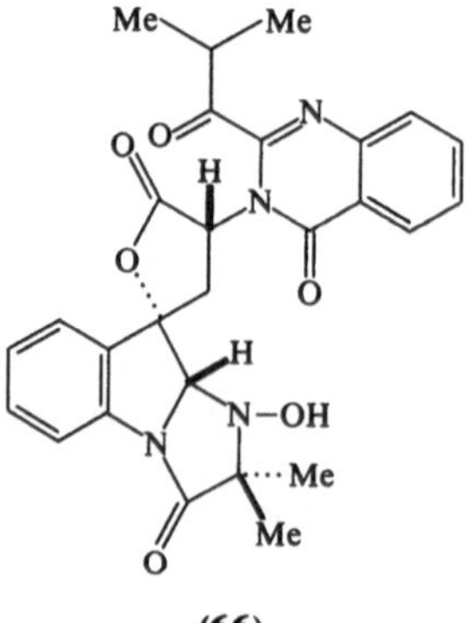

(66)

When nortryptoquivaline (**59**), which has the 12*S* configuration, is treated with 0.1% potassium hydroxide in methanol, a desacetyl product with the 12*R* configuration is obtained. This product was easily acetylated with acetic anhydride and acetic acid to afford a mixture of a monoacetate, $C_{28}H_{28}N_4O_7$, m.p. 135—139° C, $[\alpha]_D^{23}$ − 161°, and a diacetate, $C_{30}H_{30}N_4O_8$, $[\alpha]_D^{21}$ − 148°. The latter compound was identical with tryptoquivaline M acetate obtained from tryptoquivaline M on acetylation. Tryptoquivaline M, $C_{28}H_{28}N_4O_7$, m.p. 157—164° C, $[\alpha]_D^{24}$ − 154°, was isolated from cultures of *A. fumigatus* by YAMAZAKI *et al.* and it is evident from the above evidence that the metabolite is the 12*R* epimer of nortryptoquivaline (**59**) (*122*).

YAMAZAKI *et al.* (*122*) have shown that tryptoquivaline N is identical with deoxynortryptoquivalone (**62**).

7.7 Synthesis of the Tryptoquivalines

7.7.1 Tryptoquivaline A (**48**)

The total synthesis of (+)-tryptoquivaline A (**48**) by a biomimetic oxidative double cyclization was reported by NAKAGAWA *et al.* (*127*) (see Scheme 6). The protected amide (**67**) which served as starting material in the synthesis can be prepared by the reaction of D-tryptophan benzyl ester with isatoic anhydride in benzene (*128*). An alternative method used by BÜCHI *et al.* (*129*) involved condensation of L-tryptophan with o-nitrobenzoyl-chloride to give an amide which was esterified with phenyldiazomethane to the benzyl ester. Reduction of the benzyl ester with iron and hydrochloric acid in ethanol gave (**67**) with the L-configuration.

Condensation of (*S*)-α-acetoxyisovaleraldehyde with the protected amide (**67**) in the presence of *p*-toluenesulphonic acid and 4 Å molecular sieves gave a mixture of two diastereomers (**68**). Acylation of the indole nitrogen in (**68**) with the protected amino acid *N*-trichloroethoxy-carbonylmethylalanine *p*-nitrophenyl ester by the method developed by NAKAGAWA *et al.* (*128*) proceeded in 57% yield to give (**69**) as a mixture of two diastereomers. Subsequent dehydrogenation of (**69**) with DDQ gave the quinazolinone (**70**) (91%) which was quantitatively debenzylated to give acid (**71**). The oxidative double cyclization of (**71**) was executed by addition of *N*-iodosuccinimide to a solution of (**71**) in trifluoroacetic acid to give a mixture of spirolactones (**72**) and (**73**). In preliminary studies on the formation of the imidazoindole spirolactone ring system NAKAGAWA *et al.* (*128*) established the stereochemistry of the product formed in the double cyclization reaction by *X*-ray crystallography: the C-2 protected nitrogen atom and the C-3 oxygen atom have the *cis* orientation.

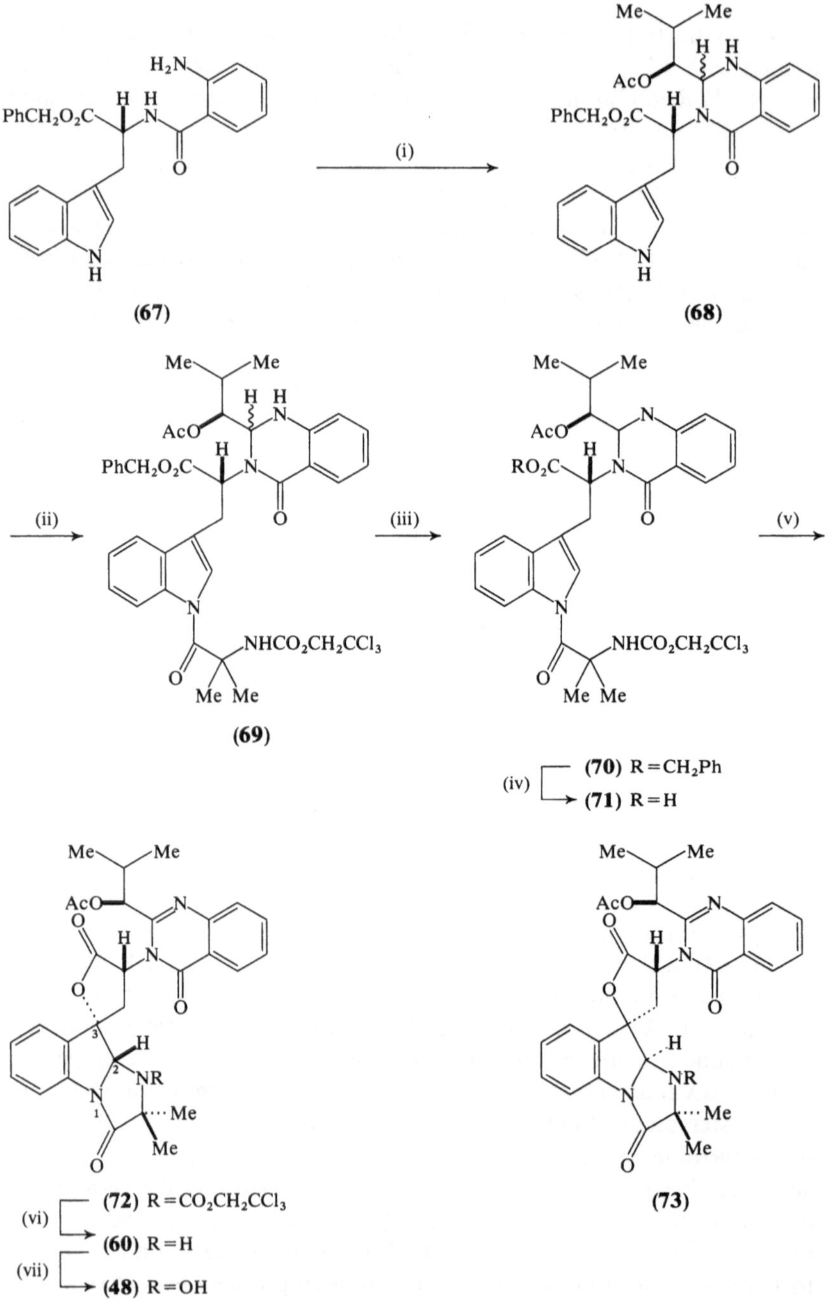

(67)   **(68)**

(69)

(iv)   **(70)** R=CH$_2$Ph

       (71) R=H

(vi)   **(72)** R=CO$_2$CH$_2$CCl$_3$

       (60) R=H

(vii)   **(48)** R=OH

               (73)

Scheme 6. Synthesis of tryptoquivaline A (**48**): (i) p-TsOH,(S)-α-acetoxyisovaleraldehyde; (ii) CCl$_3$CH$_2$O$_2$CNHCMe$_2$CO$_2$C$_6$H$_4$-p-NO$_2$; (iii) DDQ; (iv) Pd-C, H$_2$; (v) N-iodo-succinimide, CF$_3$CO$_2$H; (vi) Zn, HOAc; (vii) m-ClC$_6$H$_4$CO$_3$H

Deprotection of the spirolactone (72) with zinc in acetic acid provided the amine deoxatryptoquivaline (60) (66%) which was converted to the hydroxylamine (93%), identical with (+)-tryptoquivaline A (48), by oxidation with m-chloroperbenzoic acid.

7.7.2 Tryptoquivaline G (64)

The first total synthesis of tryptoquivaline G (64) was achieved by Büchi et al. (129). The total synthesis confirmed the proposed structure (121) and established both the relative and absolute configuration. The synthesis relied on a new method for the oxidative cyclization of N-acyltryptophans to spirolactones and the steric course of the reaction was explored with a model compound. Oxidation of N-phthalimido-L-tryptophan with 2 equivalents of trichloromethanesulphonylchloride - dimethylsulphoxide in dichloromethane at $-20°$ C gave a 65% yield of two diastereomeric lactones (74) and (75) in a ratio of 7 : 3. The major isomer (74), $[\alpha]_D^{25} -133°$ was recovered unchanged after treatment with sodium hydride or imidazole in dimethylformamide. The minor isomer (75), $[\alpha]_D -206°$ under the same conditions was converted into the enantiomer of (74), $[\alpha]_D^{25} +132°$, demonstrating different configurations at the spiro carbon atom in the two diastereomers. Since the epimerization of tryptoquivaline G (64) to tryptoquivaline L is accompanied by a large negative shift in optical rotation, the absolute configurations at C-3 and C-12 should be opposite to those of model compound (75). The major, and thermodynamically more stable, epimer (74) formed in the oxidative lactonization thus has the correct stereochemistry at the spiro centre (129).

(74)   (75)

The enantiomer of the protected amide (67) (Scheme 7) was treated with formic acid in benzene to give a formamide which on dehydration with p-toluenesulphonic acid gave the quinazoline (76). Hydrogenolysis of (76) over palladium on charcoal in ethanol gave the free acid (77). Oxidative cyclization of (77) with methanesulphonic anhydride - dimethyl sulphoxide

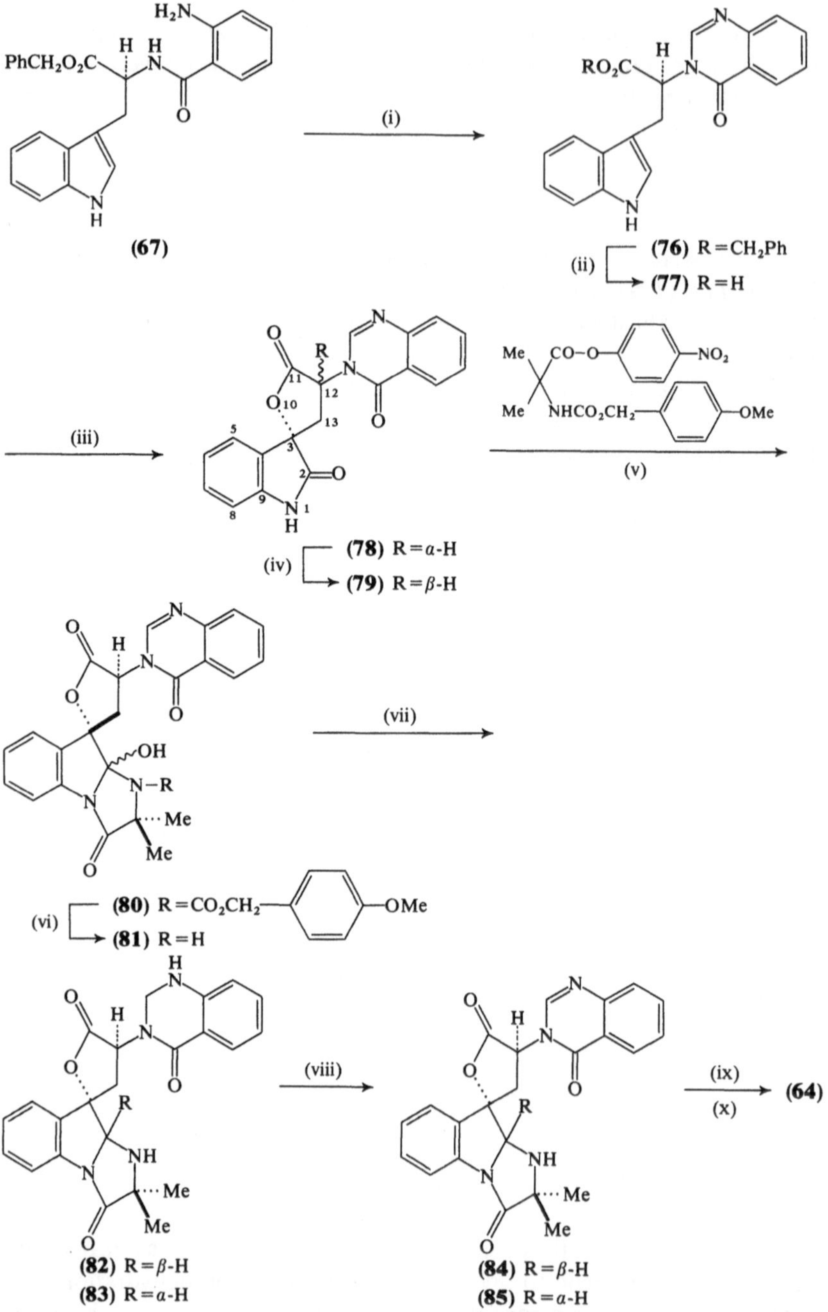

Scheme 7. Synthesis of tryptoquivaline G (**64**): (i) HCO₂H; dehydration with *p*-TsOH; (ii) Pd-C, H₂; (iii) (MeSO)₂O–Me₂SO; (iv) KH; (v) MeCON(SiMe₃)₂ (vi); CF₃CO₂H; (vii) NaBH₃CN; (viii) DDQ; (ix) *m*-ClC₆H₄CO₃H; (x) KH

in dichloromethane gave the spirolactone (78), $[\alpha]_D^{25}$ $-377°$ in addition to < 10% of the epimer differing in configuration at the spiro centre. To confirm the stereochemistry (78) was epimerized with potassium hydride to give the C-12 epimer (79), $[\alpha]_D^{25}$ $+320°$, which is the enantiomer of the minor product formed in the oxidative lactonization. Owing to the exceptional lability of (79) to base the synthesis was continued with (78).

The spirolactone (78) was silylated with bis(trimethylsilyl)acetamide and the crude product condensed with the p-nitrophenol ester of the N-(p-methoxybenzyloxycarbonyl)methylalanine to give the cyclol (80). The p-methoxybenzyloxycarbonyl protecting group was removed by treatment of (80) with trifluoroacetic acid and the deprotected cyclol (81) was reduced with sodium cyanoborohydride to give a 4:1 mixture of the dihydro-quinazolinones (82) and (83). Reoxidation to the corresponding quinaz-olinones (84) and (85), respectively was accomplished with DDQ in chloroform. The ^{1}H n.m.r. spectra of the two isomers were used to establish the configurations at C-2 and the major isomer (84) on oxidation with m-chloroperbenzoic acid gave tryptoquivaline L with spectral properties in accord with those published by YAMAZAKI et al. (122). Contra-thermodynamic epimerization, as in the conversion of (78) into (79) gave tryptoquivaline G (64) identical with material isolated from A. fumigatus (121). The tryptoquivalines are thus derived from D-tryptophan.

A total synthesis of (+)-tryptoquivaline G (64) using a biomimetic double cyclization was reported by NAKAGAWA et al. (128). The protected amide (67) when heated in benzene with triethylorthoformate in the presence of a catalytic amount of p-toluenesulphonic acid gave the quinazolinone (76). The subsequent steps in the elaboration of the quinazolinone (86) to tryptoquivaline G were analogous to those described for tryptoquivaline A (see earlier).

A formal synthesis of (±)-tryptoquivaline G has been described by OHNUMA et al. (130).

7.8 Biosynthesis of the Tryptoquivalines

The novel structure of the tryptoquivalines may be biogenetically derived from four amino acids: tryptophan, anthranilic acid, valine and alanine. Deoxynortryptoquivaline (62) could be the first compound formed in the tryptoquivaline biosynthesis. Oxidation of the secondary amine to a hydroxylamine gives nortryptoquivalone (58). Subsequent reduction of the side-chain carbonyl group would lead to the formation of nortrypto-quivaline (59). The geminal dimethyl group at C-15 may be formed by the incorporation of a C_1-unit into deoxynortryptoquivalone (62) or by the direct participation of methylalanine instead of alanine in the first step of the biosynthesis. The actual incorporation of ^{14}C-labelled methylalanine into tryptoquivaline A (**48**) has been demonstrated by Yamazaki (*131*).

References

1. Steyn, P. S. (ed.): The Biosynthesis of Mycotoxins — a Study in Secondary Metabolism. New York: Academic Press. 1980.
2. Rodricks, J. V., C. W. Hesseltine, and M. A. Mehlman (eds.): Mycotoxins in Human and Animal Health. Park Forest South, Illinois: Pathotox Publishers Inc. 1977.
3. Betina, V. (ed.): Mycotoxins — Production, Isolation, Separation and Purification. Amsterdam: Elsevier. 1984.
4. Floss, H. G., and J. A. Anderson: Biosynthesis of Ergot Toxins. In: The Biosynthesis of Mycotoxins — a Study in Secondary Metabolism (P. S. Steyn, ed.), 17. New York: Academic Press. 1980.
5. Sakabe, N., T. Goto, and Y. Hirata: The Structure of Citreoviridin, a Toxic Compound Produced by *P. citreoviride* Molded in Rice. Tetrahedron Lett. 1825 (1964).
6. Steyn, P. S., and R. Vleggaar: Biosynthesis of the Aurovertins B and D. The Role of Methionine and Propionate in the Simultaneous Operation of Two Independent Biosynthetic Pathways. J. Chem. Soc. Perkin Trans. I, 1298 (1981).
7. Kruger, G. J., P. S. Steyn, and R. Vleggaar: X-Ray Crystal Structure of Asteltoxin, a Novel Mycotoxin from *Aspergillus stellatus* Curzi. J. Chem. Soc. Chem. Commun. 441 (1979).
8. Burka, L. T., M. Ganguli, and B. J. Wilson: Verrucosidin, a Tremorgen from *Penicillium verrucosum* var. *cyclopium*. J. Chem. Soc. Chem. Commun. 544 (1983).
9. Heu, T. H., C. K. Yang, K. H. Ling, C. J. Wang, and C. P. Tang: (4aR,6aR,12aS,12bS)-4a,6,6a,12,12,12b-hexahydro-4a,12a,dihydroxy-4,4,6a,12b-tetramethyl-9-(3,4,5-trimethoxyphenyl)-4H, 11H-naphtho (2,1-b) pyrano (3,4-*e*) pyran-1,11(5H)-dione, territrem B, $C_{29}H_{34}O_9$. Cryst. Struct. Comm. **11**, 199 (1982).
10. Cole, R. J.: Tremorgenic Mycotoxins: An Update. In: Antinutrients and Natural Toxicants in Foods (R. L. Ory, ed.), 17. Westport, CT, Food and Nutrion Press. 1980.
11. — Tremorgenic Mycotoxins. In: Mycotoxins in Human and Animal Health (J. V. Rodricks, C. W. Hesseltine, and M. A. Mehlman, eds.), 583. Park Forest South, Illinois: Pathotox Publishers Inc. 1977.
12. — Fungal Tremorgens. J. Food Protec. **44**, 715 (1981).
13. Ciegler, A., R. F. Vesonder, and R. J. Cole: Tremorgenic Mycotoxins. In: Mycotoxins and Other Fungal Related Food Problems. Adv. Chem. Series. **49**, 163, American Chemical Society, Washington, D. C., 1976.
14. Mantle, P. G., and R. H. C. Penny: Tremorgenic Mycotoxins and Neurological Disorders — A Review. The Veterinary Journal. **21**, 51 (1981).
15. Shreeve, B. J., D. S. P. Patterson, B. A. Roberts, and S. M. Macdonald: Tremorgenic Fungal Toxins. Vet. Res. Commun. **7**, 155 (1983).
16. Betina, V.: Indole-derived Tremorgenic Toxins. In: Mycotoxins, Production, Isolation, Separation and Purification (V. Betina, ed.), 415. Amsterdam: Elsevier. 1984.
17. Cysewski, S. J.: Chemistry of the Tremorgenic Mycotoxins (T. D. Wylie and L. G. Morehose, eds.), 357. New York: Marcel Dekker Inc. 1977.
18. Wilson, B. J., C. H. Wilson, and A. W. Hayes: Tremorgenic Toxin from *Penicillium cyclopium* grown in Food Materials. Nature **220**, 77 (1968).
19. Ciegler, A.: A Tremorgenic Toxin from *Penicillium palitans*. Appl. Microbiol. **18**, 128 (1969).

20. WILSON, B. J., and C. H. WILSON: Toxin from *Aspergillus flavus:* Production and Food Materials of a Substance Causing Tremors in Mice. Science **144**, 177 (1964).

21. HOU, C T., A. CIEGLER, and C. W. HESSELTINE: Tremorgenic Toxins from Penicillia. A New Tremorgenic Toxin, Tremortin B, from *Penicillium palitans*. Can. J. Microbiol. **17**, 599 (1971).

22. CIEGLER, A., and J. I. PITT: Survey of the Genus Penicillium for Tremorgenic Toxin Production. Mycopathol. Mycol. Appl. **42**, 119 (1970).

23. PITT, J. I.: *Penicillium crustosum* and *P. simplicissimum,* the Correct Names of Two Common Species Producing Tremorgenic Mycotoxins. Mycologia **71**, 1166 (1979).

24. FRISVAD, J. C.: Physiological Criteria and Mycotoxin Production as Aids in Identification of Common Asymmetric Penicillia. Appl. Environ. Microbiol. **41**, 568 (1981).

25. DE JESUS, A. E., P. S. STEYN, F. R. VAN HEERDEN, R. VLEGGAAR, P. L. WESSELS, and W. E. HULL: Structure and Biosynthesis of the Penitrems A − F. Six Novel Tremorgenic Mycotoxins from *Penicillium crustosum*. J. Chem. Soc. Chem. Commun. 289 (1981).

26. DE JESUS, A. E., W. E. HULL, P. S. STEYN, F. R. VAN HEERDEN, R. VLEGGAAR, and P. L. WESSELS: High-field ^{13}C N.M.R. Evidence for the Formation of [1,2-^{13}C]acetate from [2-^{13}C]acetate During the Biosynthesis of Penitrem A by *Penicillium crustosum*. J. Chem. Soc. Chem. Commun. 837 (1982).

27. DE JESUS, A. E., P. S. STEYN, F. R. VAN HEERDEN, R. VLEGGAAR, P. L. WESSELS, and W. E. HULL: Tremorgenic Mycotoxins from *Penicillium crustosum:* Isolation of Penitrems A − F and the Structure Elucidation and Absolute Configuration of Penitrem A. J. Chem. Soc. Perkin Trans. I. 1847 (1983).

28. − − − − − − Tremorgenic Mycotoxins from *Penicillium crustosum*. Structure Elucidation and Absolute Configuration of Penitrems B − F. J. Chem. Soc. Perkin Trans. I. 1857 (1983).

29. DE JESUS, A. E., C. P. GORST-ALLMAN, P. S. STEYN, F. R. VAN HEERDEN, R. VLEGGAAR, P. L. WESSELS, and W. E. HULL: Tremorgenic Mycotoxins from *Penicillium crustosum*. Biosynthesis of Penitrem A. J. Chem. Soc. Perkin Trans. I. 1863 (1983).

30. PATTERSON, D. S. P., B. A. ROBERTS, B. J. SHREEVE, S. M. MACDONALD, and A. W. HAYES: Tremorgenic Toxins Produced by Soil Fungi. Appl. Environ. Microbiol. **37**, 172 (1979).

31. WAGENER, R. E., N. D. DAVIS, and U. L. DIENER: Penitrem A and Roquefortine Production by *Penicillium commune*. Appl. Environ. Microbiol. **39**, 882 (1980).

32. VESONDER, R. F., L. TJARKS, W. ROHWEDDER, and D. O. KIESWETTER: Indole Metabolites from *Penicillium cyclopium* NRRL 6093. Experientia **36**, 1303 (1980).

33. KYRIAKIDIS, N., E. S. WAIGHT, J. B. DAY, and P. G. MANTLE: Novel Metabolites from *Penicillium crustosum,* including Penitrem E, a Tremorgenic Mycotoxin. Appl. Environ. Microbiol. **42**, 61 (1981).

34. DORNER, J. W., R. J. COLE, and R. A. HILL: Tremorgenic Mycotoxins Produced by *Aspergillus fumigatus* and *Penicillium crustosum* from Molded Corn Implicated in a Natural Intoxication of Cattle. J. Agric. Food Chem. **32**, 411 (1984).

35. GORST-ALLMAN, C. P., and P. S. STEYN: Screening Methods for the Detection of Thirteen Common Mycotoxins. J. Chromatogr. **175**, 325 (1979).

36. GIMENO, A.: Thin Layer Chromatographic Determination of Aflatoxins, Ochratoxins, Sterigmatocystin, Zearalenone, Citrinin, T-2 Toxin, Diacetoxyscirpenol, Penicillic Acid, Patulin, and Penitrem. J. Assoc. Off. Anal. Chem. **62**, 579 (1979).

37. MAES, C. M., P. S. STEYN, and F. R. VAN HEERDEN: High-performance Liquid Chromatography and Thin-layer Chromatography of Penitrems A − F, Tremorgenic Metabolites from *Penicillium crustosum*. J. Chromatogr. **234**, 489 (1982).

38. NEWMARK, R. A., and J. R. HILL: Use of ^{13}C Isotope Shifts to Assign Carbon Atoms α and β (and *para*) to a Hydroxy Group in Alkyl Substituted Phenols and Alcohols. Org. Magn. Reson. **13**, 40 (1980).

39. BEGTRUP, M., R. M. CLARAMUNT, and J. ELGUERO: Carbon-13 Nuclear Magnetic Resonance Study of *N*-methyl and *N*-acetyl Derivatives of Azoles and Benzazoles. J. Chem. Soc. Perkin Trans. II. 99 (1978).

40. FELLOWS, P. A., N. KYRIAKIDIS, P. G. MANTLE, and E. S. WAIGHT: Electron Impact Mass Spectra of Penitrem A, Some Derivatives and Its Analogues. Org. Mass Spec. **16**, 403 (1981).

41. SPRINGER, J. P., and J. CLARDY: Paspaline and Paspalicine, Two Indolemevalonate Metabolites from *Claviceps paspali*. Tetrahedron Lett. 231 (1980).

42. HOREAU, A.: Principe et Applications d'une Nouvelle Méthode de Détermination des Configurations dite "Par Dédoublement Partiel". Tetrahedron Lett. 506 (1961).

43. LANIGAN, G. W., A. L. PAYNE, and P. A. COCKRUM: Production of Tremorgenic Toxins by *Penicillium janthinellum* Biourge: a Possible Aetiological Factor in Ryegrass Staggers. AJEBAK **57**, 31 (1979).

44. GALLAGHER, R. T., G. C. M. LATCH, and R. K. KEOGH: The Janthitrems: Fluorescent Tremorgenic Toxins Produced by *Penicillium janthinellum* Isolates from Ryegrass Pastures. Appl. Environ. Microbiol. **39**, 272 (1980).

45. DE JESUS, A. E., P. S. STEYN, F. R. VAN HEERDEN, and R. VLEGGAAR: Structure Elucidation of the Janthitrems, Novel Tremorgenic Mycotoxins from *Penicillium janthinellum*. J. Chem. Soc. Perkin Trans. I. 697 (1984).

46. LAUREN, D. S., and R. T. GALLAGHER: High-performance Liquid Chromatography of the Janthitrems: Fluorescent Tremorgenic Mycotoxins produced by *Penicillium janthinellum*. J. Chromatogr. **150**, 248 (1982).

47. PACHLER, K. G. R., and P. L. WESSELS: Sensitivity Gain in a Progressive-saturation Selective Population Inversion NMR Experiment. J. Magn. Reson. **28**, 53 (1977).

48. MORTIMER, P. H.: Perennial Ryegrass Staggers in New Zealand. In: Effects of Poisonous Plants on Livestock (R. F. KEELER, K. R. VAN KAMPEN, and L. F. JAMES, eds.), 353. New York: Academic Press. 1978.

49. LATCH, G. C. M.: Endophytes and Ryegrass Staggers. International Mycotoxin Symposium, Abstract 66, Sydney, Australia, August 1983.

50. GALLAGHER, R. T., A. D. HAWKES, P. S. STEYN, and R. VLEGGAAR: Tremorgenic Neurotoxins from Perennial Ryegrass Causing Ryegrass Staggers Disorder of Livestock: Structure Elucidation of lolitrem B. J. Chem. Soc. Chem. Commun. 614 (1984).

51. GALLAGHER, R. T., J. CLARDY, and B. J. WILSON: Aflatrem, a Tremorgenic Toxin from *Aspergillus flavus*. Tetrahedron Lett. 239 (1980).

52. DODDRELL, D. M., D. T. PEGG, and M. R. BENDALL: Distortionless Enhancement of NMR Signals by Polarization Transfer. J. Magn. Reson. **48**, 323 (1982).

53. BAX, A., and G. A. MORRIS: Am Improved Method for Heteronuclear Chemical Shift Correlation by Two-dimensional NMR. J. Magn. Reson. **42**, 501 (1981).

54. GALLAGHER, R T., and B. J. WILSON: Aflatrem, the Tremorgenic Mycotoxin from *Aspergillus flavus*. Mycopathol. **66**, 183 (1978).

55. LUK, K. C., B. KOBBE, and J. M. TOWNSEND: Production of Cyclopiazonic Acid by *Aspergillus flavus* Link. Appl. Environ. Microbiol. **33**, 212 (1977).

56. GALLAGHER, R. T., J. CLARDY, and B. J. WILSON: Aflatrem, a Tremorgenic Toxin from *Aspergillus flavus*. Tetrahedron Lett. 239 (1980).

57. COLE, R. J., J. W. DORNER, J. P. SPRINGER, and R. H. COX: Indole Metabolites from a Strain of *Aspergillus flavus*. J. Agric. Food. Chem. **29**, 293 (1981).

58. GALLAGHER, R. T., T. McCABE, K. HIROTSU, and J. CLARDY: Aflavinine, a Novel Indole-mevalonate Metabolite from Tremorgen-producing *Aspergillus flavus* Species. Tetrahedron Lett. 243 (1980).

59. COLE, R. J., J. W. KIRKSEY, and J. M. WELLS: A New Tremorgenic Metabolite from *Penicillium paxilli*. Can. J. Microbiol. **20**, 1159 (1974).

60. SPRINGER, J. P., J. CLARDY, J. M. WELLS, R. J. COLE, and J. W. KIRKSEY: The Structure of Paxilline, a Tremorgenic Metabolite of *Penicillium paxilli* Bainier. Tetrahedron Lett. 2531 (1975).

61. GERMAIN, G., P. MAIN, and M. M. WOOLFSON: On the Application of Phase Relationships to Complex Structures. II. Getting a Good Start. Acta Cryst. **B 24**, 274 (1970).

62. FEHR, TH., and W. ACKLIN: Die Isolierung zweier neuartiger Indol-Derivate aus dem Mycel von *Claviceps paspali* Stevens et Hall. Helv. Chim. Acta **49**, 1907 (1966).

63. GYSI, P. G.: Die Konstitution des Paspalins. Dissertation ETH Nr. 4990 (1973).

64. LEUTWEILER, A.: Die Konstitution des Paspalicins. Dissertation ETH Nr. 5163 (1973).

65. COLE, R. J., J. W. DORNER, J. A. LAMSDEN, R. H. COX, C. PAPE, B. CUNFER, S. S. NICHOLSON, and D. M. BEDELL: Paspalum Staggers: Isolation and Identification of Tremorgenic Metabolites from Sclerotia of *Claviceps paspali*. Agric. Food Chem. **25**, 1197 (1977).

66. GALLAGHER, R. T., J. FINER, J. CLARDY, A. LEUTWEILER, F. WEIBEL, W. ACKLIN, and D. ARIGONI: Paspalinine, a Tremorgenic Metabolite from *Claviceps paspali* Stevens et Hall. Tetrahedron Lett. 235 (1980).

67. STEYN, P. S., F. R. VAN HEERDEN, and R. VLEGGAAR: Unpublished results (1984).

68. WEIBEL, F.: Diss. ETH Zürich.

69. ACKLIN, W., F. WEIBEL, and D. ARIGONI: Zur Biosynthese von Paspalin und verwandten Metaboliten aus *Claviceps paspali*. Chimia **31**, 63 (1977).

70. TANABE, M.: Abstract 8C 206, 26th IUPAC Symposium, Tokyo 1977.

71. PETERSON, D. W., R. H. C. PENNY, J. B. DAY, and P. G. MANTLE: A Comparative Study of Sheep and Pigs given the Tremorgenic Mycotoxins Verrucologen and Penitrem A. Res. Vet. Sci. **33**, 183 (1977).

72. HAYES, A. W., and R. D. HOOD: Effects of Prenatal Administration of Penicillic Acid and Penitrem A to Mice. Toxicon **16**, 92 (1978).

73. HAYES, A. W., D. B. PRESLEY, and J. A. NEVILLE: Acute Toxicity of Penitrem A in Dogs. Toxicol. Appl. Pharmacol. **35**, 311 (1976).

74. HAYES, A. W., R. D. PHILLIPS, and L. C. WALLACE: Effect of Penitrem A on Mouse Liver Composition. Toxicon **15**, 293 (1977).

75. SOBOTKA, T. J., R. E. BRODIE, and S. L. SPAID: Neurobehavioral Studies of Tremorgenic Mycotoxins Verrucologen and Penitrem A. Pharmacol. **16**, 287 (1978).

76. MANTLE, P. G., P. H. MORTIMER, and E. P. WHITE: Mycotoxic Tremorgens of *Claviceps paspali* and *Penicillium cyclopium*: A Comparative Study of Effects on Sheep and Cattle in Relation to Natural Staggers Syndromes. Res. Vet. Sci. **24**, 49 (1977).

77. STERN, P.: Pharmacological Analysis of the Tremor induced by Cyclopium Toxin. Jugoslav. Physiol. Pharmacol. Acta **7**, 187 (1971).

78. NORRIS, P. J., C. C. T. SMITH, J. DE BELLEROCHE, H. F. BRADFORD, P. G. MANTLE, A. J. THOMAS, and R. H. C. PENNY: Action of Tremorgenic Fungal Toxins on Neurotransmitter Release. J. Neurochem. **34**, 33 (1980).

79. GALLAGHER, R. T.: Neurotoxins and Ryegrass Staggers. Proceedings of the VIIIth Congress International Society for Human and Animal Mycology (M. BAXTER, ed.). Massey University, New Zealand. 1982.

80. CUNNINGHAM, I. J., and W. J. HARTLEY: Ryegrass Staggers. N. Z. Vet. J. **7**, 1 (1959).

81. SHREEVE, B. J., D. S. P. PATTERSON, B. A. ROBERTS, S. M. MACDONALD, and E. N. WOOD: Isolation of Potentially Tremorgenic Fungi from Pasture associated with a Condition Resembling Ryegrass Staggers. Vet. Res. **103**, 209 (1978).

82. GALLAGHER, R. T., A. G. CAMPBELL, A. D. HAWKES, P. T. HOLLAND, D. A. McGAVESTON, E. A. PANSIER, and I. C. HARVEY: Ryegrass Staggers: The Presence of Lolitrem Neurotoxins in Perennial Ryegrass Seed. N. Z. Vet. J. **30**, 183 (1982).

83. Gallagher, R. T., G. S. Smith, M. E. Di Menna, and P. W. Young: Some Observations on Neurotoxin Production in Perennial Ryegrass. N. Z. Vet. J. **30**, 203 (1982).

84. Di Menna, M. E., and P. G. Mantle: The Role of Penicillia in Ryegrass Staggers. Res. Vet. Sci. **24**, 347 (1978).

85. Porter, J. K., C. W. Bacon, and J. D. Robbins: Major Alkaloids of a *Claviceps* Isolated from Toxic Bermuda Grass. J. Agric. Food Chem. **22**, 838 (1974).

86. Tung, S. S., K. H. Ling, S. E. Tsai, C. H. Chung, J. J. Wang, and T. C. Tung: Study on Fungi of the Stored, Unhulled Rice of Taiwan. (1) Mycological Survey of the Stored Unhulled Rice. J. Formosan Med. Assoc. **70**, 251 (1971).

87. Chung, C. H., K. H. Ling, S. S. Tung, and T. C. Tung: Study on Fungi of the Stored Unhulled Rice of Taiwan. (2) Aflatoxin B_1 like Compounds from the Culture of *Aspergillus* Genus. J. Formosan Med. Assoc. **70**, 258 (1971).

88. Ling, K. H., and M. T. Huang: Study on Mycotoxins Contaminated in Food in Taiwan. (1) Study on Pseudo-aflatoxin B_2 from *Aspergillus terreus*. Proc. Natl. Sci. Council, Republic of China, no. 8, part 2, 65 (1975).

89. Ling, K. H.: Study on Mycotoxins contaminated in Food in Taiwan. (2) Tremor inducing Compounds from *Aspergillus terreus*. Proc. Natl. Sci. Council, Republic of China, no. 9, part 2, 121 (1976).

90. Ling, K. H., C. Yang, and F. Peng: Territrems, Tremorgenic Mycotoxins of *Aspergillus terreus*. Appl. Environ. Microbiol. **37**, 365 (1979).

91. Ling, K. H., C. Yang, C. Kuo, and M. Kuo: Solvent Systems for Improved Isolation and Separation of Territrems A and B. Appl. Environ. Microbiol. **44**, 860 (1982).

92. Ling, K. H., H. Lion, C. Yang, and C. Yang: Isolation, Chemical Structure, Acute Toxicity and some Physicochemical Properties of Territrem C from *Aspergillus terreus*. Appl. Environ. Microbiol. **47**, 98 (1984).

93. Ling, K. H., C. Yang, and H. Huang: Differentiation of Aflatoxins from Territrems. Appl. Environ. Microbiol. **37**, 358 (1979).

94. Wilson, B. J., C. S. Byerly, and L. T. Burka: Neurologic Disease of Fungal Origin in Three Herds of Cattle. J. Am. Vet. Med. Assoc. **179**, 480 (1981).

95. Cole, R. J., J. W. Kirksey, and G. Morgan-Jones: Verruculotoxin, a New Mycotoxin from *Penicillium verruculosum*. Toxicol. Appl. Pharmacol. **31**, 465 (1975).

96. Macmillan, J. G., J. P. Springer, J. Clardy, R. J. Cole, and J. W. Kirksey: Structure and Synthesis of Verruculotoxin, a New Mycotoxin from *Penicillium verruculosum* Peyronel. J. Am. Chem. Soc. **98**, 246 (1976).

97. Cole, R. J., J. W. Kirksey, J. H. Moore, B. R. Blankenship, U. L. Diener, and N. D. Davis: Tremorgenic Toxin from *Penicillium verruculosum*. Appl. Microbiol. **24**, 248 (1972).

98. Schroeder, H. W., R. J. Cole, H. Hein, and J. W. Kirksey: Tremorgenic Mycotoxins from *Aspergillus caespitosus*. Appl. Microbiol. **29**, 857 (1975).

99. Yoshizawa, T., N. Marooka, Y. Sawada, and S.-I. Udagawa: Tremorgenic Mycotoxin from *Penicillium paraherquei*. Appl. Environ. Microbiol. **32**, 441 (1976).

100. Gallagher, R. T., and G. C. M. Latch: Production of the Tremorgenic Mycotoxins Verruculogen and Fumitremorgin B by *Penicillium piscarium* Westling. Appl. Environ. Microbiol. **33**, 730 (1977).

101. Cockrum, P. A., C. C. J. Culvenor, J. A. Edgar, and A. L. Payne: Chemically Different Tremorgenic Mycotoxins in Isolates of *Penicillium paxilli* from Australia and North America. J. Nat. Prods. **42**, 534 (1979).

102. Cole, R. J., J. W. Kirksey, J. W. Dorner, D. M. Wilson, J. C. Johnson, A. N. Johnson, D. M. Bedell, J. P. Springer, K. K. Chexal, J. C. Clardy, and R. H. Cox: Mycotoxins Produced by *Aspergillus fumigatus* Species Isolated from Molded Silage. J. Agric. Food Chem. **25**, 826 (1977).

103. YAMAZAKI, M., S. SUZUKI, and K. MIYAKI: Tremorgenic Toxins from *Aspergillus fumigatus* Fres. Chem. Pharm. Bull. (Tokyo) **19**, 1739 (1971).

104. DIX, D. T., J. MARTIN, and C. E. MOPPETT: Molecular Structure of the Metabolite Lanosulin. J. Chem. Soc. Chem. Commun. 1168 (1972).

105. YAMAZAKI, M., K. SASAGO, and K. MIYAKI: The Structure of Fumitremorgin B (FTB), a Tremorgenic Toxin from *Aspergillus fumigatus* Fres. J. Chem. Soc. Chem. Commun. 408 (1974).

106. COLE, R. J., and J. W. KIRKSEY: The Mycotoxin Verruculogen: A 6-*O*-methylindole. J. Agric. Food Chem. **21**, 927 (1973).

107. FAYOS, J., D. LOKENSGARD, J. CLARDY, R. J. COLE, and J. W. KIRKSEY: Structure of Verruculogen, a Tremor Producing Peroxide from *Penicillium verruculosum*. J. Am. Chem. Soc. **96**, 6785 (1974).

108. COLE, R. J., and R. H. COX: Handbook of Toxic Fungal Metabolites. New York: Academic Press. 1981.

109. COLE, R. J., J. W. KIRKSEY, R. H. COX, and J. CLARDY: Structure of the Tremor-producing Indole, TR-2. J. Agric. Food Chem. **23**, 1015 (1975).

110. TANABE, M.: Proc. US-Jpn Joint Seminar on Biosynthesis of Natural Products, Honolulu, Hawaii (1976).

111. YAMAZAKI, M., H. FUJIMOTO, T. AKIYAMA, U. SANKAWA, and Y. IITAKA: Crystal Structure and Absolute Configuration of Fumitremorgin B, a Tremorgenic Toxin from *Aspergillus fumigatus* Fres. Tetrahedron Lett. 27 (1975).

112. YAMAZAKI, M., H. FUJIMOTO, and T. KAWASAKI: The Structure of a Tremorgenic Metabolite from *Aspergillus fumigatus* Fres., Fumitremorgin A. Tetrahedron Lett. 1241 (1975).

113. EICKMAN, N., J. CLARDY, R. J. COLE, and J. W. KIRKSEY: The Structure of Fumitremorgin A. Tetrahedron Lett. 1051 (1975).

114. COLE, R. J.: Proc. US-Jpn Conf. on Mycotoxins in Human and Animal Health, University of Maryland, College Park.

115. YAMAZAKI, M., H. FUJIMOTO, T. KAWASAKI, E. OKUYAMA, and T. KUGA: Proc. 19th Symp. on Chemistry of Natural Products, Hiroshima, Japan (1975).

116. URAMOTO, M., L. CARY, M. TANABE, K. HIROTSU, and J. CLARDY: Proc. Ann. Meeting, Kanto Branch. Agric. Chem. Soc. (Japan), Tokyo, Japan (1977).

117. GLINSUKON, T., S. YUAN, R. WIGHTMAN, Y. KITAURA, G. BÜCHI, R. C. SHANK, G. N. WOGAN, and C. M. CHRISTENSEN: Isolation and Purification of Cytochalasin E and Two Tremorgens from *Aspergillus clavatus*. Plant Foods for Man **1**, 113 (1974).

118. CLARDY, J., J. P. SPRINGER, G. BÜCHI, K. MATSUO, and R. WIGHTMAN: Tryptoquivaline and Tryptoquivalone, Two Tremorgenic Metabolites of *Aspergillus clavatus*. J. Am. Chem. Soc. **97**, 663 (1975).

119. YAMAZAKI, M., H. FUJIMOTO, and E. OKUYAMA: Structure Determination of Six Tryptoquivaline-related Metabolites from *Aspergillus fumigatus*. Tetrahedron Lett. 2861 (1976).

120. YAMAZAKI, M., H. FUJIMOTO, and E. OKUYAMA: Structure of Tryptoquivaline C (FTC) and D (FTD). Novel Fungal Metabolites from *Aspergillus fumigatus*. Chem. Pharm. Bull. (Tokyo) **25**, 2554 (1977).

121. YAMAZAKI, M., H. FUJIMOTO, and E. OKUYAMA: Structure Determination of Six Fungal Metabolites, Tryptoquivaline E, F, G, H, I and J from *Aspergillus fumigatus*. Chem. Pharm. Bull. (Tokyo) **26**, 111 (1978).

122. YAMAZAKI, M., E. OKUYAMA, and Y. MAEBAYASHI: Isolation of Some New Tryptoquivaline-related Metabolites from *Aspergillus fumigatus*. Chem. Pharm. Bull. (Tokyo) **27**, 1611 (1979).

123. DEMAIN, A. L., N. A. HUNT, V. MALIK, B. KOBBE, H. HAWKINS, K. MATSUO, and G. N. WOGAN: Improved for Procedure for Production of Cytochalasin E and Tremorgenic Mycotoxins by *Aspergillus clavatus*. Appl. Environ. Microbiol. **31**, 138 (1976).

124. Hesseltine, C. W.: Solid State Fermentations. Biotechnol. Bioeng. **14,** 517 (1972).
125. Büchi, G., K. C. Luk, B. Kobbe, and J. M. Townsend: Four New Mycotoxins of *Aspergillus clavatus* Related to Tryptoquivaline. J. Org. Chem. **42,** 244 (1977).
126. Springer, J. P.: The Absolute Configuration of Nortryptoquivaline. Tetrahedron Lett. 339 (1979).
127. Nakagawa, M., M. Ito, Y. Hasegawa, S. Akashi, and T. Hino: Total Synthesis of (+)-Tryptoquivaline. Tetrahedron Lett. **25,** 3865 (1984).
128. Nakagawa, M., M. Taniguchi, M. Sodeoka, M. Ito, K. Yamaguchi, and T. Hino: Total Synthesis of (+)- and (−)-Tryptoquivaline G by Biomimetic Double Cyclization. J. Am. Chem. Soc. **105,** 3709 (1983).
129. Büchi, G., P. R. Deshong, S. Katsumura, and Y. Sugimura: Total Synthesis of Tryptoquivaline G. J. Am. Chem. Soc. **101,** 5084 (1979).
130. Ohnuma, T., Y. Kimura, and Y. Ban: Synthetic Studies on Oxindole Spirolactones with Thallium(III) Nitrate: A Formal Total Synthesis of (±)-Tryptoquivaline G. Tetrahedron Lett. **22,** 4969 (1981).
131. Yamazaki, M.: The Biosynthesis of Neurotropic Mycotoxins. In: The Biosynthesis of Mycotoxins — A Study in Secondary Metabolism (P. S. Steyn, ed.), 193. New York: Academic Press. 1980.

(Received December 7, 1984)

Structure of Palytoxin

By R. E. MOORE, Department of Chemistry, University of Hawaii, Honolulu, Hawaii, U.S.A.

With 10 Figures

Contents

I. Introduction

Palytoxin (**1**) is an extremely poisonous substance associated with marine coelenterates (zoanthids) of the genus *Palythoa* (*1*). Its intravenous lethality (LD_{50}), which ranges from 0.025 µg/kg in the rabbit to 0.45 µg/kg in the mouse (*2*), is exceeded only by certain proteins and polypeptides (*3*). The toxicity of *Palythoa* was probably noticed for the first time by the Hawaiians who used exudates of a rare, but very toxic species, *P. toxica* (*4*), to poison spear tips for warfare. Investigators at the University of Hawaii were led to *P. toxica* when they followed up the lead provided by the entry *limu-make-o-Hana* (the deadly seaweed of Hana) in the Hawaiian-English Dictionary (*5*). Japanese researchers, on the other hand, were led to *P. tuberculosa* when they traced the dietary origin of a water-soluble toxin that had been found in the digestive tract of some toxic filefish, *Alutera scripta* (*6, 7*).

Using reverse-phase chromatography and successive anion- and cation-exchange gel filtration chromatography, the Hawaii group was able to isolate palytoxin for the first time in mid 1963 (*1, 8*). Attempts in the next decade to elucidate its structure or the structures of degradation products were largely unsuccessful. The structure elucidation of palytoxin posed a tremendous challenge to the organic chemist of the 1960s, as the toxin appeared to have a molecular weight of about 3000 daltons but totally lacked repeating units, such as amino acid, sugar, and fatty acid residues, commonly found in biomolecules of this size. Significant progress on the determination of its structure began only in the mid 1970s with the advent of better separation techniques and analytical methods, in particular high-pressure liquid chromatography to separate the complex mixtures that palytoxin produced on chemical degradation and high frequency nuclear magnetic resonance spectroscopy and field desorption mass spectrometry to determine the structures of the numerous degradation products.

References, pp. 199—202

(1)

II. Isolation

The procedure used by MOORE and SCHEUER to isolate palytoxin from *P. toxica* is outlined in Scheme I (*1*). The toxin could be completely extracted from the unground wet animal with 70% ethanol-water. Reverse-phase chromatography of the defatted extract on powdered polyethylene separated the toxin from inorganic salts and very polar organic material. Palytoxin was one of the few substances to be absorbed onto polyethylene when the extract was passed through a column of this material with water. The toxin, however, could be readily removed from the absorbent with 50% aqueous ethanol. Successive ion exchange gel filtration of the toxic fraction, first on DEAE-Sephadex at pH 7 and then on CM-Sephadex at pH 4.5 − 5, resulted in pure palytoxin in 0.027% yield based on the wet weight of the animal. On the anion exchange Sephadex, palytoxin passed through the column shortly after the void volume and was separated from compounds of lower molecular weight and acidic contaminants in this step. On the cation Sephadex, however, palytoxin was retarded as it was weakly basic, resulting in separation of the toxin from non-basic substances of high molecular weight.

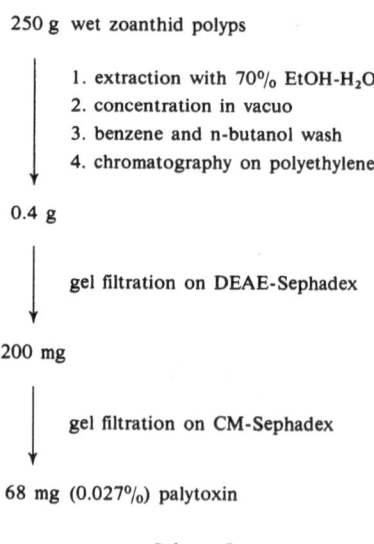

250 g wet zoanthid polyps

1. extraction with 70% EtOH-H_2O
2. concentration in vacuo
3. benzene and n-butanol wash
4. chromatography on polyethylene

0.4 g

gel filtration on DEAE-Sephadex

200 mg

gel filtration on CM-Sephadex

68 mg (0.027%) palytoxin

Scheme I

A similar procedure was used by Japanese investigators to isolate palytoxin from *P. tuberculosa* (*9*). The reverse-phase chromatography, however, was carried out on polystyrene gel instead of powdered polyethylene.

The same toxin has also been isolated from *P. vestitus* (*10*), *P. mammilosa* (*11*), *P. caribaeorum* (*11, 12*), and a Tahitian *Palythoa* sp. (*13*).

III. Characterization

Palytoxin is a white, amorphous, hygroscopic solid. Attempts to crystallize the toxin and various derivatives have so far failed. The palytoxin from *P. toxica* was found to have an optical rotation of $[\alpha]_D + 26°$ in water; its optical rotatory dispersion (ORD) curve exhibited a positive Cotton effect with $[\alpha]_{250} + 700°$ and $[\alpha]_{215} - 600°$ (*1*).

Based on combustion analysis and integration of the proton nmr spectrum, MOORE and SCHEUER estimated palytoxin's molecular weight to be roughly 3300 daltons (*1*). Its exact molecular weight, 2678.5 daltons, and elemental composition, $C_{129}H_{223}N_3O_{54}$, however, were not to be known unambiguously until after the gross structure had been determined. Shortly before completion of the gross structure work, MACFARLANE was able to

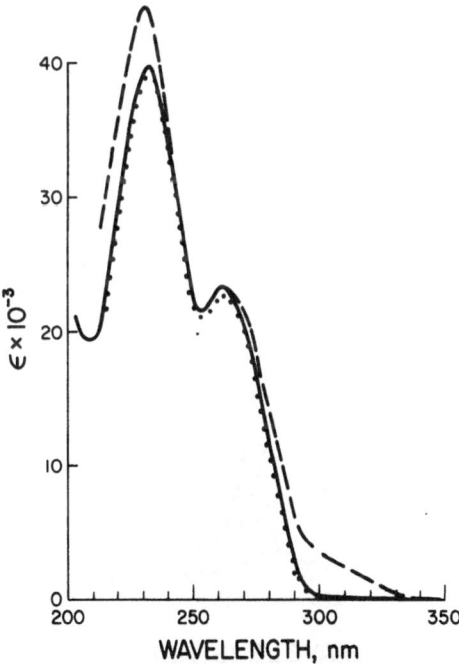

Fig. 1. The ultraviolet absorption spectrum of palytoxin in water (——), aqueous 0.05 N hydrochloric acid (·····), and aqueous 0.05 N sodium hydroxide (---)

calculate the molecular weight as 2681.1 by ^{252}Cf plasma desorption mass spectrometry (14). Since the ^{13}C satellites could not be resolved, MACFARLANE's measurement actually represented the average mass of the molecular ion cluster. A better resolved spectrum and a more precise value, 2678.9 daltons, were ultimately obtained by fast-atom bombardment mass spectrometry (15).

The ultraviolet absorption spectrum in water (Fig. 1) showed two intense peaks at 233 nm (ε 40,500) and 263 nm (ε 23,600) (1). No change was observed in the spectrum after the toxin was treated with sodium borohydride. The spectrum was completely obliterated, however, when palytoxin was hydrogenated catalytically. The chromophore accounting for the 263 nm absorption was very labile to acid and base, the λ 263 peak disappearing with a half-life of 85 minutes in methanolic 0.05 N hydrochloric acid and a half-life of 55 minutes in aqueous 0.05 N sodium hydroxide (Fig. 2) (1). The λ 233 peak, however, was unaffected by this mild acid or base treatment.

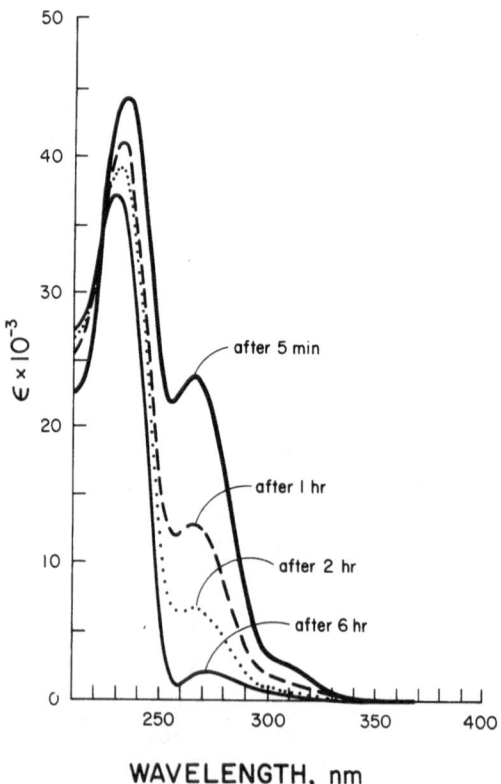

Fig. 2. Effect of aqueous 0.05 N sodium hydroxide on the ultraviolet spectrum of palytoxin

References, pp. 199—202

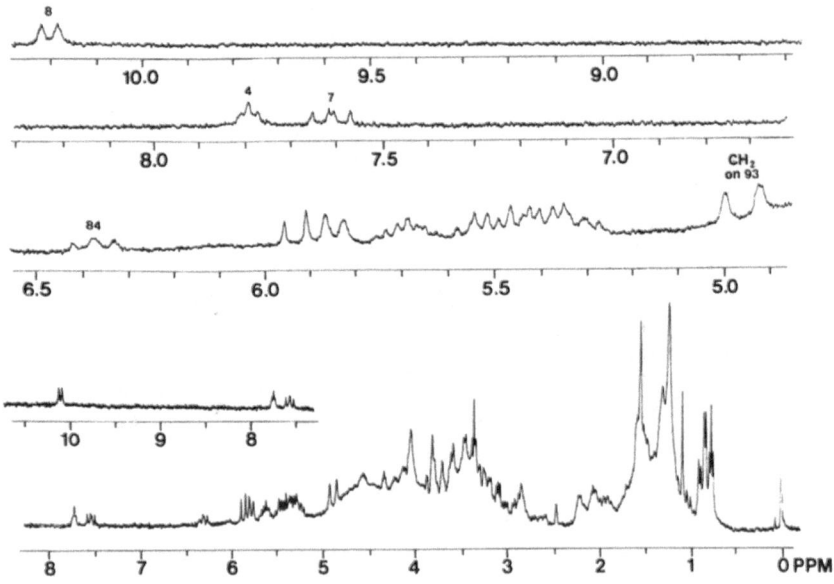

Fig. 3. The 360 MHz ^{1}H nmr spectrum of palytoxin in dimethyl sulfoxide-d$_6$

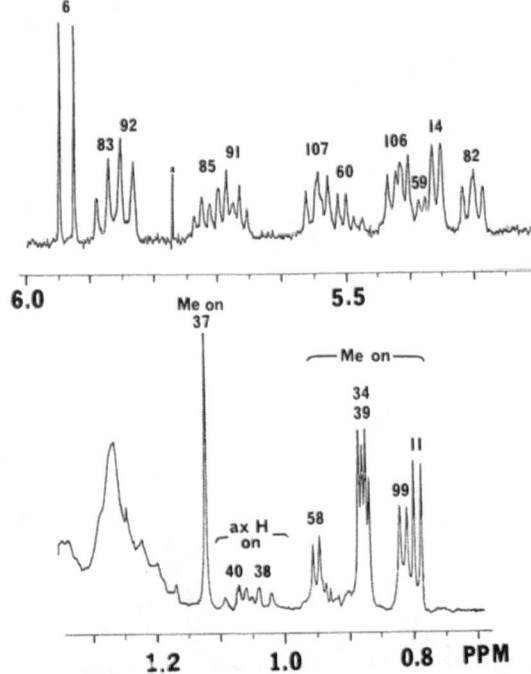

Fig. 4. The 5.3–6.0 ppm and 0.7–1.35 ppm regions of the 600 MHz ^{1}H nmr spectrum of palytoxin in dimethyl sulfoxide-d$_6$. Numbers refer to carbon position of protons and methyl groups

R. E. MOORE:

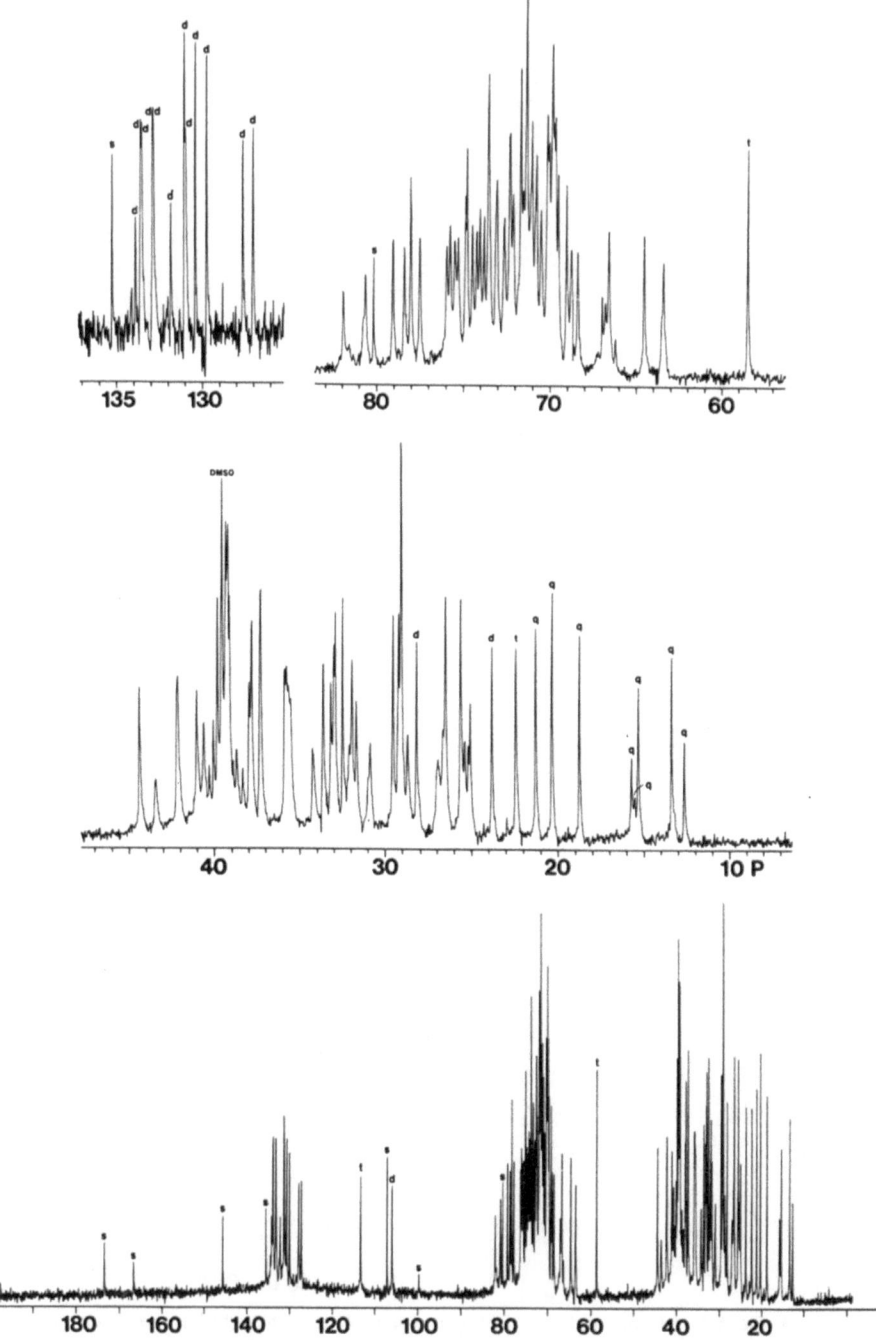

Fig. 5. The 90 MHz ^{13}C nmr spectrum of palytoxin in dimethyl sulfoxide-d$_6$ at 55°

References, pp. 199—202

The infrared spectrum showed a band at $1670 \, \text{cm}^{-1}$ which was assigned to amide carbonyl (*1*). The high frequency ^1H nmr spectrum (Fig. 3 and 4) and ^{13}C nmr spectrum (Fig. 5) indicated the presence of two secondary amide groups, eight olefinic double bonds, and seven methyl groups (Table 1). Proton-proton decoupling experiments showed that the two amide groups and one of the carbon-carbon double bonds were in a N-(3′-hydroxypropyl)-*trans*-3-aminoacrylamide unit (**1a**) (*16*), a unit which also accounted for the uv absorption at 263 nm. The ^1H und ^{13}C chemical shifts associated with unit (**1a**) were essentially identical with those of synthetic N-(3′-hydroxypropyl)-*trans*-3-acetamidoacrylamide (**2**) (*17, 18*). In (**1a**) the primary alcohol group had to be free and not masked in an acid-labile functionality, since the methylene chemical shifts of the N-(3′-hydroxy-propyl) group were exactly the same for the toxin and model compound (**2**). Furthermore the C-1 carbon of palytoxin exhibited a longer relaxation time (T_1) than any other proton-bearing carbon.

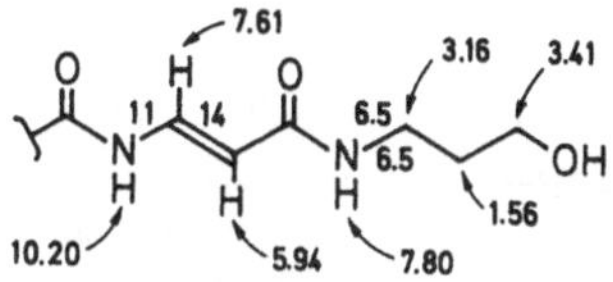

(**1a**)

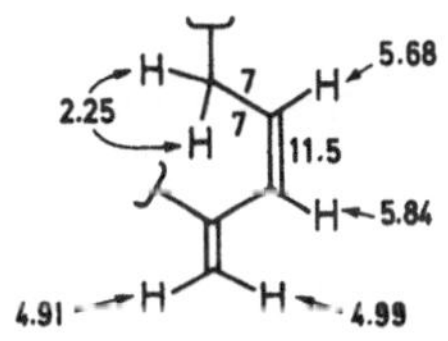

(**1b**)

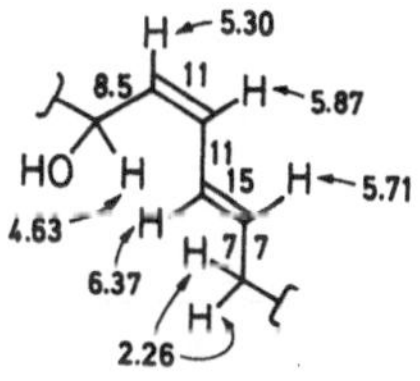

(**1c**)

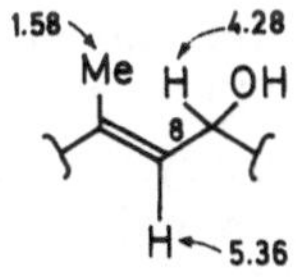

(**1d**)

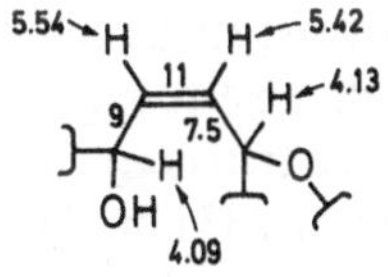

(**1e**)

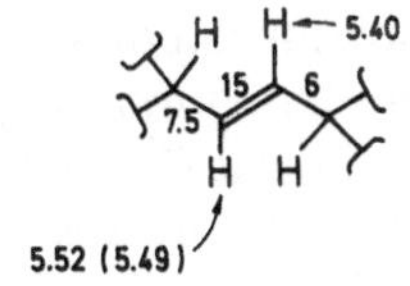

(**1f**)

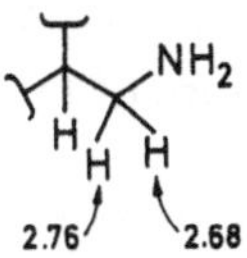

(**1g**)

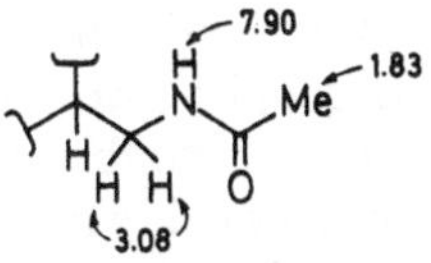

N-acetylpalytoxin

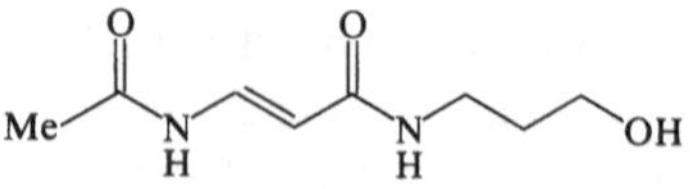

(2)

Table 1. ^{13}C Chemical Shifts in ppm from TMS of Palytoxin in Dimethyl Sulfoxide-d_6 at 55°

Types of Carbons	Chemical Shifts*
2 amide C=O	singlets at 166.6, 173.5
5 CH₃—CH—	quartets at 12.70, 13.45, 15.38, 15.76, 18.81, 20.40, 21.35
7 CH₃ 1 CH₃—C—	
1 CH₃—C=	
6 —CH=CH—	triplet at 113.06
1 —C=CH—	doublets at 105.78, 127.07, 127.69, 129.82, 130.47, 131.06, 131.11, 131.91, 132.90, 132.96, 133.60, 133.66, 133.97
8 C=C	
1 —C=CH₂	singlets at 135.33, 145.51
3 —C—	singlets at 80.17, 99.8, 106.93
1 —CH₂—O—	triplet at 58.52

* Assignments are based on single frequency off resonance decoupled spectra and partially relaxed Fourier transform spectra. Data obtained at 90 MHz.

Extensive nmr studies strongly suggested that four other olefinic double bonds were in a *trans,cis*-1,4-disubstituted diene (**1b**) and a *cis*-1,3-disubstituted diene (**1c**) unit. These two conjugated diene systems accounted for the intense uv absorption at 233 nm. The three remaining olefinic double bonds were isolated. One was trisubstituted and had a methyl group on it; since the ^{13}C chemical shift of the methyl group was 12.70 ppm, the geometry of this double bond had to be *E* (**1d**). The two other isolated double bonds were *cis*-1,2-disubstituted (**1e**) and *trans*-1,2-disubstituted (**1f**). Detailed decoupling work established the nature of the carbons that were attached to each double bond.

Evidence for the weakly basic nitrogen of palytoxin was obtained when the toxin was treated with *p*-nitrophenyl acetate (*9, 14*), acetic ethylcarbonic anhydride (*19*), or N-succinimidyl acetate (*19*). An N-acetylpalytoxin derivative was formed. ^{1}H nmr analysis indicated that a primary amino group had been acetylated and that palytoxin had partial structure (**1g**). The ^{1}H nmr spectrum of palytoxin showed two doublets of doublets at 2.68 and 2.76 ppm for the protons of a methylene bearing a primary amino

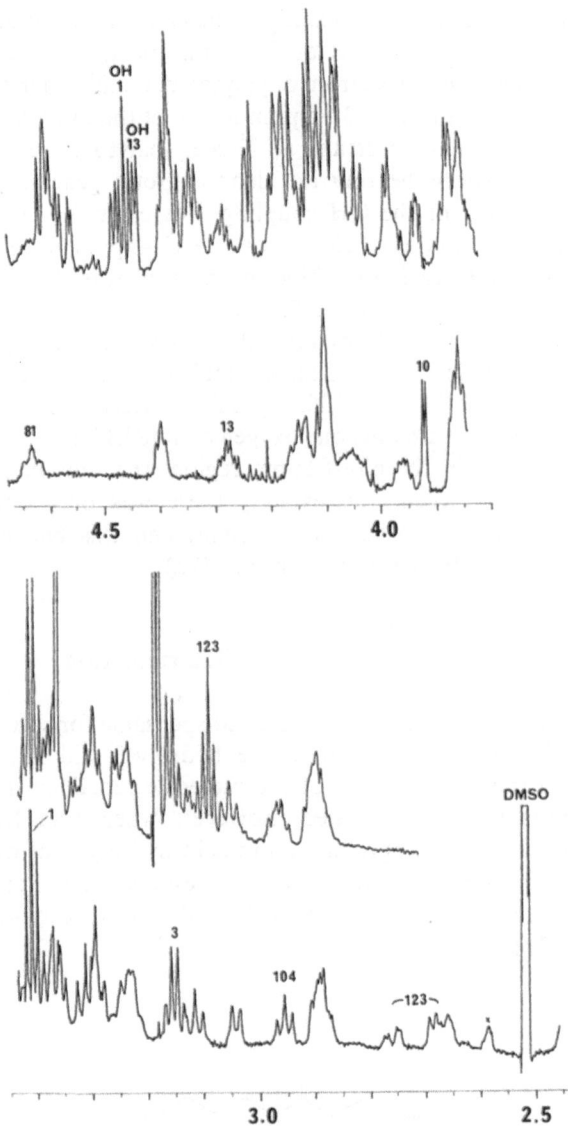

Fig. 6. The 2.45 – 3.45 ppm and 3.8 – 4.7 ppm regions of the 600 MHz ^{1}H nmr spectra of palytoxin (lower traces) and N-acetylpalytoxin (upper traces) in dimethylsulfoxide-d$_6$

group. After acetylation this methylene signal had shifted to 3.08 ppm and was a triplet. Shown also in the ^{1}H nmr spectrum of N-acetylpalytoxin was a broad triplet at 7.90 ppm for a secondary amide proton and a singlet at 1.83 ppm for acetyl protons.

Interestingly the 600 MHz ^{1}H nmr spectrum of N-acetylpalytoxin in dimethyl sulfoxide-d$_6$ exhibited well resolved signals for many of the hydroxyl protons (Fig. 6). For example, the OH protons in units (**1a**) and (**1d**) could be seen as a triplet at 4.45 ppm and a doublet at 4.43 ppm respectively. The ^{1}H nmr spectrum of palytoxin, on the other hand, showed a single, very broad signal at 3 – 5 ppm for all of the hydroxyl protons. In the 600 MHz ^{1}H nmr spectrum of N-acetylpalytoxin, each signal for methine or methylene bearing an alcohol group generally showed an additional splitting for the OH coupling. For example, the OH bearing methylene in unit (**1a**) was a 1:3:3:1 quartet (3.41 ppm) instead of a 1:2:1 triplet and the OH bearing methine in unit (**1d**) was a triplet of triplets (4.28 ppm) instead of a triplet of doublets.

The ^{13}C nmr spectrum showed that there were only three sp^3 quaternary carbon signals (Table 1). Two of these signals were for quaternary carbons bearing two oxygens (99.8 and 106.93 ppm) whereas the third signal was for a quaternary carbon with only one oxygen on it (80.17 ppm). All but one of the carbon signals in the 63 – 82 ppm region were for methine carbons bearing a single oxygen. Interestingly there was only one signal for methylene attached to oxygen (58.52 ppm) and this had already been assigned to the $-CH_2OH$ carbon in unit (**1a**).

IV. Gross Structure Determination

Chemical degradation studies were indispensable for solving the gross structure of palytoxin. Acid and base hydrolysis did not yield much structural information as most of the molecule remained intact. Partial structure (**1a**) was corroborated, however, when acid hydrolysis of palytoxin yielded 3-aminopropanol and acid hydrolysis of hydrogenated palytoxin (hexadecahydropalytoxin) (*14*) resulted in a 1:1 mixture of β-alanine and 3-aminopropanol or in N-(3'-hydroxypropyl)-3-aminopropionamide (**3**) (*16*).

(**3**)

Two reactions, periodate oxidation and ozonolysis, proved to be the most useful for degradation. Each reaction led to a complex mixture of products, the structures of which defined structual units in the toxin. The results of the two degradations allowed the MOORE and HIRATA groups to assemble the various units into a total gross structure.

In order to isolate degradation products containing the primary amino group, palytoxin had to be first converted to an N-acyl derivative (Scheme II). The choice of a N-*p*-bromobenzoyl group as the protective group (rather than an N-acetyl group) facilitated the purification of the oxidation products possessing the primary amino group, since the basic nitrogen was tagged with a uv absorbing chromophore.

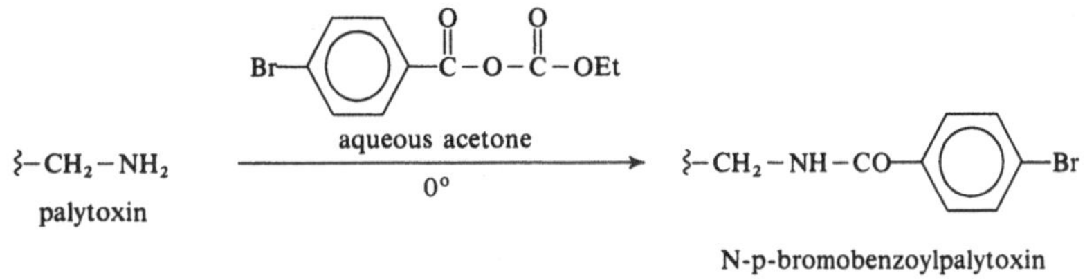

Scheme II

The isolation of the actual oxidation products, however, in particular the very reactive aldehydes from periodate oxidation, proved to be exceedingly difficult and frequently unfruitful. It was found best to add excess sodium borohydride directly to the reaction mixture to reduce the labile oxidation products to stable alcohols and to destroy excess oxidizing agent. Use of borohydride allowed instant termination of the periodate oxidation and therefore complete control of the reaction. Separation of the alcohols could be achieved by reverse-phase HPLC, but it was found to be much easier to separate the corresponding acetates by HPLC on silica.

It could be reasoned from ^{13}C nmr data that periodate oxidation of palytoxin would lead only to aldehydic products and not ketonic ones. There was only one signal, a singlet at 80.17 ppm, for a quaternary carbon bearing a single oxygen atom (Table 1). Since this oxygen was not in a hydroxyl group, ketones could not possibly be formed on periodate oxidation. This meant that all of the $-CH_2OAc$ groups in the various degradation products, except the one coming from unit (**1a**), had to arise from aldehydic groups in the periodate products (Scheme III). All of the CHOAc groups, therefore, had to arise from secondary alcohol functionalities present originally in the toxin.

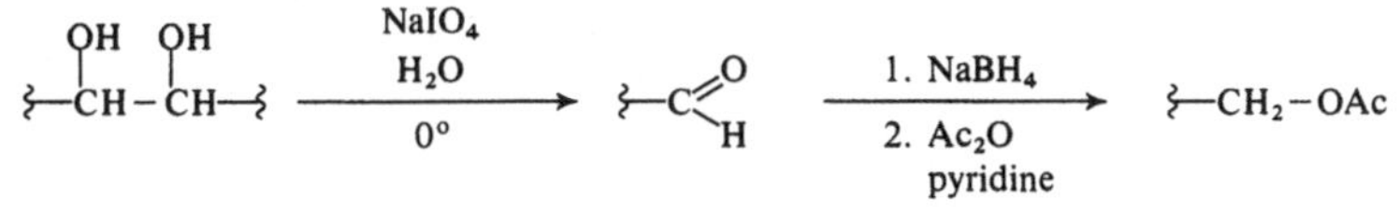

Scheme III

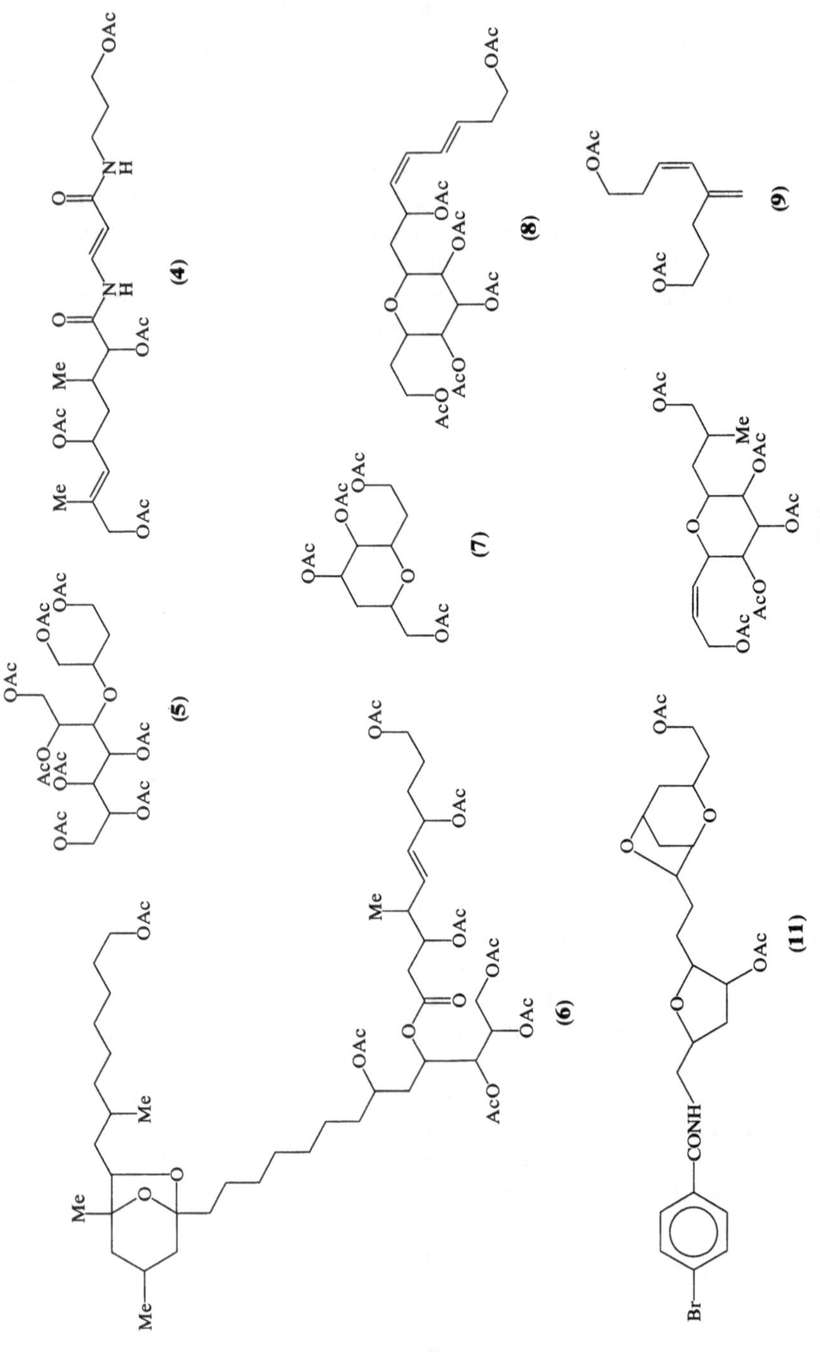

Scheme IV. Periodate oxidation products of N-p-bromobenzoylpalytoxin

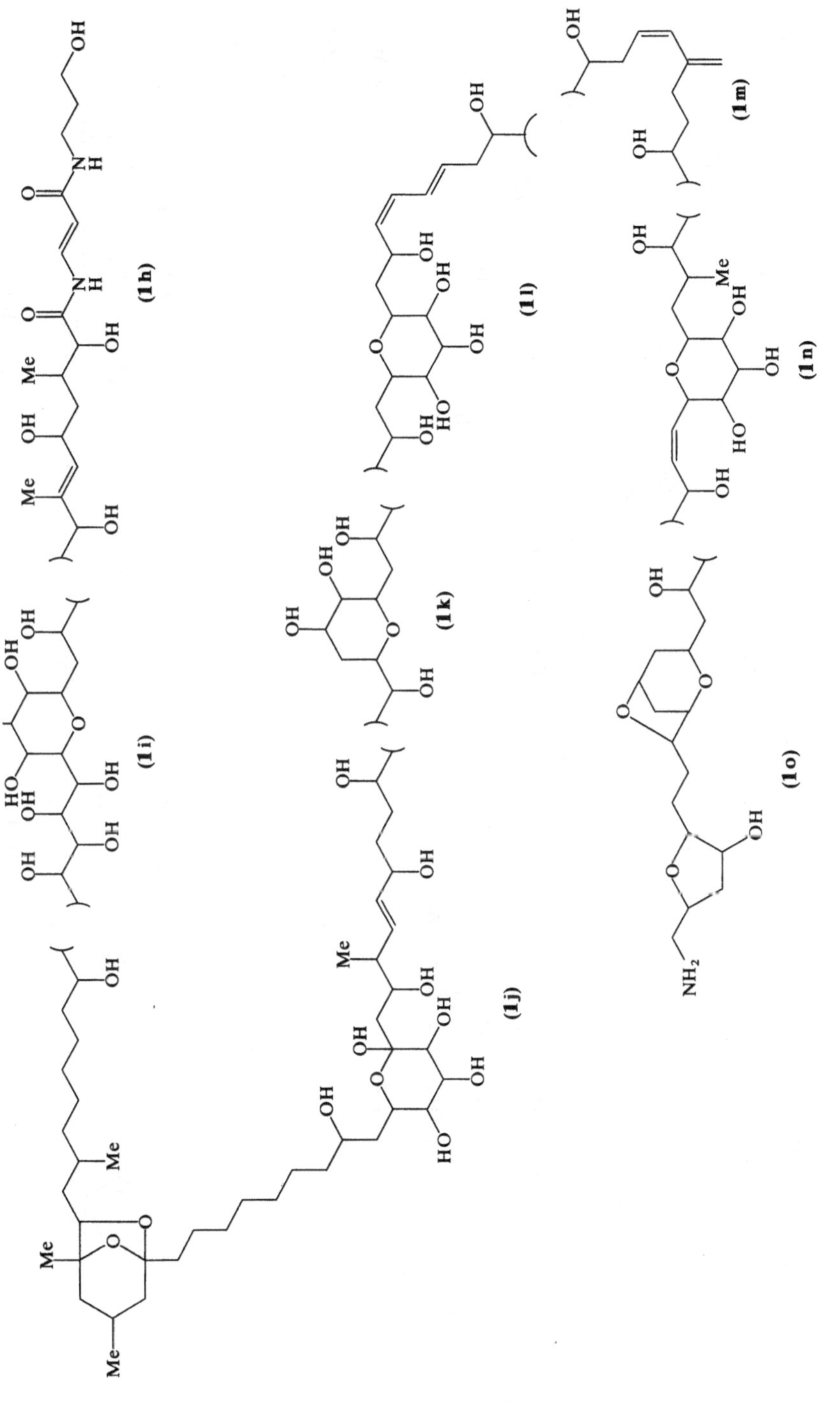

Scheme V. Structural units in palytoxin implied from periodate oxidation products

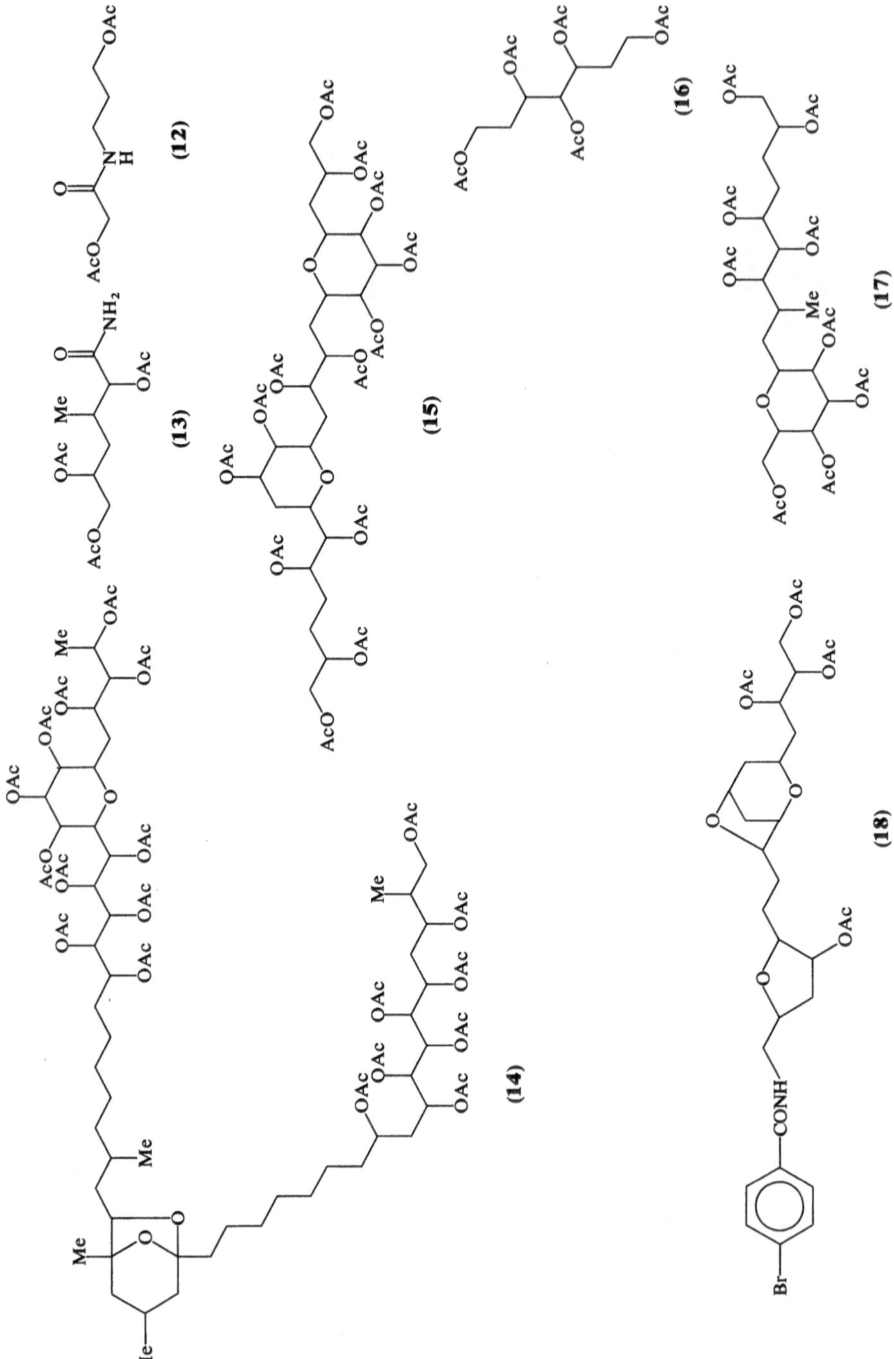

Scheme VI. Ozonolysis products of N-p-bromobenzoylpalytoxin

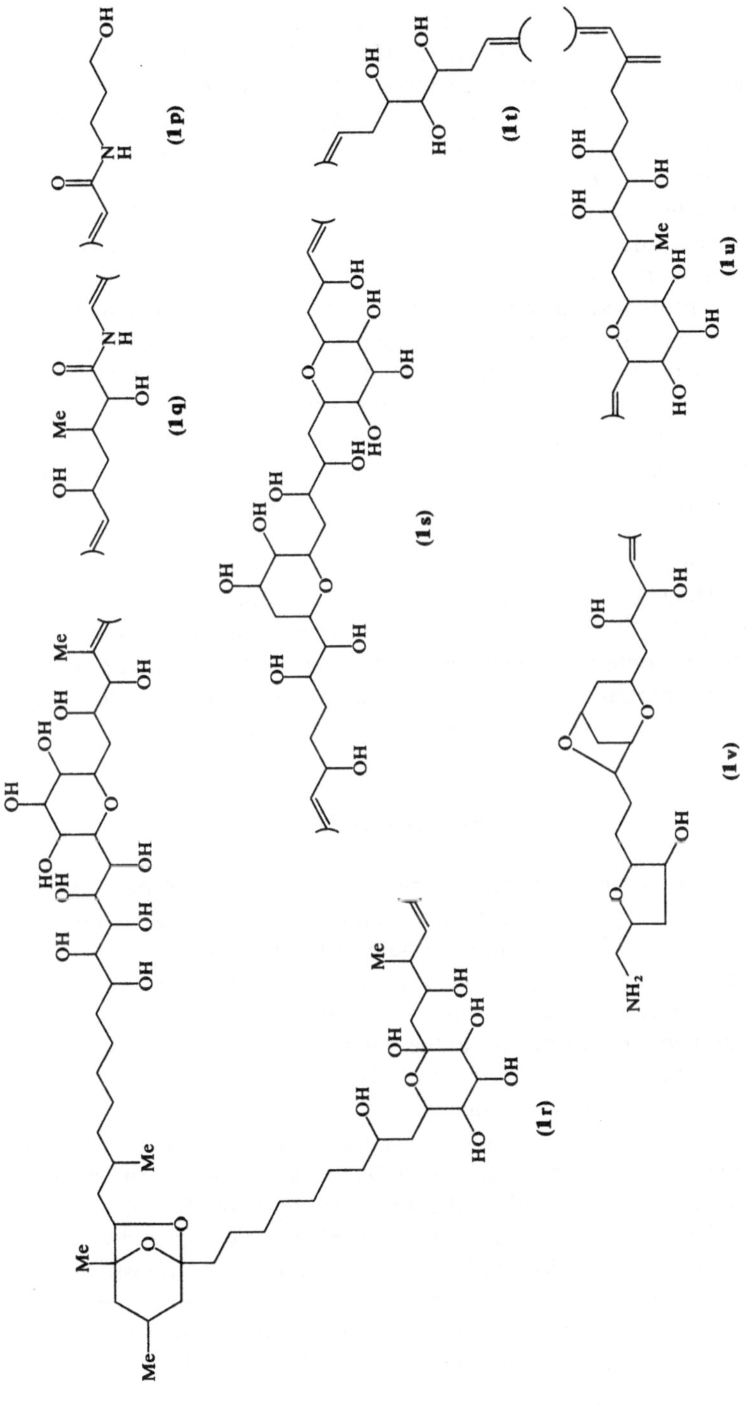

Scheme VII. Structural units in palytoxin implied from ozonolysis products

The most important products from periodate oxidation of N-*p*-bromobenzoylpalytoxin were **(4)**, **(5)**, **(6)**, **(7)**, **(8)**, **(9)**, **(10)** and **(11)**, shown in Scheme IV. The structures of these compounds indicated that partial structures **(1h-o)** (Scheme V) were present in palytoxin (*17, 19, 20*). Subunits **(1a-g)** and the other functional groups summarized in Table 1 were all accounted for by **(1h-o)**.

Complete ozonolysis of N-*p*-bromobenzoylpalytoxin, followed by borohydride reduction and acetylation, gave **(12)**, **(13)**, **(14)**, **(15)**, **(16)**, **(17)**, and **(18)** shown in Scheme VI. Three points in the palytoxin molecule, viz. the trisubstituted double bond in unit **(1h)**, the terminal methylene group in unit **(1m)**, and the hemiketal functionality in unit **(1j)**, had to lead to stereoisomers on borohydride workup of N-*p*-bromobenzoylpalytoxin octaozonide. Two of these degradation products, **(14)** and **(17)**, were actually mixtures of four and two diastereomers, respectively. The gross structures of these ozonolysis products, however, implied that units **(1p-v)** (Scheme VII) were present in palytoxin. These data allowed the MOORE group to assemble units **(1h-o)** and two CHOH residues into a gross structure for palytoxin (*20*).

The HIRATA group concluded that palytoxin had the same gross structure (*21, 22, 23, 24*). In their work, however, more complex ozonolysis products were obtained which were further degraded by periodate or ozone. The various partial structures that were elucidated from these degradative studies were finally reconstructed into a total gross structure.

IV.1 Periodate Oxidation

The MOORE group carried out periodate oxidations of palytoxin and N-*p*-bromobenzoylpalytoxin with 25 molar equivalents of sodium periodate in water at 0° for ≥ 3 hours (long-term) or ≤ 15 minutes (short-term). To isolate the aldehydic products from long-term oxidation, the reaction mixture was first extracted with chloroform to obtain the relatively nonpolar oxidation products and the water-soluble material that remained was then subjected to countercurrent distribution between *n*-butanol and water to separate the more polar oxidation products.

The aldehydic products, however, were generally not isolated. Usually excess sodium borohydride was added to the reaction mixture after 3 hours or 15 minutes to reduce all of the aldehydes to alcohols and to destroy unreacted periodate and iodate. Excess borohydride was then decomposed by acidifying the solution with sodium monohydrogen phosphate. The mixture was freeze-dried and the residue treated with acetic anhydride and pyridine. Normal workup gave a mixture of lipophilic acetates which were separated by HPLC on silica with dichloromethane-ethyl acetate, EtOAc, and ethanol-EtOAc. On one occasion the excess borohydride was decomposed by acidifying (pH 6) the mixture with hydrochloric acid, but this procedure led to the formation of artifacts in the acetylation step (25).

The HIRATA group carried out their periodate oxidations in a slightly different manner (9). The toxin or N-acylated toxin was first absorbed onto polystyrene and then treated with excess aqueous sodium periodate. After removal of periodate and formic acid from the absorbent with a water wash, the aldehydic oxidation products were eluted with a water-ethanol gradient.

a) Degradation Products Implying C(1) − C(16) Segment

Compound (4), which indicated the presence of unit (1h) in palytoxin, was obtained by short term periodate oxidation of palytoxin and N-*p*-bromobenzoylpalytoxin followed by sodium borohydride reduction of the intermediate α,β-unsaturated aldehyde (19) (9) and acetylation. Its structure was determined in a straightforward manner by ^{1}H nmr analysis. Compound (4) was always accompanied by the Z isomer (20), which was eluted from silica before (4). The coupling constant for the two *trans* olefinic protons in the β-amidoacrylamide moiety of (4) and palytoxin (1a) was 14 Hz whereas it was 8.5 Hz for the *cis* olefinic protons of (20). Isomerization of (4) to (20) probably occurred during the acetylation step.

Long term oxidation of palytoxin led to (23) (17), presumably by addition of water to (19) (9) followed by further periodate oxidation of the resulting tetrol (21) to give (22), which after borohydride reduction and acetylation resulted in (23). Aldehyde (22), however, existed predominantly in the cyclic lactam form (24), since an aldehyde signal could not be seen in its ^{1}H nmr spectrum and because treatment with MeOH readily gave (25) and acetylation with acetic anhydride and pyridine produced (26) (17).

b) Degradation Products Implying C(17) − C(27) and C(65) − C(72) Segments

Tetrahydropyrans (7), (27), and (28) were obtained in low yield by short-term oxidation of palytoxin and N-*p*-bromobenzoylpalytoxin. Compound (7), however, was isolated in sufficient quantity for complete characterization by ^{1}H nmr spectroscopy (19). Compounds (27) and (28), on the other hand, were identified solely by gas chromatography-mass spectrometry as minor constituents in a silica gel HPLC fraction that was predominately a complex mixture of related acyclic ethers (29 − 38) (Fig. 6). The mass spectra of (27) and (28) showed a base peak at m/z 123 ($C_7H_7O_2$) and several fragment ion peaks for losses of side chain, acetic acid, and ketene; for example (28) fragmented as outlined in Scheme IX.

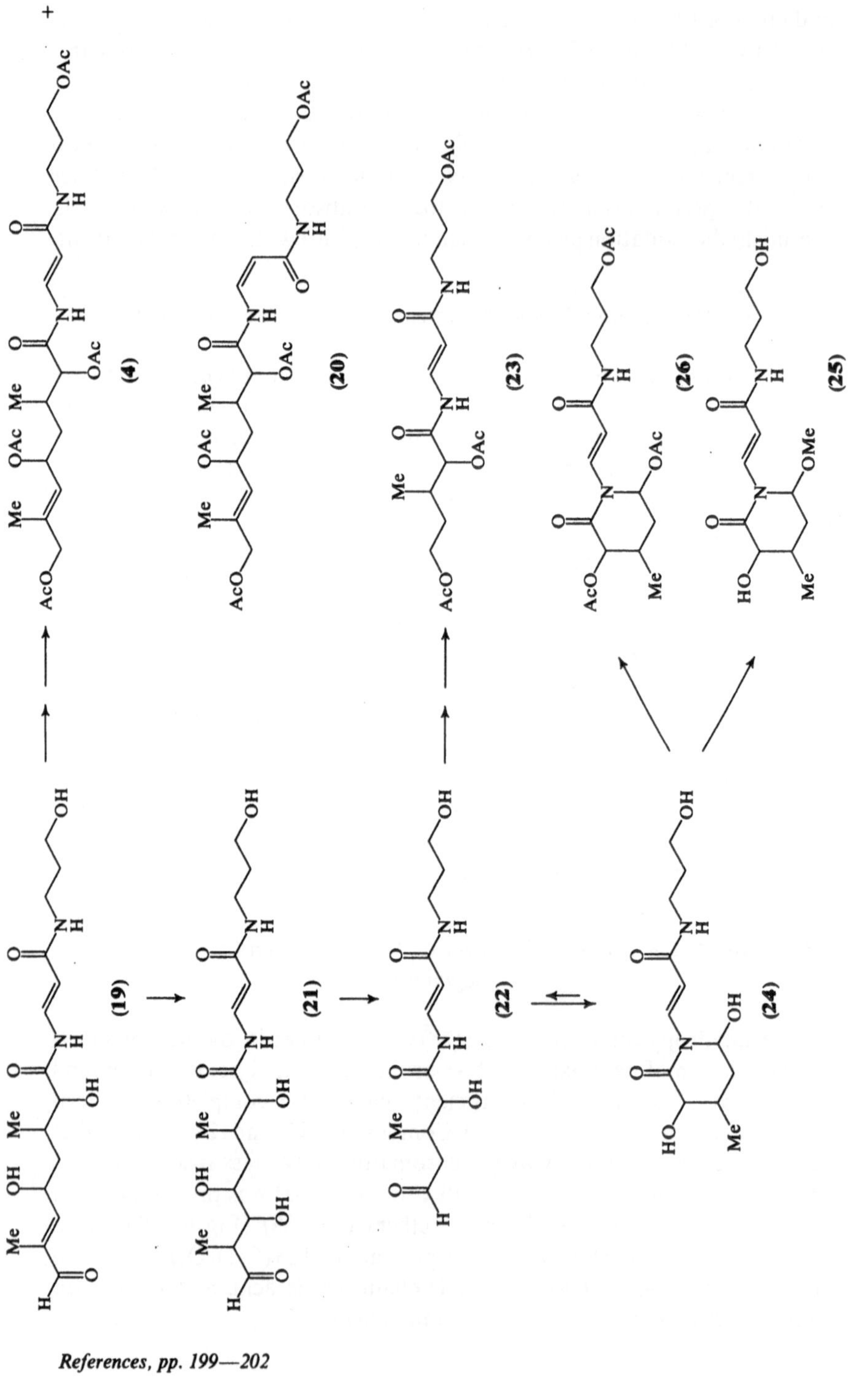

Scheme VIII

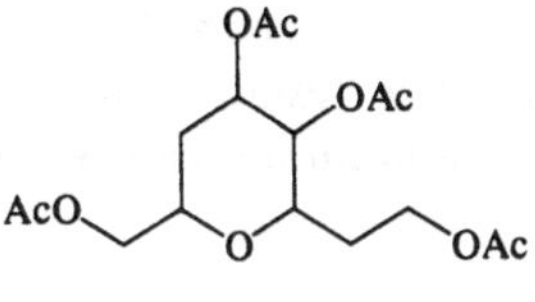
OAc
OAc
AcO
O
OAc

(7)

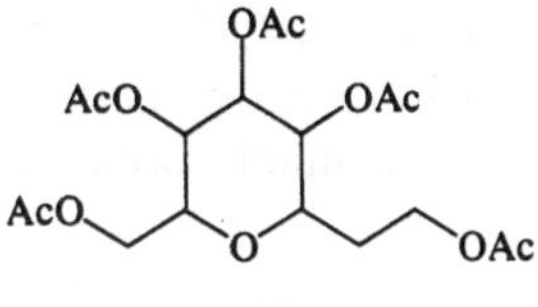
OAc
AcO
OAc
AcO
O
OAc

(27)

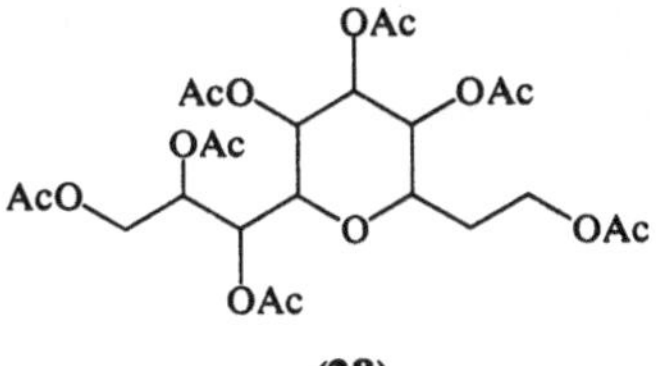
OAc
AcO
OAc
OAc
AcO
O
OAc
OAc

(28)

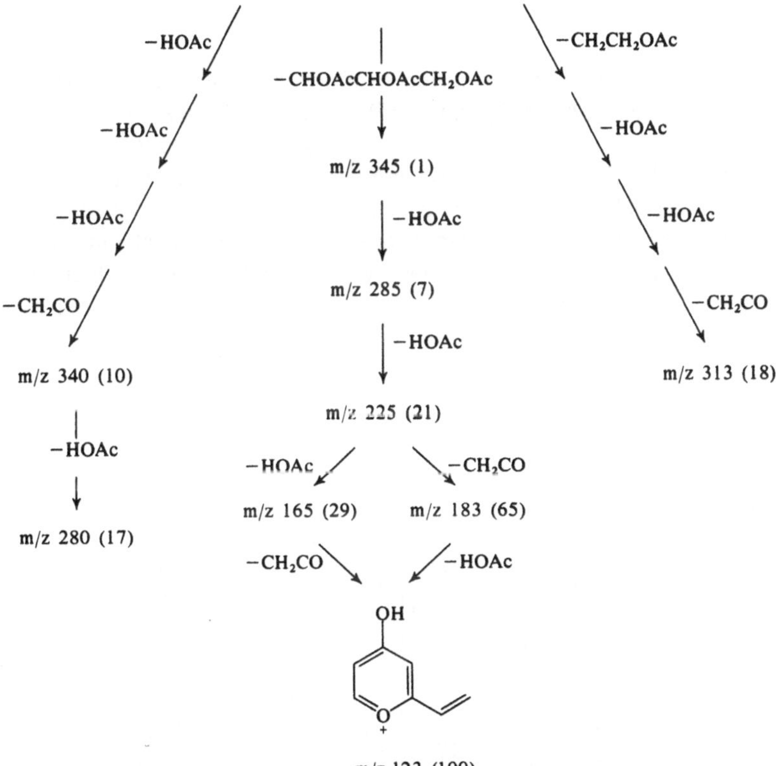
−HOAc
−HOAc
−HOAc
−CH₂CO
m/z 340 (10)
−HOAc
m/z 280 (17)
−CHOAcCHOAcCH₂OAc
m/z 345 (1)
−HOAc
m/z 285 (7)
−HOAc
m/z 225 (21)
−HOAc
−CH₂CO
m/z 165 (29)
m/z 183 (65)
−CH₂CO
−HOAc
OH
O
+
m/z 123 (100)
−CH₂CH₂OAc
−HOAc
−HOAc
−CH₂CO
m/z 313 (18)

Scheme IX

The major short-term oxidation product representing unit (**1k**) was the acyclic ether (**29**). The compound was identified by ^{1}H and ^{13}C nmr analysis and its characteristic mass spectrum. Unlike most acyclic ethers which cleave principally at the C-C bonds adjacent to the ether oxygen, the most intense peak, m/z 173, in the mass spectrum of (29) results from fission of the C-O bond, presumably because a stable oxonium ion can be formed.

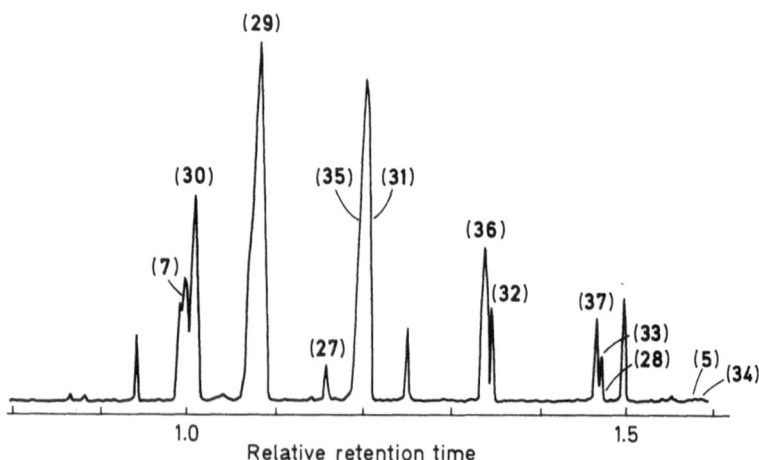

The major short-term oxidation products representing unit (1i) were the acyclic ethers (30), (31), and (35). Compounds (30) and (31) were characterized by ¹H nmr spectroscopy. Progressively smaller amounts of (36), (32), (37), (33), (34), and (5) were also produced. All of these acyclic ethers were easily identified by gas chromatography-mass spectrometry (Fig. 7), since the most intense peaks in the mass spectrum of each compound again resulted from cleavage of the ether bonds, which led to fragment ions similar to the ones from (29). Ethers (34) and (5) were formed in only trace amounts.

Fig. 7. Gas chromatography-mass spectrometry of a silica HPLC fraction containing compounds (5), (7) and (27—37) from short-term oxidation of palytoxin on a 0.325 mm × 32 m column of 0.2% SE-52 at 150°

References, pp. 199—202

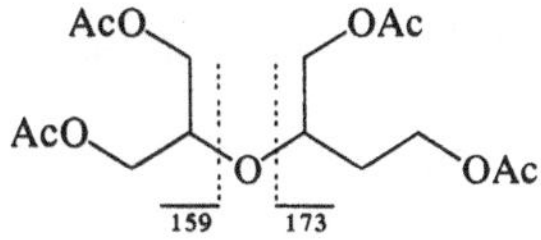

(30)

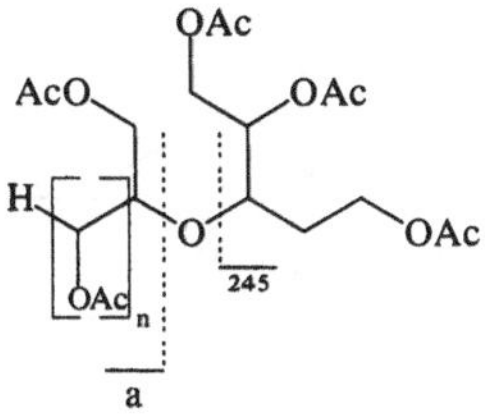

a

(31) n=1, a=159
(32) n=2, a=231
(33) n=3, a=303
(34) n=4, a=375

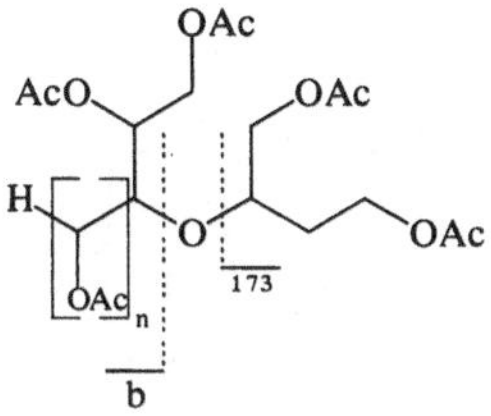

b

(35) n=1, b=231
(36) n=2, b=303
(37) n=3, b=375
(5) n=4, b=447

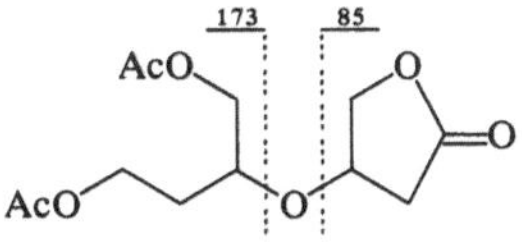

(38)

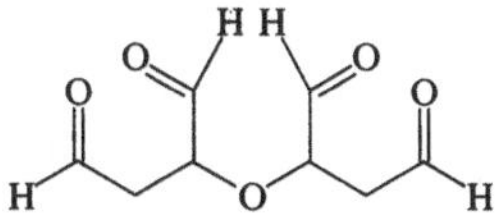

(39)

Long term periodate oxidation yielded **(7)** and **(29)** and appreciable amounts of γ-lactones such as **(38)**. Apparently one of the aldehydic groups in the intermediate tetral **(39)** had oxidized during the 3-hr period, resulting in **(38)** after borohydride reduction and acetylation.

c) Degradation Products Implying C(73) − C(87) and C(98) − C(108) Segments

Tetrahydropyrans **(8)** and **(10)**, which indicated the presence of units **(1l)** and **(1n)** in the toxin, were formed in high yield on short-term oxidation of palytoxin and N-p-bromobenzoylpalytoxin *(19)*. Acyclic ethers such as **(40)** and **(41)** were produced in very low yield.

As expected long-term oxidation resulted in the formation of acyclic ethers **(42)** and **(43)** *(19, 23)*. Also formed were the *trans, trans* isomer **(44)** and the α,β-unsaturated lactone **(45)**. Presumably **(45)** was produced from the intermediate tetral **(46)** in the same manner as **(38)** from **(39)**.

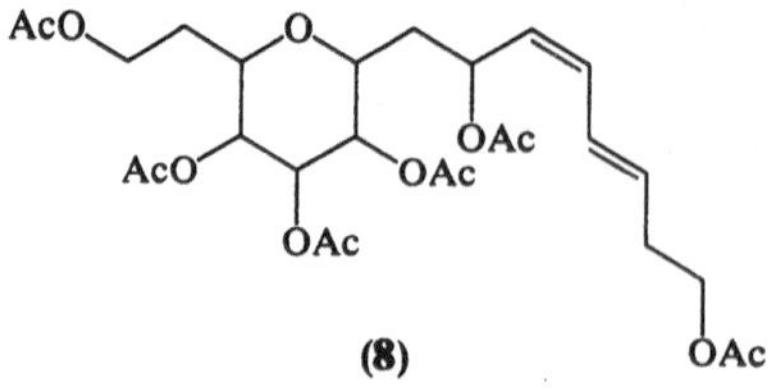

(8)

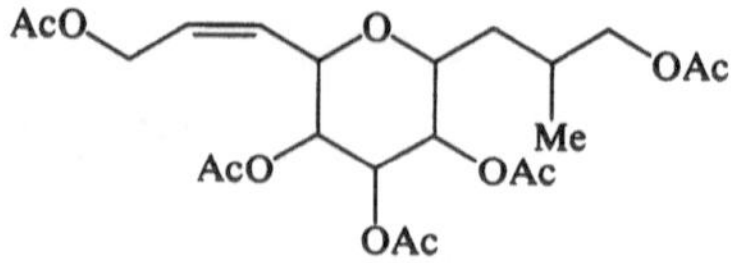

(10)

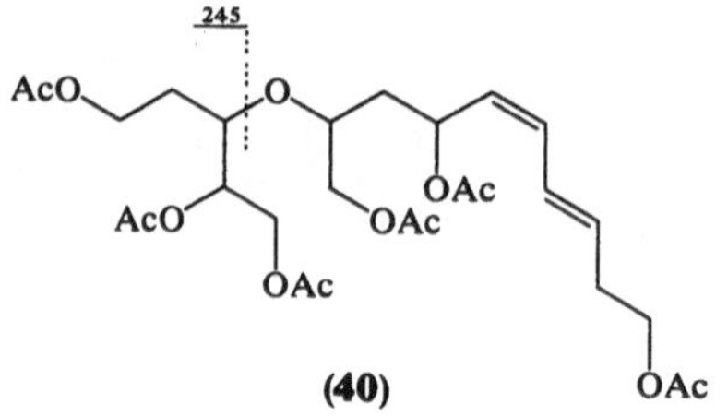

(40)

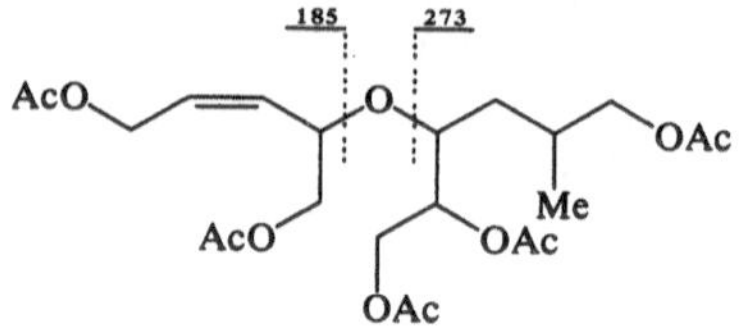

(41)

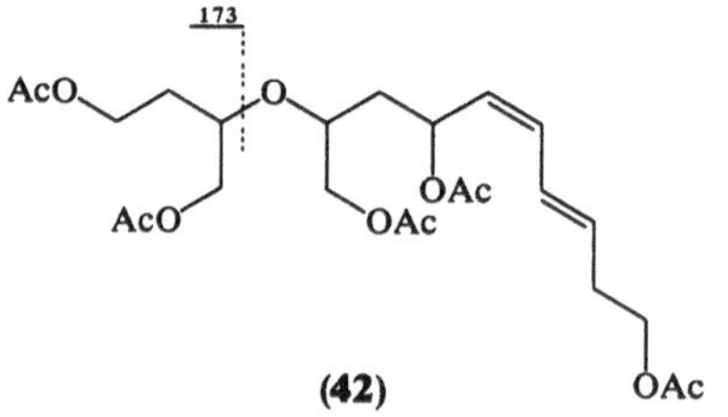

(42)

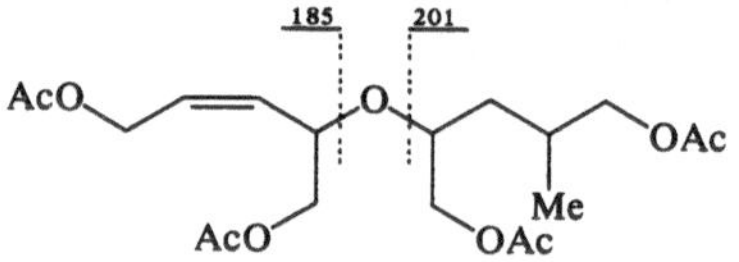

(43)

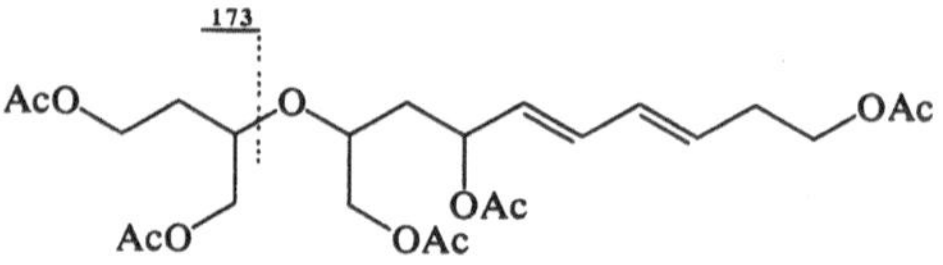

(44)

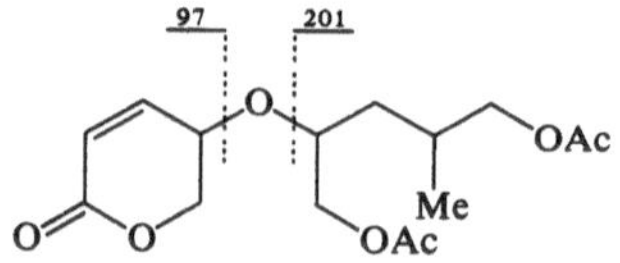

(45)

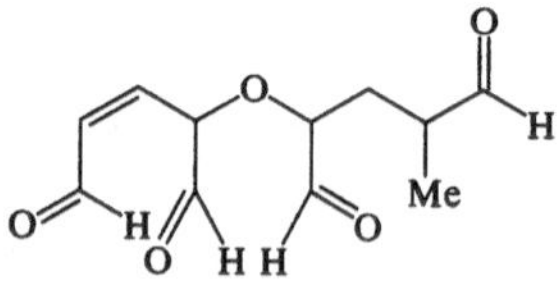

(46)

d) Degradation Products Implying C(28) — C(64) Segment

Long-term oxidation of palytoxin or N-p-bromobenzoylpalytoxin led to a complex mixture of lipophilic aldehydes and masked aldehydes which precipitated during the reaction as a gummy solid. Further oxidation of this mixture with periodate-permanganate (Lemieux-von Rudloff oxidation) resulted in a single product, palyoic acid (**47**) (20). Its molecular weight, 440

Table 2. ^{13}C *Chemical Shifts of Palyoic Acid in Dimethyl Sulfoxide-d_6 at 25 MHz*

^{13}C Chemical Shift[a]	Assignment[b]
18.76 q	Me on 34
20.34 q	Me on 37
21.31 q	Me on 39
22.46 t	43
23.87 d	39
24.57(2) t	30 and 47
26.16 t	32
28.09 d	34
28.8(4) t	31, 44, 45, and 46
33.64(2) t	29 and 48
37.50 t ⎫	
38.15 t[c] ⎬	33, 35, and 42
39.56 t[c] ⎭	
42.46 t[c]	40
44.30 t	38
77.94 d	36
80.05 s	37
106.74 s	41
179.3(2) s[c]	28 and 49

[a] In ppm using solvent peak as internal reference, 39.60.

[b] Using palytoxin numbering system in structure (**1**).

[c] In chloroform-d; signals obscured by DMSO-d_6 peaks or not clearly observed.

daltons, and elemental composition, $C_{25}H_{44}O_6$, were determined by mass spectrometry. Palyoic acid was a decarboxylic acid since it formed a dimethyl ester (**48**) on treatment with diazomethane. The ^{13}C nmr spectral data for (**47**) shown in Table 2 indicated the presence of 3 methyl, 15 methylene, 3 methine, and 4 quaternary carbons. Two of the quaternary carbons and four of the oxygens were accounted for by the two carboxylic acid functionalities. The two CO_2H groups had to be connected to methylene chains that were longer than three carbons, since the ^{13}C chemical shifts associated with the two $CH_2CH_2CO_2H$ groups were identical. The remaining two oxygens were connected to a quaternary carbon (δ 106.74). One of these oxygens was also connected to a second quaternary carbon (δ 80.05) and the other oxygen was attached to a methine (δ 77.94). Palyoic acid therefore possessed subunit (**47a**), which had to be present in a bicyclic system since P/HI reduction led to a mixture of C_{25} dicarboxylic acids (*26*). The five methylene signals in the 37.50 – 44.30 ppm region indicated that there were five methylenes connected to branched carbon atoms and the two methylene signals at δ 22.46 and 26.12 suggested that there were only two methylenes β to branched carbons.

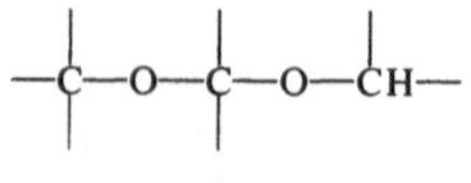

(47 a)

The environments of the three methyl groups were determined from the 600 MHz ^{1}H nmr spectrum of palyoic acid in dimethyl sulfoxide-d_6, which showed a singlet at δ 1.120 for a methyl group on a quaternary carbon and doublets at δ 0.876 (J = 7 Hz) and 0.870 (J = 7 Hz) for methyl groups attached to methine carbons. Decoupling experiments showed that the two secondary methyl groups were attached to two different methines (δ 1.64 and 2.00, respectively) and that one of these methyl groups was present in subunit (47 b). Since no further coupling of the methylenes in (47 b) with other protons could be observed, (47 b) had to be joined to the quaternary carbons of (47 a). Subunits (47 a) and (47 b) were therefore in a six-membered ring and this was supported by the coupling constants associated with (47 b). Since the oxygen-bearing methine of (47 a) (δ 3.87 in dimethyl sulfoxide-d_6) was a doublet of doublets (J = 12 and 1 Hz), this methine had to be attached to a methylene and a quaternary carbon. Spin-spin decoupling experiments with the aid of difference spectroscopy and in benzene-d_6 as the solvent showed that the latter methylene was connected to the methine bearing the other secondary methyl group as shown in (47 c). Palyoic acid was therefore a 6,8-dioxabicyclo[3.2.1] octane.

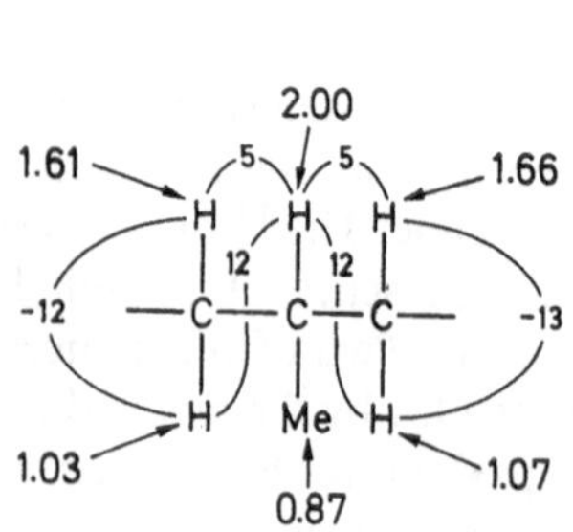

(47 b)

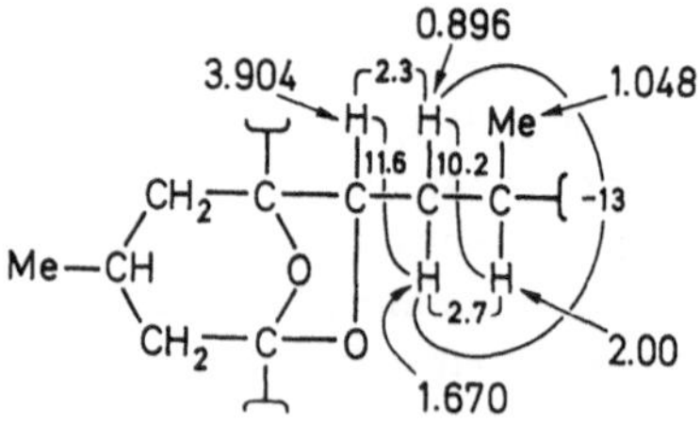

(47 c)

To complete the gross structure of palyoic acid, a methyl group had to be placed on one of the quaternary carbons and two methylene chains that were at least four carbons long had to be situated between the two carboxylic acid groups and the remaining two positions of (47 c). The position of the methyl group and the lengths of the methylene chains were determined by mass spectrometry. Palyoic acid and its dimethyl ester exhibited mass spectral fragmentation similar to that shown by *exo*-brevicomin (*27*). In the mass spectrum of *exo*-brevicomin (Scheme X) the three most intense fragment ion peaks were at m/z 43 for $CH_3C\equiv O^+$, 85 for $CH_2=CHC(OH^+)CH_2CH_3$ or $M-\text{ketene}-C_2H_5$, and 114 for $CH_3CH_2CH_2CH=C(OH^+)CH_2CH_3$ or $M-\text{ketene}$. In the mass spectrum of (**48**) the three most intense fragment ion peaks were at m/z 185 (ion c), 284 (ion d), and 241 (ion e). Ions c, d, and e corresponded to $CH_3C\equiv O^+$, $[M-CH_2CO]^+$, **and** $[M-CH_2CO-Et]^+$, respectively, from *exo*-brevicomin.

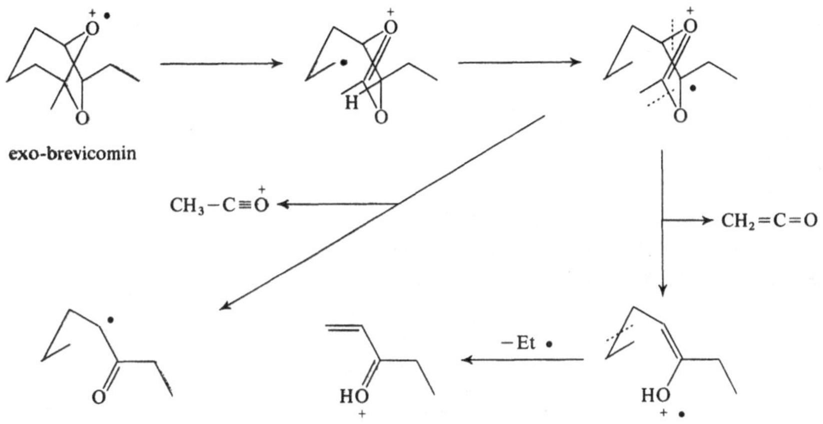

Scheme X

One of the aldehydes in the lipophilic gummy precipitate from long-term oxidation of palytoxin was the formate ester (**49**) which eliminated formic acid on sublimation to give the α,β-unsaturated aldehyde (**50**). Reduction of (**49**) with $NaBH_4$ gave a triol (**51**) which was fully characterized as the triacetate (**52**). ^1H nmr analysis of the gummy precipitate indicated that the other products were also formate esters. The other aldehydes in the gummy precipitate were not fully characterized, but reduction with borohydride led to comparable amounts of (**51**) and the tetrol (**53**). Tetrol (**53**) was also fully characterized as the tetraacetate (**54**).

Short-term oxidation of palytoxin or N-(*p*-bromobenzoyl)palytoxin followed by sodium borohydride reduction and acetylation led to (**52**), (**54**), and three new compounds (**55**), (**6**), and (**56**). Interestingly the trideuterated compounds (**57**) and (**58**) were formed when palytoxin was treated with D_2O at 55° prior to periodate oxidation. These degradation products pinpointed the location of the hemiketal functionality and implied that unit (**1j**) was present in palytoxin.

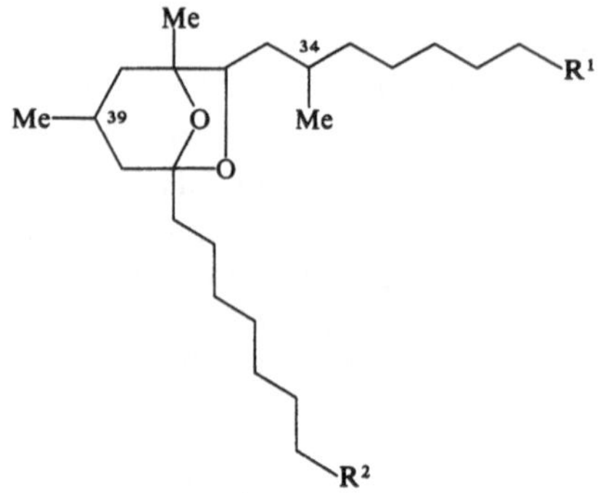

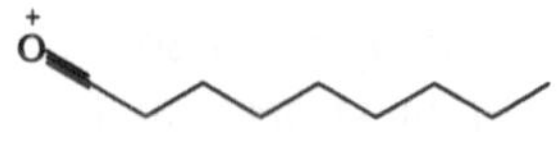

(47) $R^1 = R^2 = CO_2H$

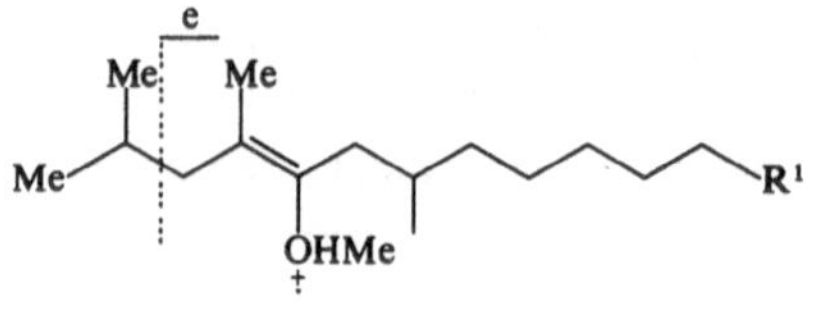

(48) $R^1 = R^2 = CO_2Me$

(49) $R^1 = CHO$; $R^2 = \overset{\displaystyle OCHO}{\underset{|}{CH}} - CH_2 - CHO$

(50) $R^1 = CHO$; $R^2 = CH \doteq CH - CHO$

(51) $R^1 = CH_2OH$; $R^2 = CHOH - CH_2CH_2OH$

(52) $R^1 = CH_2OAc$; $R^2 = CHOAcCH_2CH_2OAc$

(53) $R^1 = CH_2OH$; $R^2 = CHOHCH_2CHOHCH_2OH$

(54) $R^1 = CH_2OAc$; $R^2 = CHOAcCH_2CHOAcCH_2OAc$

(55) $R^1 = CH_2OAc$; $R^2 = CHOAcCH_2CHOAcCHOAcCH_2OAc$

(56) $R^1 = CH_2OAc$; $R^2 = CHOAcCH_2CHOAc\underset{|}{C}HCHOHOAcCH_2OAc$

$OCOCH_2CHOAcCHMeCH \doteq CHCHOAcCH_2CH_2CH_2OAc$

(57) $R^1 = CH_2OAc$; $R^2 = CHOAcCH_2\underset{|}{C}HCHOHOAcCHOAcCHDOAc$

$OCOCD_2CHOAcCHMeCH \doteq CHCHOAcCH_2CH_2CH_2OAc$

(58) $R^1 = CH_2OAc$; $R^2 = CHOAcCH_2CHOAc\underset{|}{C}HCHOHOAcCHDOAc$

$OCOCD_2CHOAcCHMeCH \doteq CHCHOAcCH_2CH_2CH_2OAc$

(59) $R^1 = CH_2OCO$—⟨benzene ring⟩—Br $R^2 = CHOHCH_2CH\langle\begin{smallmatrix}O - CH_2 \\ O - CH_2\end{smallmatrix}$

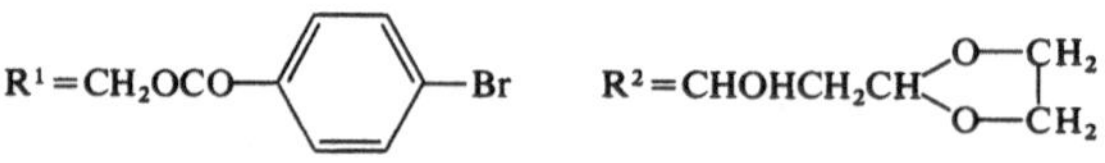

The structures of all of the **(1j)** related periodate products isolated by the MOORE group (20) were determined by extensive 600 MHz ^{1}H nmr and mass spectral studies. The electron-impact mass spectra always showed intense peaks for c, d, and e type ions.

The HIRATA group only isolated **(51)** and **(53)** from periodate oxidation of N-(p-bromobenzoyl)palytoxin, but were able to convert **(53)** to a crystalline derivative **(59)**, m. p. $60 - 62°$, and to establish the structure of **(59)** by X-ray crystallography (22).

The HIRATA group had been able to isolate tetraacetate **(60)** from periodate oxidation of palytoxin absorbed on polystyrene followed by borohydride reduction and acetylation. Even though isolation of this degradation product did not establish the position of the hemiketal functionality, it did identify the skeleton of the C(55) – C(64) segment for the first time.

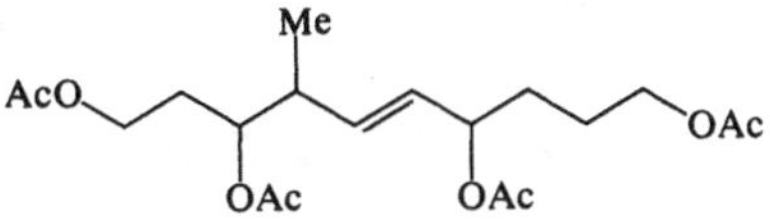

(60)

e) Degradation Products Implying C(89) – C(96) Segment

Periodate oxidation of palytoxin or N-(p-bromobenzoyl)palytoxin led to dialdehyde (61) but all attempts to isolate this substance failed as it was readily isomerized to the conjugated dienals (62a) and (62b) (9, 26).

Interestingly neither (61) nor the conjugated dienals were produced when the periodate oxidation was carried out by vigorously stirring the aqueous reaction mixture with dichloromethane. The ultraviolet spectrum of the dichloromethane layer indicated that dienals were not present. Borohydride reduction and acetylation of the dichloromethane extract led unexpectedly to a high yield of tetrahydropyran (64) (26). As shown in Scheme XI compound (64) appeared to have been formed by first an air-induced epoxidation of the diene system in (61) or the toxin before periodate cleavage, followed by borohydride reduction of the resulting epoxydial-dehyde (63) and subsequent acetylation of the intermediate ether diol.

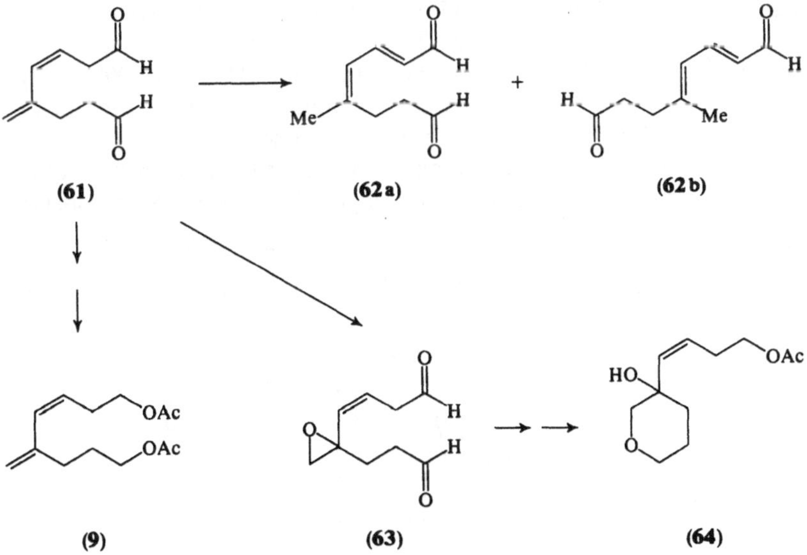

Scheme XI

Normal periodate oxidation followed by $NaBH_4$ reduction and acet-ylation gave diacetate (9). The structure of (9) was deduced from spectral data and supported by synthesis (17).

f) Degradation Products Implying C(109) − C(123) Segment

Short-term oxidation of N-(p-bromobenzoyl)palytoxin followed by reduction and acetylation led to a crystalline diacetate (11), m.p. 114−115.5°, which possessed the N-p-bromobenzoyl group and had a molecular formula $C_{26}H_{34}O_8NBr$ by mass spectrometry (19). Detailed ^1H nmr studies at 600 MHz in various solvents established two partial structures (11a) and (11b). All of the methines in (11a) and (11b) had to be connected to oxygen. One of these methines had an acetoxyl group on it (δ 5.320) whereas the remaining six methines were attached to ether oxygen (δ 3.991 − 4.488). Even though no coupling could be seen between the protons absorbing at 4.191 ppm and 4.115 ppm, (11a) and (11b) still had to be joined to each other by a carbon-carbon bond, since (11) did not exhibit the properties of an acetal. To complete the gross structure, the six ethereal methines had to be connected to three oxygen atoms to make three ether rings. Fifteen structures could be generated, but only one structure (11) was consonant with both nmr and mass spectral data. The coupling constants associated with the protons of unit (11b) indicated that all three methines in (11b) were located in a six-membered ring (11c). When the central methine in (11b) and the terminal methine of (11a) were joined in an ether ring and the remainder of (11a) was *exo* on the resulting 2,6-dioxabicyclo[3.2.1]octane system (11c), the dihedral angle between the protons on the carbons connecting (11a) and (11b) was 90°, in full agreement with the observed absence of coupling. This meant that the two remaining methines in (11a) had to be connected so as to form a tetrahydrofuran (11d). The coupling constants for the ring protons agreed with those seen in the trisubstituted tetrahydrofuran ring of monensin (19).

(11)

(11a)

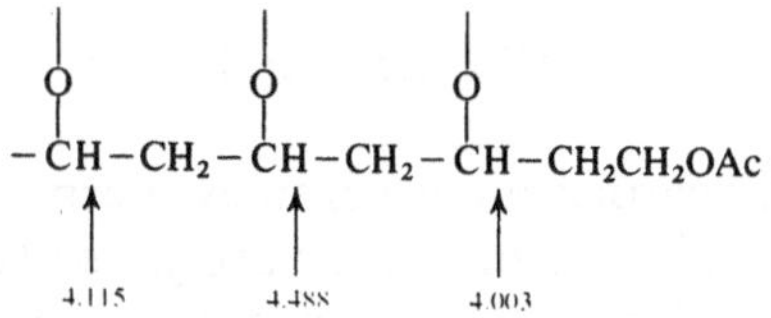

(11b)

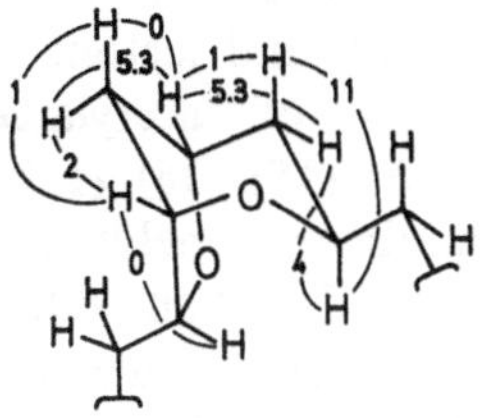

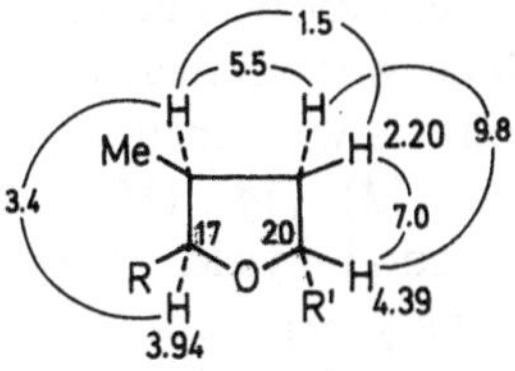

monensin

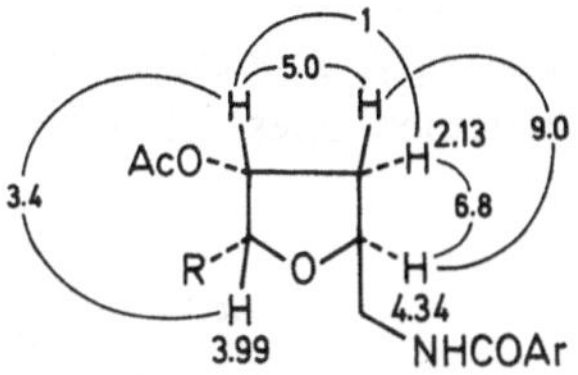

(11d)

(11c)

On periodate oxidation of palytoxin, the HIRATA group was able to isolate a ninhydrin positive substance which contained an aldehyde and a primary amino group and formed a N-acetyl derivative on treatment with *p*-nitrophenyl acetate (*9*). The same N-acetyl derivative was produced from periodate oxidation of N-acetylpalytoxin. Oxidation of N-*p*-bromo-benzoylpalytoxin led to aldehyde (**65**) which the HIRATA group was able to convert to the ethylene acetal (**66**) and finally to a crystalline acetate (**67**), m.p. 172–173°. The structure of (**67**) was determined by x-ray crystallographic analysis (*22*).

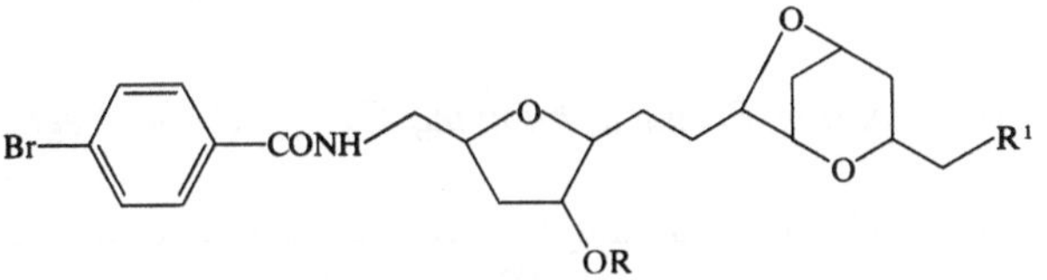

(65⁻67)

(65) R=H   R¹=CHO

(66) R=H   R¹=CH

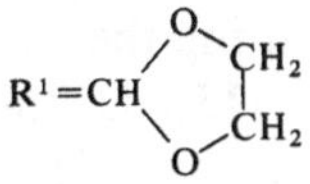

(67) R=Ac   R¹=CH

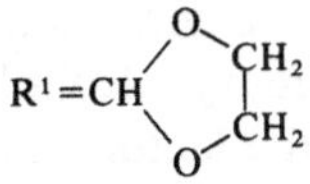

IV.2 Ozonolysis

The Moore group carried out two ozonolysis experiments, one on N-(p-bromobenzoyl)palytoxin and another on acetylated N-(p-bromobenzoyl)-palytoxin. Each ozonolysis was conducted in aqueous ethanol at 0°. The resulting ozonides were decomposed with sodium borohydride to a mixture of polyols which was acetylated with acetic anhydride in pyridine. The acetylated products were then separated by HPLC on silica.

Compounds (12), (14), (15), (16), (17), (18), and lactone (68) were obtained in good yield from ozonolysis of acetylated N-(p-bromobenzoyl)palytoxin. Compounds (12), (13), (15), (16), (17), and (18) were obtained from ozonolysis of N-(p-bromobenzoyl)palytoxin, but the yields of (15), (16), and (17) were much lower and (14) and (68) were not detected; interestingly, moderate amounts of partially acetylated compounds, in particular ones related to (14), (15), (16), and (17), were produced.

(68)

The Hirata group carried out ozonolyses of palytoxin, N-acetyl-palytoxin, and N-(p-bromobenzoyl)palytoxin. Since N-(p-bromobenzoyl)-palytoxin was soluble in methanol, ozonolysis could be readily conducted at −78° C. Decomposition of the ozonide with dimethyl sulfide followed by reduction with sodium borohydride or reduction of the ozonide by sodium borohydride alone resulted in a mixture of polyols, some of which could be separated by reverse-phase chromatography (23, 24). Separation of most of the polyols, however, was best achieved after acetylation.

a) Degradation Products Implying C(7) − C(14) Segment

Compound (13) was a major product on ozonolysis of N-(p-bromobenzoyl)palytoxin, whereas (68) was a major product on ozonolysis of acetylated N-(p-bromobenzoyl)palytoxin (Scheme XII). In the former case the imide (69), which was produced first, presumably rearranged to the formate ester (70) during the borohydride reduction step. Acetylation of (70) then produced (13). Ozonolysis of N-(p-bromobenzoyl)palytoxin, however, produced an imide (71) which on borohydride reduction gave (68) and formamide. In the conversion of (71) to (68), an acetyl group had to migrate to the primary alcohol group from the oxygen atom which eventually became the lactone oxygen.

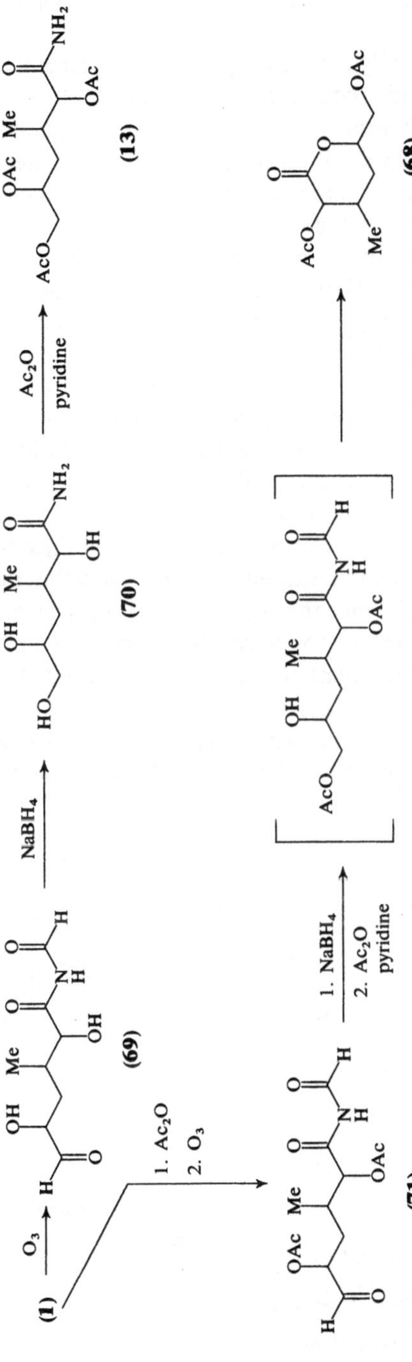

Scheme XII

b) Degradation Products Implying C(15) − C(59) Segment

Compound (14) was actually a mixture of four diastereoisomers which had resulted from borohydride reduction of the hemiketal functionality in palytoxin and the methyl ketone group generated by ozonolysis of the trisubstituted double bond. In addition to the four stereoisomers of gross structure (14), however, smaller amounts of four stereoisomers of gross structure (72) were also present in the same sample (20). Since the amount of the sample was small, no attempt was made by the MOORE group to separate the eight components by further HPLC. Instead the gross structures of each of the four stereoisomers (14) and the four stereoisomers (72) were elucidated by extensive ^{1}H nmr analysis of the mixture at 600 MHz.

The ^{1}H nmr analysis of the eight components initially presented a formidable and a seemingly insoluble problem. Actually the analytical problem was much simpler than anticipated and with the aid of double resonance difference spectra it was eventually possible to identify all of the signals for each component. Analysis of the C(15)−C(28) (palytoxin numbering system) region was less complex than expected since the proton signals were only doubled and not multiplied by eight. Only the stereochemistry at C(15) appeared to influence the chemical shifts of the signals in the C(15)−C(28) region, while the C(55) stereochemistry or the presence or absence of the C(57)−C(59) segment did not noticeably affect the chemical shifts of the protons on C(15)−C(28). Conversely there was no noticeable effect of C(15) stereochemistry on the chemical shifts in the C(49)−C(59) region. The doublet components for the C(15)−C(28) segment were of equal intensity while those for C(49)−C(59) in (14) and C(49)−C(56) in

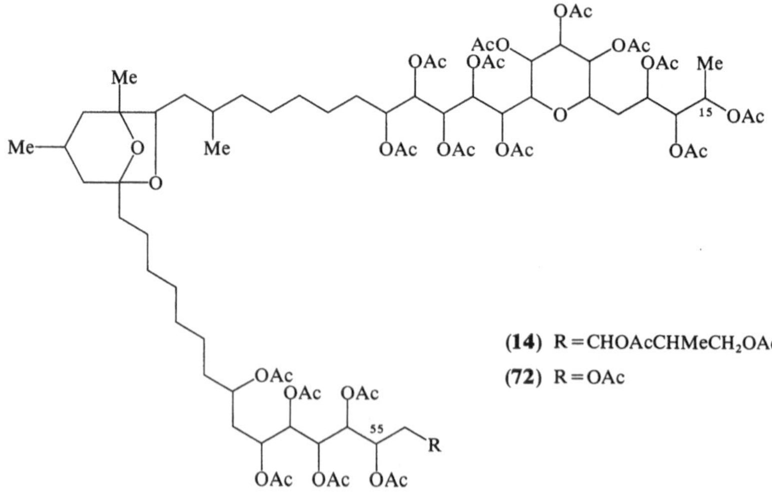

(14) R=CHOAcCHMeCH$_2$OAc

(72) R=OAc

(72) were not. This indicated that reduction of the methyl ketone group had not been stereoselective, but that there had apparently been appreciable stereoselectivity in the reduction of the hemiketal group.

The structures of **(14)** and **(72)** were supported by mass spectrometry. Molecular ions were observed in the field desorption mass spectra and characteristic brevicomin-type fragment ions were seen in the electron-impact mass spectra (20). Also seen in the EI mass spectra was an important fragment ion at m/z 503.

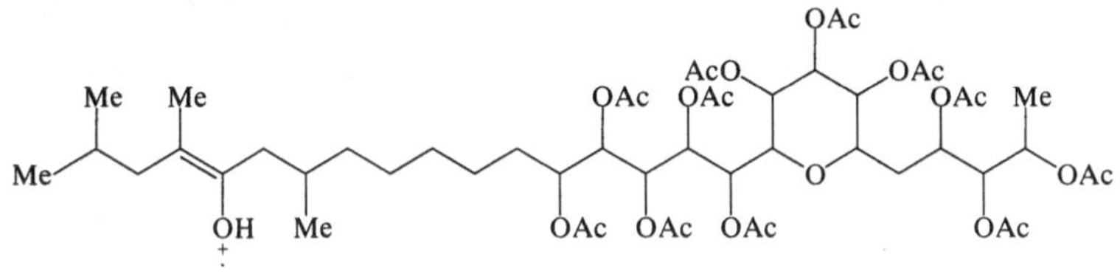

m/z 1088

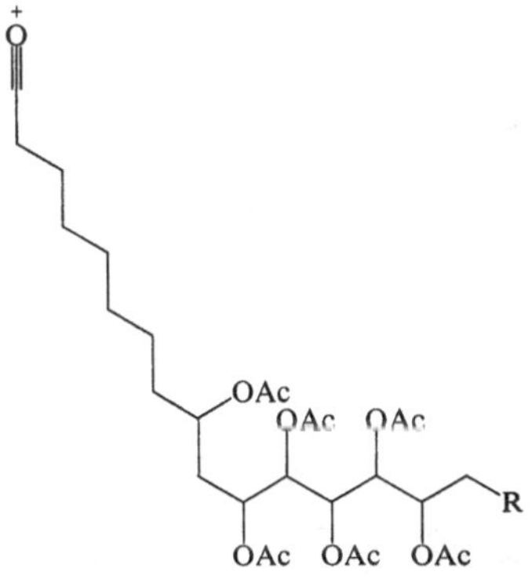

m/z 503

m/z 587 R=OAc
m/z 759 R=CHOAcCHMeCH$_2$OAc

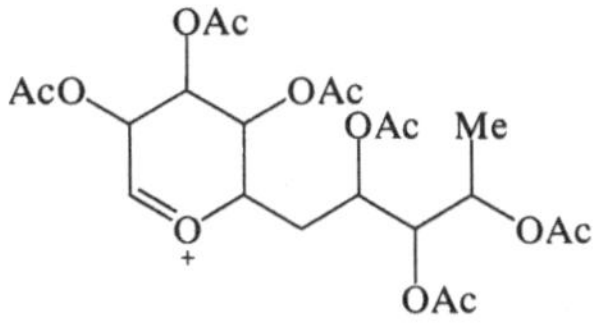

The HIRATA group was able to degrade N-(p-bromobenzoyl)palytoxin to a mixture of two epimeric polyols **(73)** by ozonolysis followed by incomplete reduction with sodium borohydride (24). Polyols **(73)** were epimeric at C(15). Attempted acetylation of the epimeric mixture **(73)** led to a low yield of two epimeric α,β-unsaturated ketones **(74)** instead of the corresponding polyacetates of **(73)**. Treatment of **(73)** with 50% aqueous acetic acid, on the other hand, led to two epimeric spiro ketals **(75)**, which could be acetylated to give the hexadecaacetates **(76)** in good yield. All of these compounds were characterized by field desorption mass spectrometry and high frequency ^{1}H nmr spectroscopy. To solve the structure of the C(15) to C(28) portion of these degradation products, **(73)** was oxidized gently with sodium periodate to give, after reduction with sodium borohydride, two decaols (77) (24). ^{1}H nmr and mass spectral analysis of the corresponding decaacetates (78) established the gross structures.

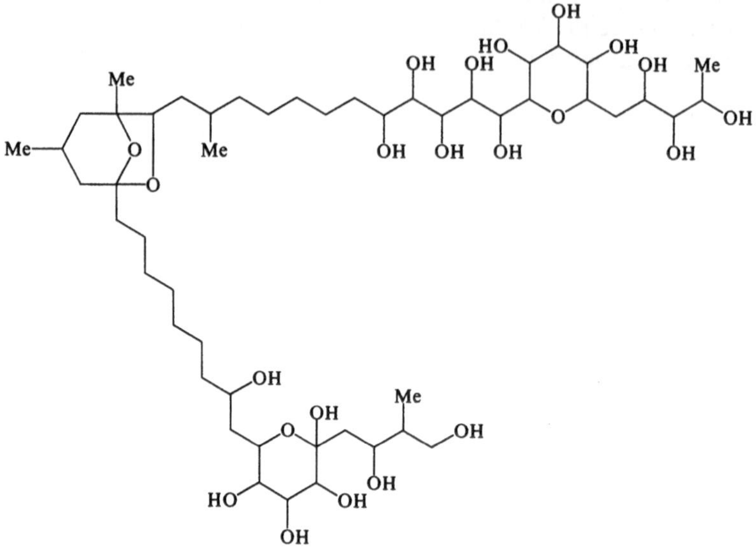

(73)

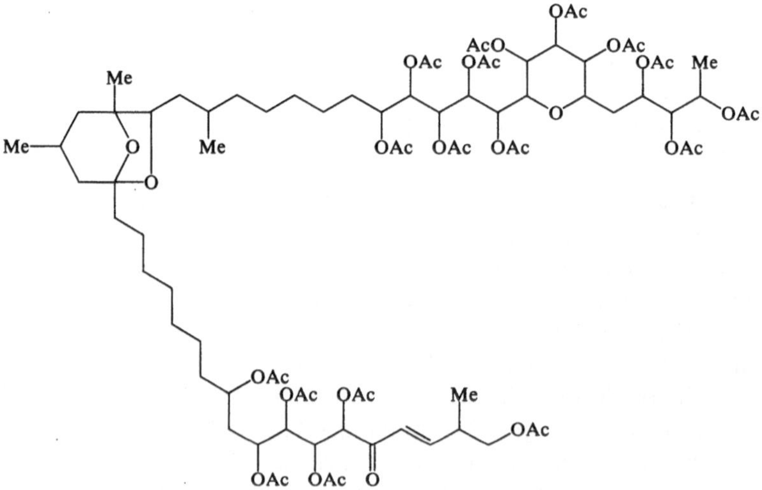

(74)

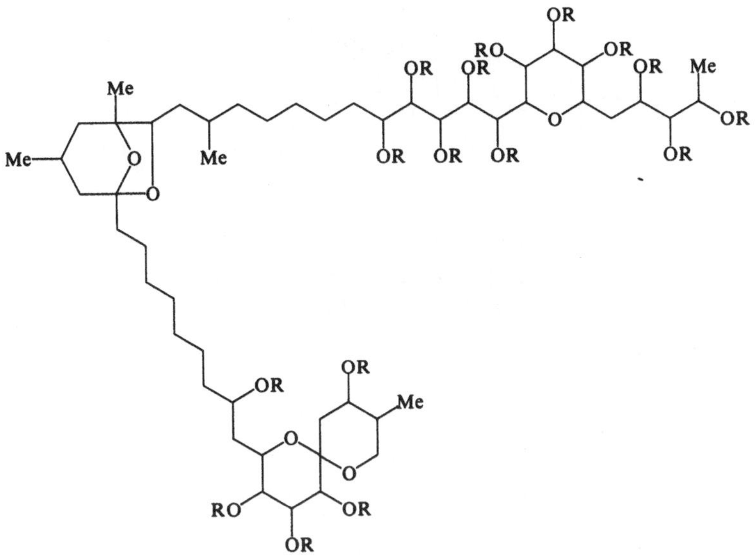

(75) R=H
(76) R=Ac

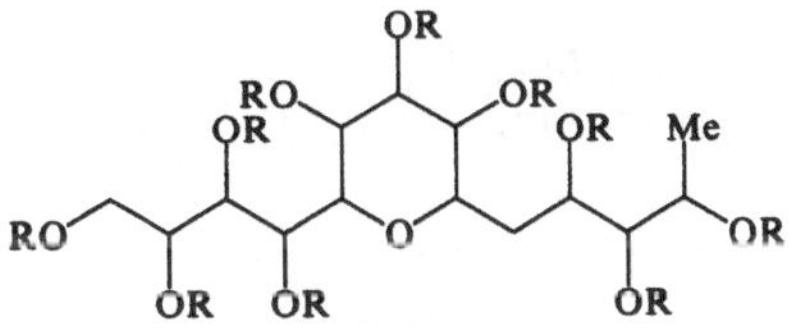

(77) R=H
(78) R=Ac

c) Degradation Products Implying C(60) – C(84) Segment

With the aid of proton-proton decoupled difference spectra, the MOORE group was able to elucidate the structure of compound (15) (20). The HIRATA group also isolated the same compound along with (79) (23, 24). Compound (79) could be converted into (15) by further ozonolysis followed by borohydride reduction and acetylation. To confirm the structure of (15), the corresponding polyol (80) was subjected to periodate oxidation to give the hexaol (81) which was characterized as the hexaacetate (82) (23, 24).

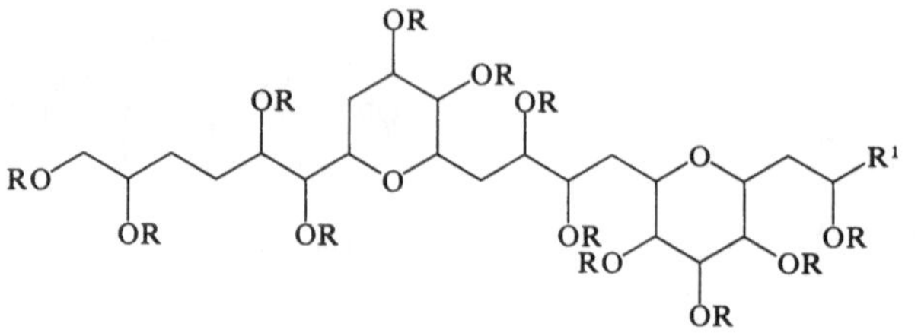

(15) R=Ac; R^1=CH$_2$OAc
(79) R=Ac; R^1=CH$\overset{c}{=}$CH–CH$_2$OAc
(80) R=H; R=CH$_2$OH

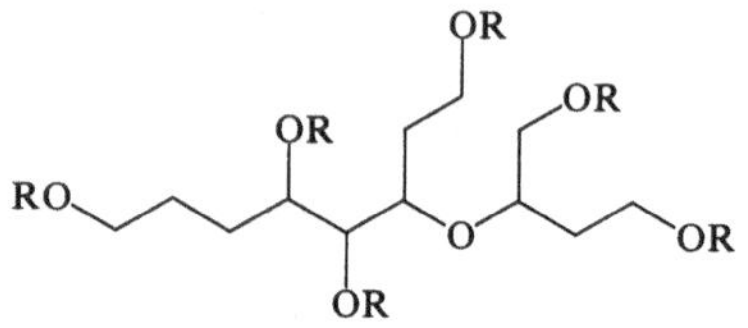

(81) R=H
(82) R=Ac

d) Degradation Products Implying C(85) – C(91), C(92) – C(106), and C(107) – C(123) Segments

The structures of (16), (17), and (18) were deduced by detailed ^{1}H nmr and mass spectral analysis (20). Actually (17) was a mixture of two epimeric compounds which had arisen from non-selective borohydride reduction of the ketone functionality produced by ozonolysis of the terminal methylene group on C(93) of palytoxin. Two C(93) epimeric compounds of gross structure (83) were obtained in moderate yield on incomplete ozonolysis of acetylated N-(p-bromobenzoyl)palytoxin followed by borohydride reduction and acetylation (28), thus confirming for the MOORE group that units (1u) and (1v) were connected to each other.

Partial ozonolysis of N-(p-bromobenzoyl)palytoxin gave after reductive workup polyols (84) and (85) (23). Compound (84) established for the HIRATA group that units (1u) and (1v) were connected to each other and (85) showed that units (1t), (1u), and (1v) were connected in sequence. Compound (85) was a mixture of the C(93) epimers. Ozonolysis of (84) and reductive workup led to tetraol (86) and two epimeric nonaols (87). Similarly ozonolysis of (85), followed by reductive workup with borohydride, led to (86), (87), and the pentaol (88).

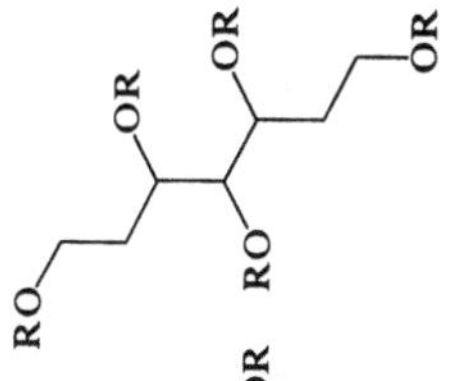

(16) R = Ac
(88) R = H

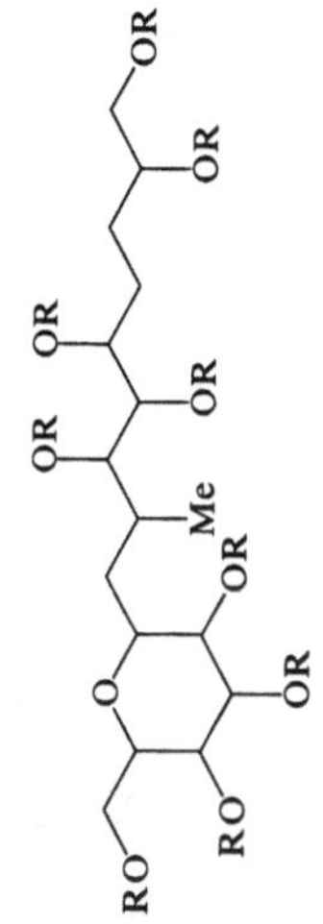

(17) R = Ac
(87) R = H

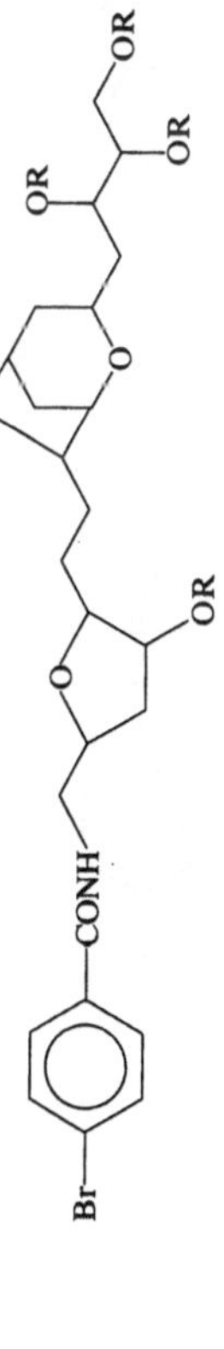

(18) R = Ac
(86) R = H

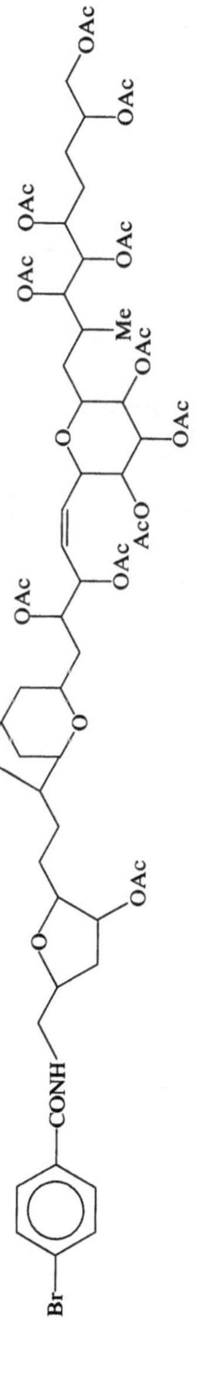

(83)

(84)

(85)

V. Stereochemistry

The absolute stereochemistry of the entire molecule, depicted in structure (1), has been unambiguously determined by HIRATA's X-ray crystallographic studies (22) and KISHI's synthetic work (15, 29, 30, 31). The absolute configurations of the 64 asymmetric carbons are 10S, 11R, 13R, 16R, 17S, 19R, 20S, 21R, 22S, 23S, 24S, 25R, 26S, 27S, 28R, 34S, 36R, 37R, 39S, 41S, 49S, 51R, 52S, 53R, 54R, 55S, 57S, 58S, 61S, 64R, 65R, 66S, 68R, 69S, 70R, 72R, 73R, 75R, 76R, 77R, 78S, 79S, 81S, 87R, 88S, 89R, 96S, 97R, 98R, 99R, 101R, 102S, 103R, 104R, 105R, 108R, 109R, 111S, 113R, 115S, 116R, 119R, 120R, 122S.

MOORE had proposed stereochemistry for 60 of the 64 asymmetric carbons (28), but twelve assignments, namely those at C(17), C(25), C(26), C(27), C(28), C(65), C(66), C(68), C(69), C(70), C(72), and C(98), were later shown to be incorrect (15). In all, seven errors led to the twelve misassignments. The MOORE group had established the correct relative stereochemistry of the six chiral centers in the C(65) − C(72) segment from ^{1}H nmr data, but unfortunately failed to relate the stereochemistry of this segment correctly to C(64) and C(73). Even though the relative chemical reactivities of the two vicinal diol groups flanking the C(66) − C(70) tetrahydropyran ring towards periodate oxidation and borate complexation strongly suggested that C(64) − C(65) was *anti* and C(72) − C(73) was *syn*, the MOORE group assigned opposite stereochemistry to these two diol functionalities on the basis of ^{1}H nmr data which had been misinterpreted. These two errors accounted for six of the misassignments.

The MOORE group had also determined the correct relative and absolute stereochemistry of the five chiral centers in the C(19) − C(23) tetrahydropyran ring, but made two errors in establishing the stereochemistries of the acyclic vicinal diol functionalities immediately adjacent to the tetrahydropyran ring, i. e. on C(16) − C(17) and C(24) − C(25), and two more errors in correlating the relative stereochemistries of the two diol groups to the tetrahydropyran ring. These four errors accounted for five more of the misassignments. One of the errors could have been avoided since the relatively sluggish reactivity of the C(16) − C(17) bond towards periodate cleavage indicated that the OH groups on C(16) and C(17) were *anti*. All four errors were the result of inadequate model studies. The information obtained from the ^{1}H nmr spectra of the fully acetylated degradation products was insufficient to distinguish among the sixteen possible C(16) C(17), C(24), C(25) stereoisomers.

The twelfth misassignment was made in the stereochemistry of C(98). The MOORE group had determined the correct relative and absolute stereochemistry for the C(101) − C(105) tetrahydropyran ring and had fortuitously related its stereochemistry to the methyl group on C(99). They

had even proposed that *syn*-1,3 substitution was present in the
C(96) — C(99) segment since the coupling constants associated with this
segment in the fully acetylated ozonolysis products were all medium-sized.
Furthermore the reactivity of the C(96), C(97), C(98) triol towards
periodate oxidation and borate complexation suggested that C(97) — C(98)
had *syn* stereochemistry. Unfortunately they misinterpreted the results of a
low temperature ^{1}H nmr study of one of the epimers of (17) and incorrectly
assigned *syn, anti, syn* stereochemistry to C(96), C(97), C(98), C(99).

Further work by the MOORE group has now shown that the absolute
stereochemistries of the palytoxins from Hawaiian *Palythoa toxica* and
P. tuberculosa are the same as that reported (15) for palytoxin from
Okinawan *P. tuberculosa* (32). New methods and procedures were developed to solve the relative and absolute stereochemistry in the acyclic
regions of the palytoxin molecule.

V.1 Methods Used for Relative Stereochemistry

a) X-Ray Crystallography

The HIRATA group was able to determine the relative stereochemistry of
13 asymmetric carbons in palytoxin, viz. C(34), C(36), C(37), C(39), C(41),
C(49), C(111), C(113), C(115), C(116), C(119), C(120), and C(122), by x-ray
crystallographic analysis of acetals (59) and (67) (22). Compound (59),
which crystallized from hexane with m.p. 60 — 62°, was formed in three
steps: periodate oxidation of tetraol (53) to aldehyde (89), acetalization of
(89) with ethylene glycol and *p*-toluenesulfonic acid in refluxing benzene to
(90), and esterification of (90) with *p*-bromobenzoyl chloride in pyridine at
50° to (59). Compound (67), which crystallized from benzene with m.p.
172 — 173°, was formed in two steps: treatment of the periodate oxidation
product (65) with ethylene glycol and *p*-toluenesulfonic acid in refluxing
benzene/1,2-dimethoxyethane to (66) followed by acetylation of (66) to
(67).

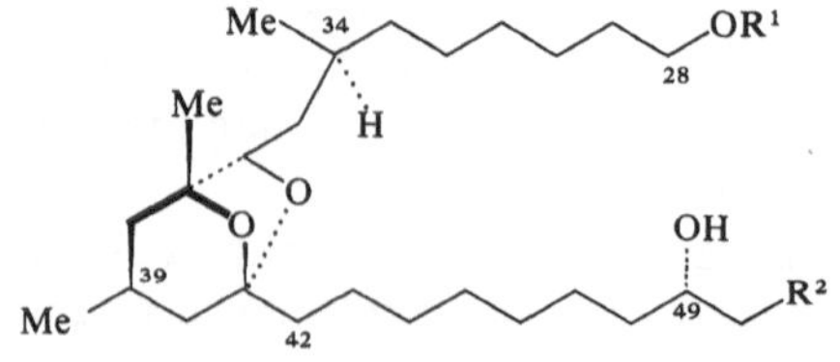

(53) R^1=H, R^2=CHOHCH$_2$OH

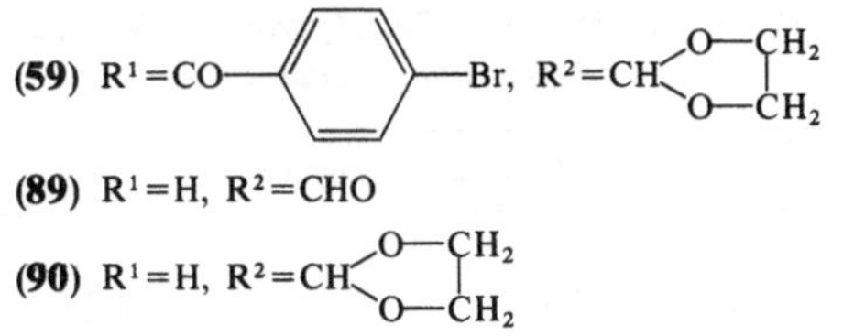

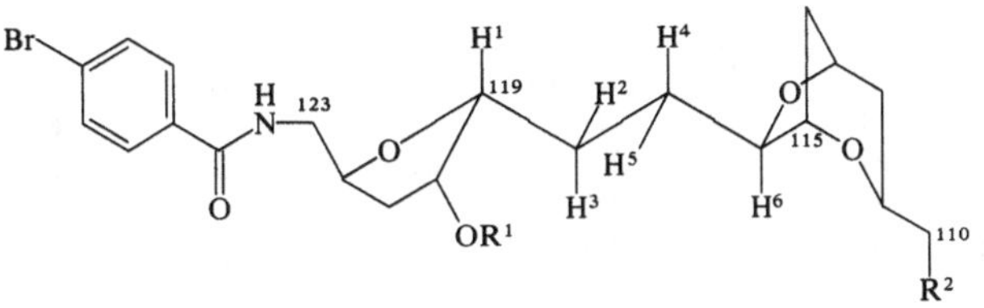

(59) R¹=CO—⟨benzene⟩—Br, R²=CH⟨O—CH₂ / O—CH₂⟩

(89) R¹=H, R²=CHO

(90) R¹=H, R²=CH⟨O—CH₂ / O—CH₂⟩

(65) R¹=H, R²=CHO

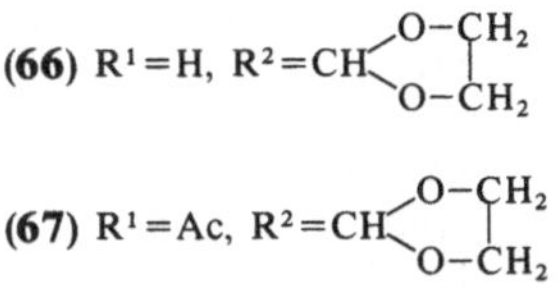

(66) R¹=H, R²=CH⟨O-CH₂ / O-CH₂⟩

(67) R¹=Ac, R²=CH⟨O-CH₂ / O-CH₂⟩

In the solid state of (59), the $C(42)-C(43)-C(44)$ ... side chain is coplanar and fully extended with $C(41)-C(40)-C(39)-CH_3$ and not coplanar with either of the $C(41)-O$ groups. Similarly the $C(33)-C(34)-C(35)$ segment of the other side chain is coplanar and fully extended with $C(36)-C(37)-C(38)$ and not with $C(36)-O$. Interestingly the $C(33)-C(32)-C(31)$ ... segment is coplanar and fully extended with $C(34)-CH_3$ and not with $C(34)-C(35)-C(36)-C(37)-C(38)$.

In the solid state of (67), $C(117)-C(118)$ orients in a planar, zig-zag manner with $C(116)-C(115)$ of the 2,6-dioxabicyclo[3.2.1]octane ring but not with $C(119)-C(120)$ of the tetrahydrofuran ring. Unlike crystal structures of other 2,5-disubstituted tetrahydrofurans in which the alkyl chains are fully extended with the C-C bonds in the ring (33), $C(117)-C(118)$ is fully extended with $C(119)-O$ in (67). 600 MHz ¹H nmr studies of (11), however, indicate that in solution $C(115)-C(116)-C(117)-C(118)-C(119)-C(120)$ is planar and fully extended (28) (see below).

b) ¹H NMR Spectroscopy

Thirty-five of the 64 asymmetric carbons in palytoxin are included in rings. The relative stereochemistries of these 35 chiral centers were deduced in a straightforward manner by ¹H nmr analysis of degradation products containing the intact rings (Table 3).

Table 3. *Coupling Constants Associated with Tetrahydropyran Methine Protons in Palytoxin Degradation Products*

Assignment	Compound	J(Hz)	References	Assignment	Compound	J(Hz)	References
19 eq, 20 ax	**(14)**	4.8	*(20, 28)*	77 ax, 78 ax	**(8)**	9.5	*(19)*
	(78)	4.5	*(24)*		**(15)**	9.6	*(20, 28)*
20 ax, 21 ax	**(14)**	8.6	*(20, 28)*			8	*(23)*[a]
	(78)	9.5	*(24)*	78 ax, 79 ax	**(8)**	9.5	*(19)*
21 ax, 22 ax	**(14)**	8.6	*(20, 28)*		**(15)**	9.6	*(20, 28)*
	(78)	9.5	*(24)*			8	*(23)*[a]
22 ax, 23 ax	**(14)**	9.8	*(20, 28)*	101 eq, 102 ax	**(10)**	5.5	*(19)*
	(78)	9.9	*(24)*			6	*(21)*[b]
51 ax, 52 eq	**(73)**	0.9	*(24)*		**(17)**	6.3[c]	*(20, 28)*
	(76)	2.0	*(31)*			8[d]	*(21)*
52 eq, 53 eq	**(73)**	3.2	*(24)*	102 ax, 103 ax	**(10)**	8.5	*(19)*
	(76)	3.6	*(31)*			8	*(21)*[b]
53 eq, 54 ax	**(73)**	3.2	*(24)*	102 ax, 103 ax	**(17)**	9.1[c]	*(20, 28)*
	(76)	4.0	*(31)*			8	*(21)*
66 eq, 67 eq	**(7)**	3.8	*(19)*	103 ax, 104 ax	**(10)**	8.5	*(19)*
	(15)	3.4	*(20, 28)*			8	*(21)*[b]
66 eq, 67 ax	**(7)**	5.0	*(19)*		**(17)**	9.1[c]	*(20, 28)*
	(15)	5.1	*(20, 28)*			8	*(21)*
68 ax, 69 ax	**(7)**	7.2	*(19)*	104 ax, 105 ax	**(10)**	8.5	*(19)*
	(15)	8.2	*(20, 28)*			8	*(21)*[b]
69 ax, 70 ax	**(7)**	7.2	*(19)*		**(17)**	9.1[c]	*(20, 28)*
	(15)	8.2	*(20, 28)*			8	*(21)*
75 ax, 76 ax	**(8)**	9.5	*(19)*				
	(15)	9.6	*(20, 28)*				
		8	*(23)*[a]				
76 ax, 77 ax	**(8)**	9.5	*(19)*				
	(15)	9.6	*(20, 28)*				
		8	*(23)*[a]				

[a] An incorrect structure for **(15)** is reported in this reference; the structure is revised in reference *(24)*.

[b] A wrong structure for **(10)** is reported in this reference.

[c] The same coupling constant is observed for both C(93) epimers.

[d] This value appears to be too large.

The Moore group was able to establish the relative stereochemistries of the four tetrahydropyran rings possessing C(19)−C(23), C(66)−C(70), C(75)−C(79), and C(101)−C(105) on the basis of coupling constant data from ozonolysis products **(14)**, **(15)**, and **(17)** and periodate oxidation products **(7)**, **(8)**, and **(10)** *(19, 20, 28)*. Sixteen of the nineteen methine protons in these rings had to be axial. Each of the sixteen protons showed at least one coupling constant of 7.2−9.5 Hz (Table 3) to neighboring protons on the ring. The remaining three methine protons, however, namely the ones on C(19), C(66), and C(101), had to be equatorial since the coupling constants to ring protons were all 4.8 – 5.5 Hz. The relative stereochemistries of the four tetrahydropyran rings in compounds (14), (15), and (17) are shown in partial structures (14a), (15a), (15b), and (17a).

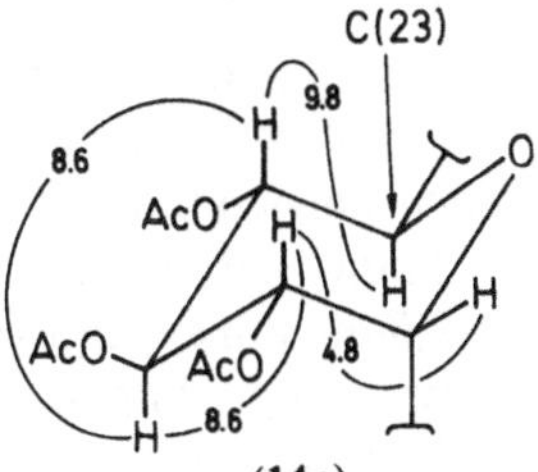

(14a)

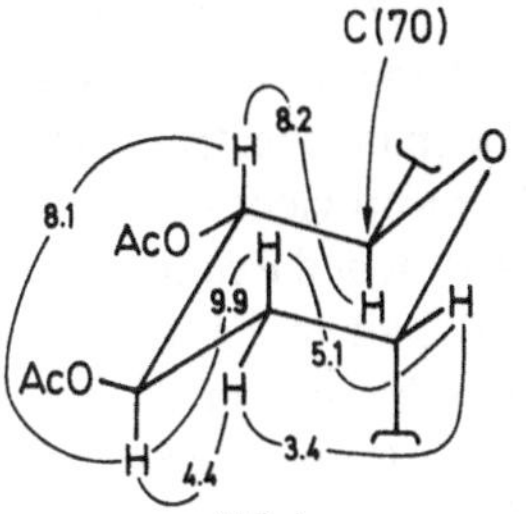

(15a)

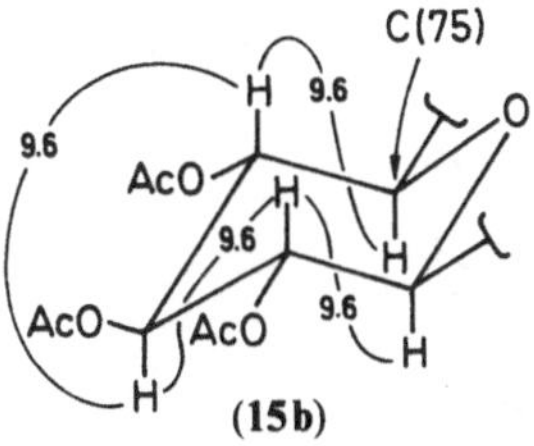

(15b)

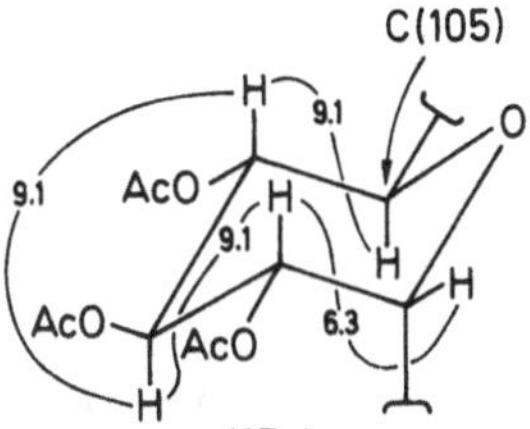

(17a)

The HIRATA group succeeded in degrading palytoxin to a compound (73) in which the hemiketal functionality was still intact (24). Coupling constants showed that the tetrahydropyran ring associated with this cyclic hemiketal had the relative stereochemistry shown in (73a). Additional evidence for this stereochemistry was provided by ^{1}H nmr analysis of (76) (31).

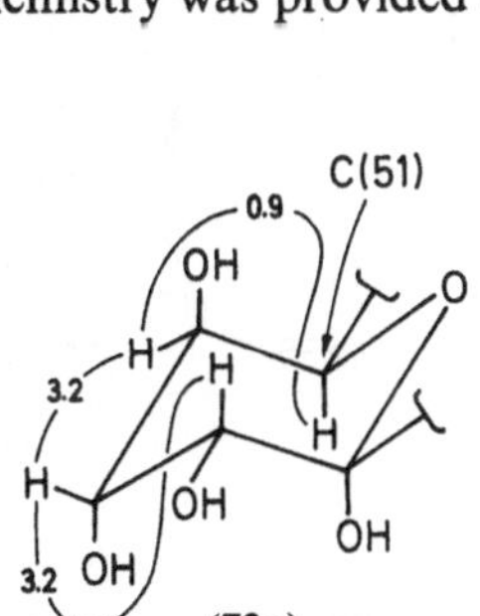

(73a)

The MOORE group was able to determine the relative stereochemistries of the two bicyclic ring systems in palytoxin from detailed spectral analysis of periodate oxidation products (47) and (11) (19, 20). The methyl group on C(39) had to be equatorial since the proton on C(39) in (47) showed large

axial-axial and small axial-equatorial couplings to the protons on C(38) and
C(40). Similarly C(110) had to be attached equatorially to C(111) since the
proton on C(111) in (11) showed a large axial-axial coupling and a small
axial-equatorial coupling to the protons on C(112). As discussed in the
gross structure determination section, ^{13}C nmr data and mass spectral data
showed that the bicyclic ring system in (47) was a 6,8-dioxabicyclo[3.2.1]-
octane. The mass spectral fragmentation pattern of (47), which was similar
to that of *exo*-brevicomin but not to that of *endo*-brevicomin, indicated that
C(35) was attached *exo* to the bicyclic ring. ^{1}H nmr analysis of (11) showed
that C(117) had to be attached *exo* to the 2,6-dioxabicyclo[3.2.1]octane
system since the proton on C(116) did not show any coupling to the proton
on C(115). With *exo* stereochemistry at C(117) the dihedral angle between
the protons on C(116) and C(115) was 90°. The relative stereochemistries of
the two bicyclic rings in (47) and (11) are shown in partial structures (47d)
and (11e).

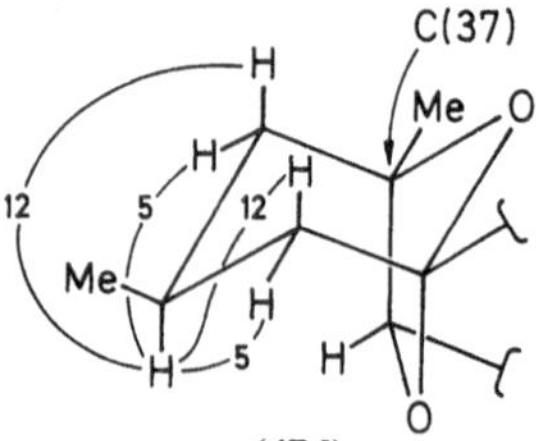

(47d)

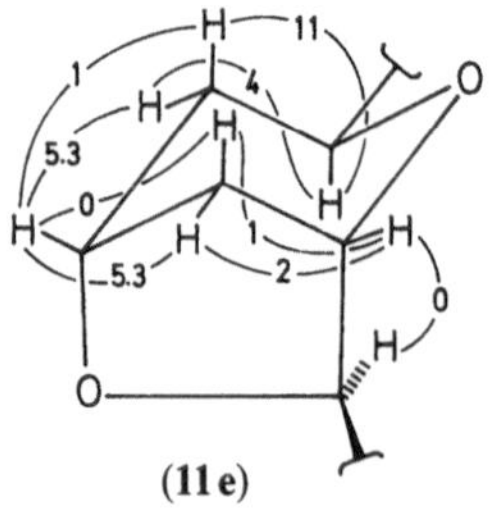

(11e) 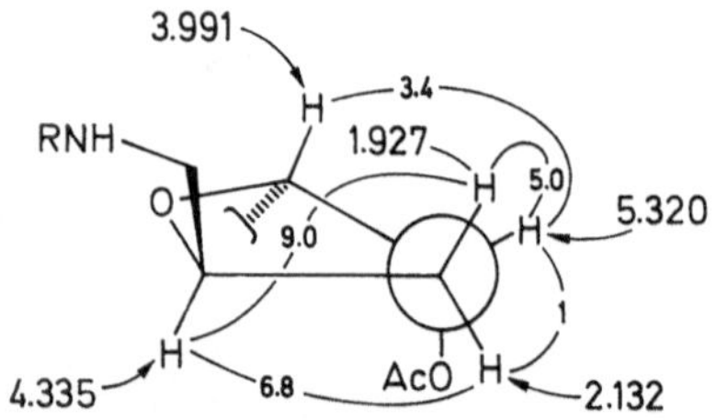   (11f)

Finally the MOORE group was able to conclude that the tenth ring, a
tetrahydrofuran ring, had the relative stereochemistry shown in partial
structure (11f) (*19*). The coupling constants and chemical shifts for the
C(119) – C(122) protons in (11) were very similar to those observed for the
protons in the tetrahydrofuran ring in monensin (19) and in (91), an
acetylated dihydro derivative of a marine algal metabolite (*33, 34*).

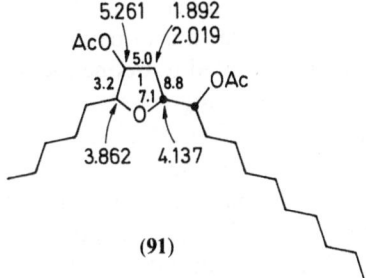

(91)

The relative stereochemistry of the three asymmetric carbons in the acyclic C(9) – C(14) segment was determined by ^{1}H nmr analysis of a cyclic degradation product (*20, 28*). Coupling constant data for the ozonolysis product (**68**) and the corresponding diol (**92**), produced by acid hydrolysis of (**68**) (*28*), indicated that these lactones had the relative stereochemistry shown. The relative stereochemistry in the related ozonolysis product (**13**) was therefore $10S^*, 11R^*, 13R^*$. The coupling constants observed for (**13**) were consistent with a fully extended conformation.

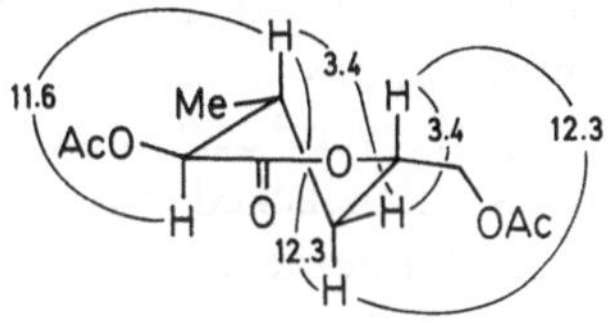

(68)

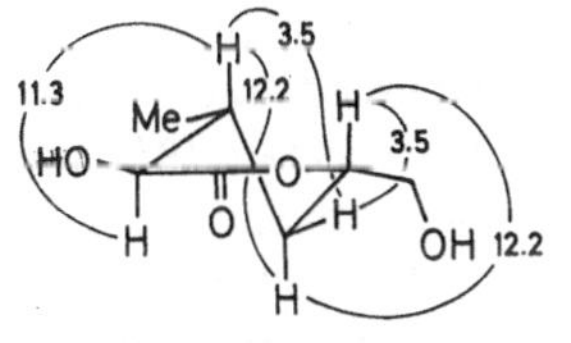

(92) 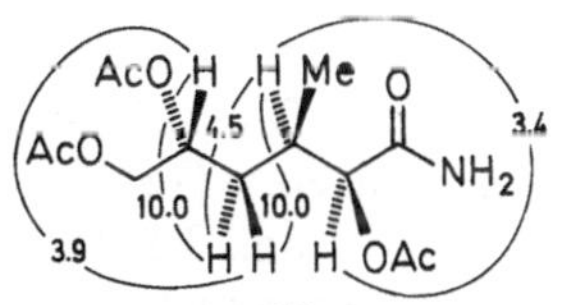   **(13)**

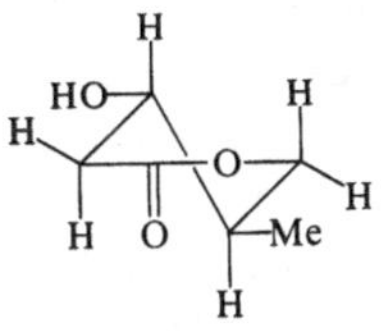

(93)

The relative stereochemistry of C(57) and C(58) could have been determined by ^{1}H nmr analysis of a cyclic degradation product such as (**93**), but this was not done. The HIRATA group had degraded palytoxin to cyclic ketal (**76**) (*24, 31*), but apparently was not able to deduce the relative stereochemistry of these two asymmetric carbons by ^{1}H nmr analysis.

Determination of relative stereochemistry of the remaining 24 asymmetric carbons in the acyclic regions of palytoxin was not straightforward.
The MOORE group had attempted to define the relative stereochemistry of
the acyclic regions from coupling constants observed in fully acetylated
degradation products, but unfortunately made several errors (28).

Assignment of correct stereochemistry from coupling constants proved
to be a very difficult task since the conformation of the acyclic segment
under examination had to be deduced first. In most cases it was not possible
to determine the preferred conformation unambiguously. The assumption
could be made that an unsubstituted acyclic carbon chain was generally
fully extended (planar, zig-zag) in the preferred conformation (22, 23). If
acetoxy substituents were present on this chain, the fully extended
conformation would still be preferred as long as none of the acetoxy groups
were syn-1,3 (35). If syn-1,3 acetoxy groups were present, however, the fully
extended conformation would no longer be preferred since in this conformation the acetoxy groups would be eclipsed. The resulting steric interaction between the two acetoxy substituents would be comparable with the
non-bonded interaction between two axially-orientated cis acetoxy groups
on a six-membered ring. Instead of a planar zig-zag conformation, one or
more non-planar bent conformations would now predominate.

In a detailed ^{1}H nmr study of several acetylated alditols (36), the MOORE
group found that compounds which lacked syn-1,3 acetoxy groups such as
arabinitol pentaacetate, mannitol hexaacetate, galactitol hexaacetate, and
perseitol heptaacetate (94) generally showed large couplings (8.4 – 10 Hz)
for anti-1,2 protons and small couplings (2 – 2.5 Hz) for syn-1,2 protons.
These coupling constants indicated that in each of these molecules the
preferred conformation was the fully extended one. There were a few
exceptions. 3-Deoxy-D-arabino-hexitol pentaacetate (95) and 3-deoxy-L-
xylo-hexitol pentaacetate (96), for example, both showed a small coupling
constant for $J_{4,5}$, 3.1 and 3.8 Hz, respectively. Compounds that possessed
syn-1,3 acetoxy groups such as 3-deoxy-L-ribo-hexitol pentaacetate (97),
3-deoxy-D-lyxo-hexitol pentaacetate (98), D-allitol hexaacetate (99),
D-iditol hexaacetate (100), sorbitol hexaacetate (101), and meso-glycero-
gulo-heptitol heptaacetate (102), however, usually showed medium-sized
coupling constants (4.5 – 7 Hz) for anti-1,2 and syn-1,2 protons, indicating
that the preferred conformations for these molecules were non-extended
ones. Again there were some exceptions. For example, D-glycero-D-
manno-heptitol heptaacetate (103), which has syn acetoxy groups on
C(2) and C(4), showed only large (8.6 – 8.8 Hz) and small (2.4 – 3.3 Hz)
coupling constants. Two or more non-extended conformations were
obviously contributing to the coupling constants for compounds (95 – 102),
but only one non-extended conformation was apparently important for
(103).

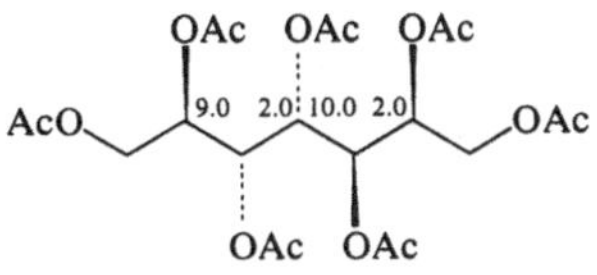

OAc OAc OAc

AcO 9.0 2.0 10.0 2.0 OAc

OAc OAc

(94)

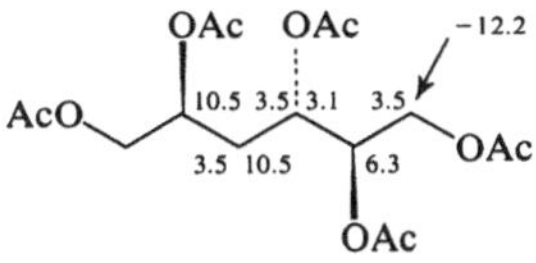

OAc OAc −12.2

AcO 10.5 3.5 3.1 3.5 OAc

3.5 10.5 6.3

OAc

(95)

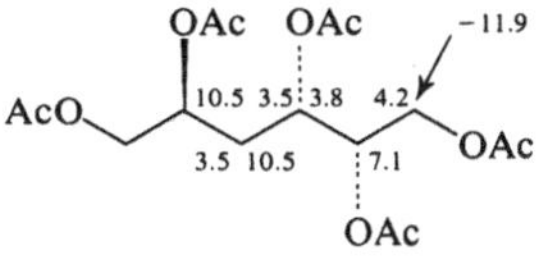

OAc OAc −11.9

AcO 10.5 3.5 3.8 4.2 OAc

3.5 10.5 7.1

OAc

(96)

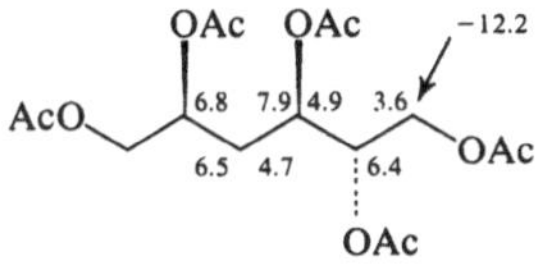

OAc OAc −12.2

AcO 6.8 7.9 4.9 3.6 OAc

6.5 4.7 6.4

OAc

(97)

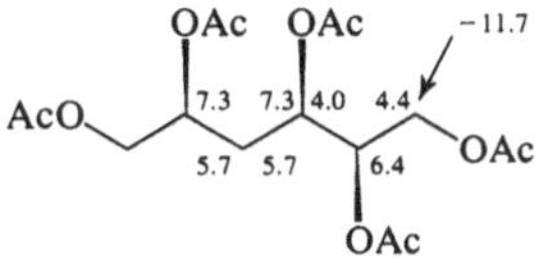

OAc OAc −11.7

AcO 7.3 7.3 4.0 4.4 OAc

5.7 5.7 6.4

OAc

(98)

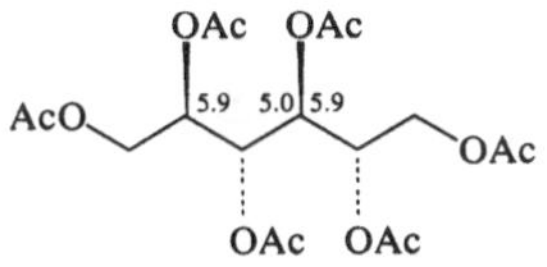

OAc OAc

AcO 5.9 5.0 5.9 OAc

OAc OAc

(99)

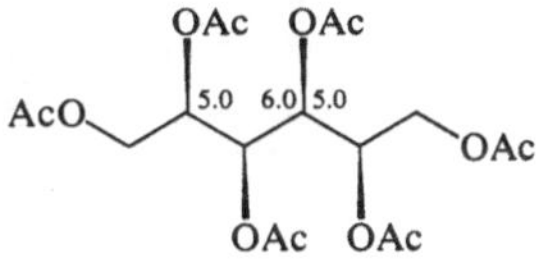

OAc OAc

AcO 5.0 6.0 5.0 OAc

OAc OAc

(100)

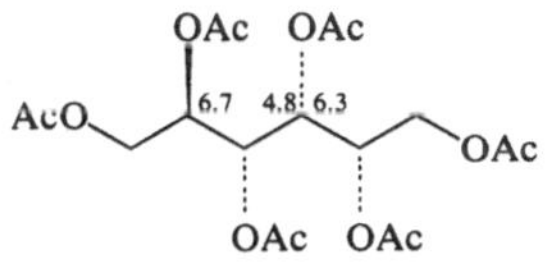

OAc OAc

AcO 6.7 4.8 6.3 OAc

OAc OAc

(101)

OAc OAc OAc

AcO 5.6 5.1 5.1 5.6 OAc

OAc OAc

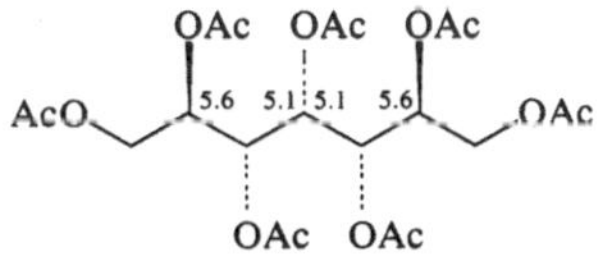

(102)

The results of this ^{1}H nmr study indicated that relative stereochemistry in the acyclic segments of palytoxin could not be determined in a straightforward manner from coupling constants observed in fully acetylated degradation products. There was one exception. The relative stereochemistry of the C(87)−C(89) segment was clearly *syn,anti* or *anti,syn*, since in the ozonolysis product (**16**) two different coupling constants were observed for $J_{87,88}$ and $J_{88,89}$. It should have been possible to differentiate between the *syn,anti* and *anti,syn* possibilities for the C(87)−C(89) segment in palytoxin by ^{1}H nmr analysis of the ozonolysis product (**85**), but this was not done (*23*).

(16)

(85)

In spite of these limitations, the MOORE group was able to deduce the relative stereochemistry of C(49) and C(51) in the periodate oxidation products (55) and (6) (28). The coupling constants associated with the C(49)−C(51) segment clearly indicated that the acetoxy groups on C(49) and C(51) were *anti* to each other (35). [1]H nmr studies of model compounds (95), (96), (97), and (98) further supported the proposed stereochemistry. [1]H nmr studies of (95) and (96) had already shown that the relative stereochemistry at C(51)−C(52) in (55) could not be determined from $J_{51,52}$; however, it was possible to decide that the stereochemistry was *syn* since the vicinal and geminal coupling constants associated with C(52)−C(53) were essentially identical with those observed in model compound (96) and not with those in (95). The medium-sized coupling constants (5.4 Hz) for $J_{51,52}$ and $J_{52,53}$ in (6), which suggested that the substituents on C(51) and C(53) were *syn* to each other, indicated that the relative stereochemistry of the C(51)−C(53) segment was *syn,syn*.

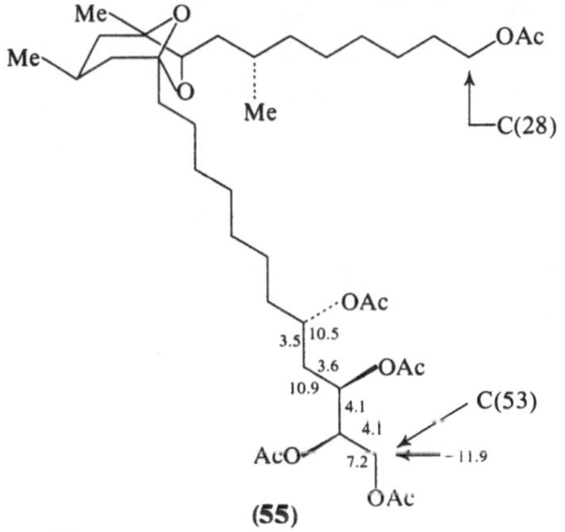

(55)

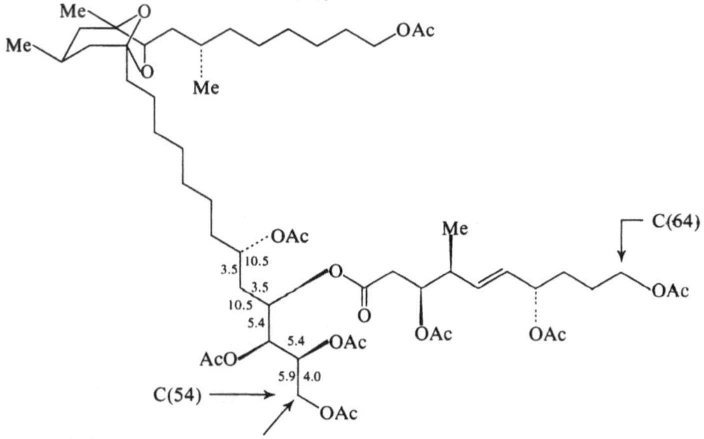

(6)

The major compound in the mixture of ozonolysis products of gross structure (14) was proposed to have the relative stereochemistry for the C(49) – C(55) segment shown in (14b) (28). Interestingly $J_{52,53}$, $J_{53,54}$, and $J_{54,55}$ were all large and small, suggesting that the relative stereochemistry of the C(51) – C(55) segment was syn,syn,anti,syn. The coupling constants for D-glycero-D-manno-hepitol heptaacetate (103), which has anti,anti,syn,anti stereochemistry, were all found to be large and small, indicating that the preferred conformation of this molecule was one in which C(2) – C(3) was bent ($J_{2,3}$ is 3.3 Hz) to relieve the steric interaction of the acetoxy groups on C(2) and C(4). The C(51) – C(52) bond was probably also bent in the preferred conformation of (14) to lessen the steric interaction between the acetoxy groups on C(51) and C(53); unfortunately, the MOORE group was unable to determine $J_{51,52}$ in (14) to verify this, as the chemical shifts of the protons on C(51) and C(52) were not separated sufficiently, even at 600 MHz, to obtain this coupling constant from decoupled difference spectra. Nevertheless the relative stereochemistry implied from (14b) was the same as the relative stereochemistry implied from (73a) for the C(51) – C(55) segment in palytoxin.

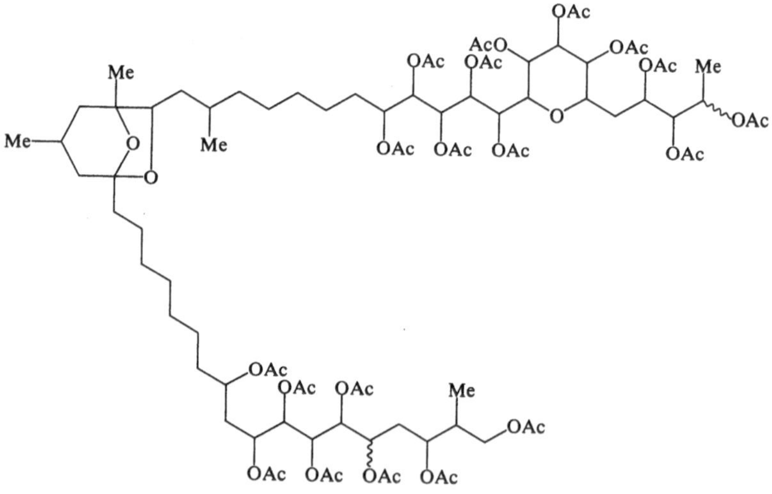

(14)

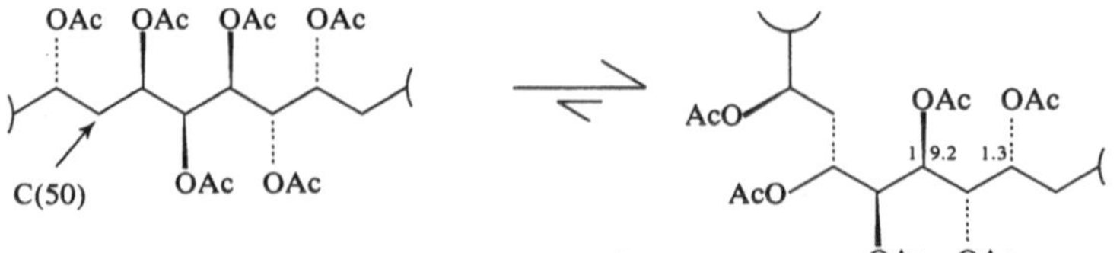

(14b)

The MOORE group was able to determine the relative stereochemistry of the C(25)−C(28) segment from the coupling constants shown by the various isomeric ozonolysis products of gross structure (14). The two major isomers, which were epimeric at C(15), showed values of 1.6, 9.5, and 1 Hz for $J_{25,26}$, $J_{26,27}$, and $J_{27,28}$, respectively, strongly suggesting a *syn,anti,syn* stereochemistry for the C(25)−C(28) segment as shown in (14c). The coupling constants were very similar to $J_{52,53}$, $J_{53,54}$, and $J_{54,55}$ in (14b) (1, 9.2, and 1.3 Hz, respectively). Since $J_{24,25}$ was 9.8 Hz in (14), the MOORE group proposed a *syn,anti,syn,anti* stereochemistry for the C(24)−C(28) segment (28). Unfortunately, the alternative *syn,anti,syn,syn* stereochemistry, which was subsequently shown to be correct by KISHI's synthetic work, was not considered. The MOORE group should have been suspicious about their assignment, however, since the HIRATA group had reported a somewhat smaller value (6.7 Hz) for $J_{24,25}$ in the degradation product (78) (24); $J_{24,25}$ would have still been 9 − 10 Hz if C(24) − C(28) had possessed *syn,anti,syn,anti* stereochemistry.

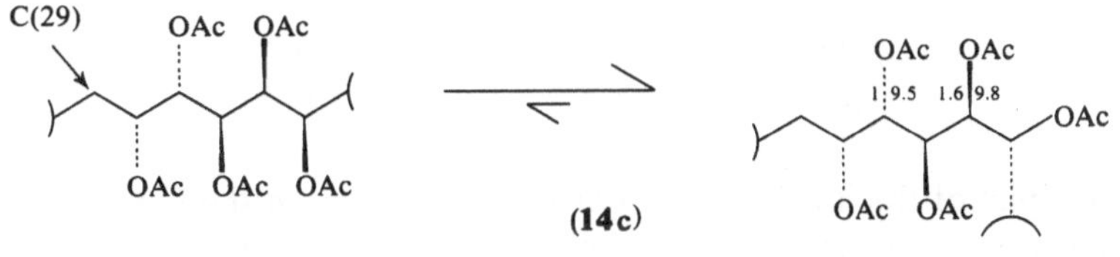

(14c)

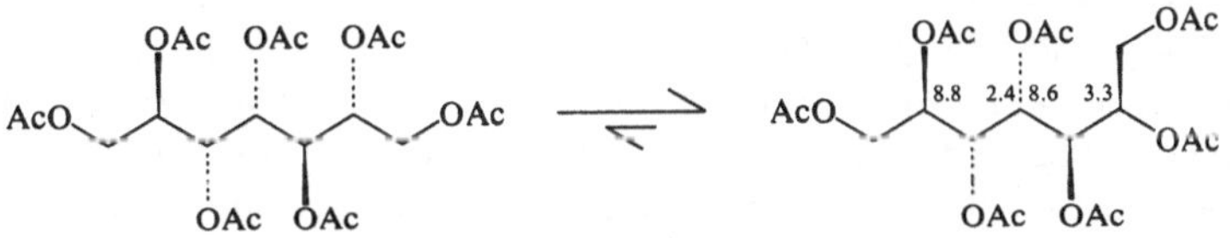

(103)

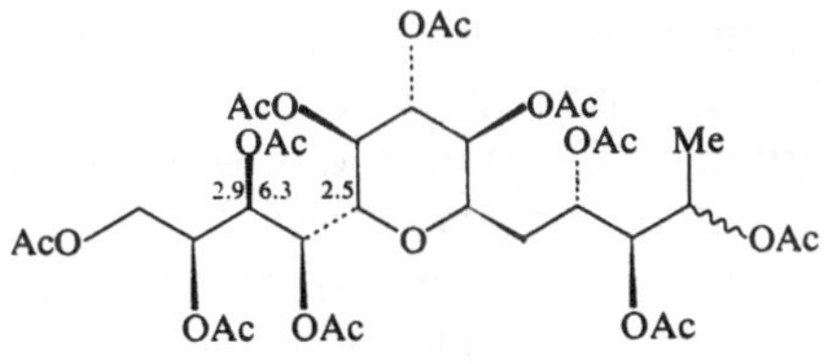

(78)

Assignment of relative stereochemistry in the C(96) − C(99) segment by ^{1}H nmr analysis of the ozonolysis product (17) was unsuccessful. Values of 5.5 Hz for $J_{96,97}$, $J_{97,98}$, and $J_{98,99}$ had been observed for one of the C(93) epimers (17A) and values of 4.3, 6.5, and 4.3 Hz for $J_{96,97}$, $J_{97,98}$, $J_{98,99}$, respectively, had been found for the other epimer (17B), indicating that syn-

1,3 substitution was present in the C(96)–C(99) segment. It was not possible, however, to differentiate among the six possibilities of stereochemistry for C(96)–C(99), viz. *anti,anti,anti, syn,syn,syn, syn,anti,anti, anti,anti,syn, syn,syn,anti,* and *anti,syn,syn,* from the coupling constants.

OAc

AcO.   .OAc

OAc   OAc   OAc

AcO.   O

AcO   OAc

OAc   Me

ÓAc ÓAc

(17)

OAc   OAc

5.5│5.5 5.5   4.3│6.5 4.3

Me   Me

ÓAc ÓAc   ÓAc ÓAc

(17A)   **(17B)**

In spite of the indication of *syn*-1,3 substitution in the C(96)–C(99) segment in **(17)** from model compound studies, the MOORE group misinterpreted the results of a low temperature study of **(17A)** and erroneously assigned a *syn,anti,syn* stereochemistry to C(96)–C(99). $J_{95,96}$, $J_{95',96}$, $J_{96,97}$, $J_{97,98}$, $J_{98,99}$, which were all 5.5 Hz at 25° C, had changed to large and small couplings, 9.3, 3, 2.9, 7.8, and 3 Hz, respectively, at −40° C (Fig. 8). MOORE felt that the change in coupling constants at −40° C reflected an increase in population of a conformer in which the C(92)–C(100) carbon chain was fully extended and that the medium-sized coupling at 25° C indicated the presence of other preferred conformers in which the C(99)–C(98)–C(97)... portion of the carbon chain was fully extended with the methyl group on C(99) instead of with C(100). This was not the case. Now that the stereochemistry of the C(96)–C(99) segment has been shown to be *syn,syn,anti* by synthesis (*29*), this change in coupling constants at −40° C can be interpreted as indicating an increase in population of a non-extended conformer, possibly **(17b)**. Curiously none of the coupling constants and chemical shifts for epimer **(17B)** showed appreciable changes at −40° C. This difference in conformational behavior of epimers **(17A)** and **(17B)** suggested that the stereochemistry at C(93), which had been generated non-selectively by the borohydride reduction, was playing an important role. The explanation of these effects needs further study. It is interesting that the values of $J_{96,97}$, $J_{97,98}$, and $J_{98,99}$ in **(17A)** at −40° C (*28*)

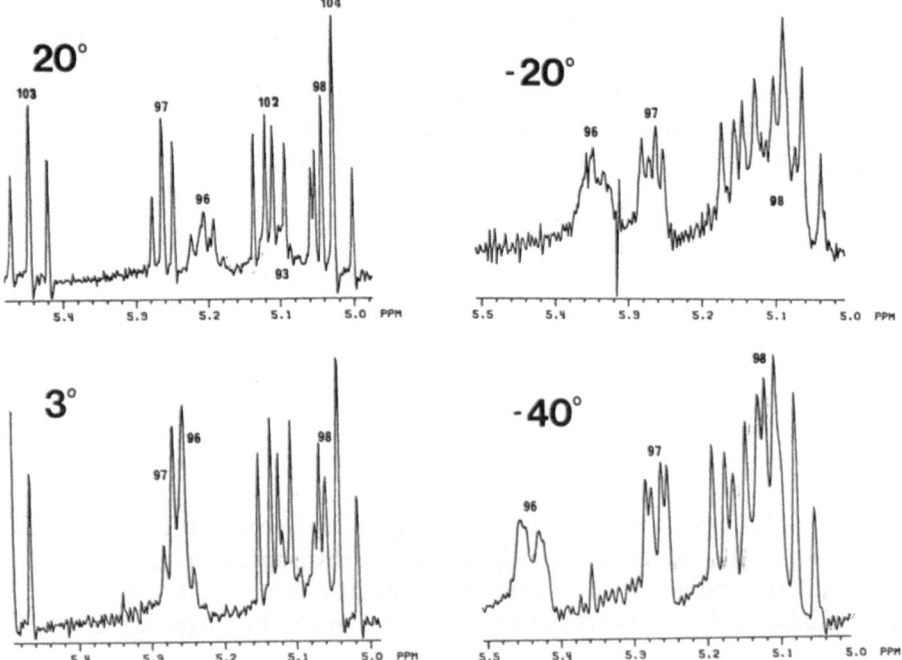

Fig. 8. The 360 MHz ¹H nmr spectrum of compound (17 A) in chloroform-d at 20°, 3°, −20°, and −40° C

are comparable to $J_{26,25}$, $J_{25,24'}$, and $J_{24,23'}$, respectively, in (78) (24). The conformations of the C(96)−C(99) segment in (17A) at −40°C and the C(26)−C(23) segment in (78) at 25°C are apparently the same.

(17b)

Recently the MOORE group has been able to determine the relative stereochemistry of the C(96) – C(99) segment by ^{1}H nmr analysis of (104), one of the major partially acetylated degradation products obtained by ozonolysis of N-(p-bromobenzoyl)palytoxin, followed by reductive workup with sodium borohydride and acetylation with acetic anhydride and pyridine (see section on borate complexation below).

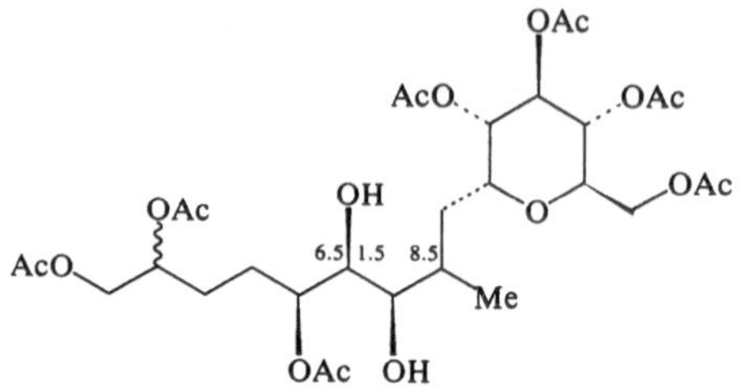

(104)

The MOORE group succeeded in relating the stereochemistries of C(72), C(73), C(81), and C(109) to neighboring tetrahydropyran rings from the coupling constants in the ozonolysis products (15) and (18) (28). X-ray crystallographic data (22) had suggested that an unsubstituted alkyl chain

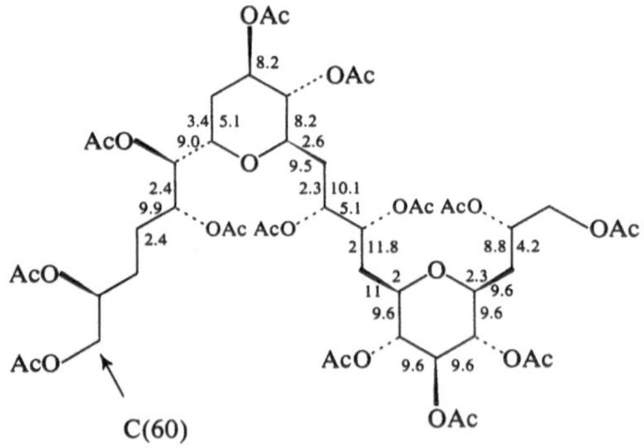

(15)

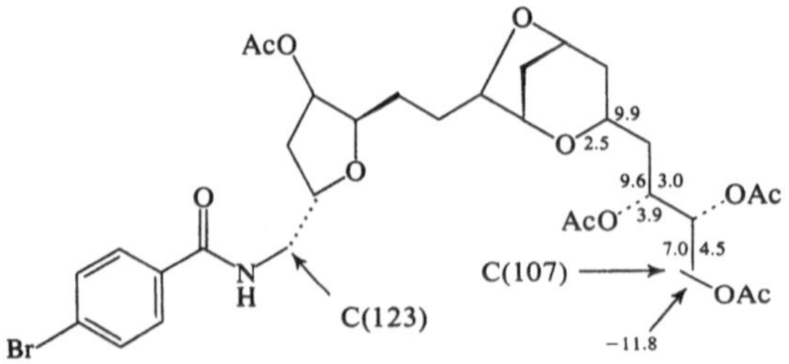

(18)

attached equatorially to C_α of a tetrahydropyran ring was preferentially oriented in a planar, zig-zag manner with $C_\alpha-C_\beta-C_\gamma$ and not with $C_\alpha-O$ in the ring. If acetoxy substituents were present on the alkyl side chain, the acyclic carbon chain was still expected to be fully extended with $C_\alpha-C_\beta-C_\gamma$ in the ring in the preferred conformation as long as no acetoxy group on $C_{\alpha'}$ or $C_{\beta'}$ in the side chain was *gauche-gauche* with an acetate group on C_β in the ring or with the cyclic ether oxygen. By studying model compounds (105) and (106), the MOORE group was able to show that stereochemistry at $C_{\beta'}$ could be easily determined from coupling constants. In the ^{1}H nmr spectrum of (105), the H-6 signal showed large and small couplings to the H-7 and H-7′ signals, which in turn showed small and large couplings, respectively, to the H-8 signal. These coupling constants were consistent with a fully extended conformation for the side chain as shown in (105a). In the ^{1}H nmr spectrum of (106), however, the H-6 signal still showed large and small couplings to the H-7 and H-7′ signals, but each signal for the C-7 protons now showed a medium-sized coupling to the H-8 signal. These coupling constants indicated that the fully extended conformer (106a) was in rapid equilibrium with the nonextended conformer (106b) and that (106b) was slightly favored over (106a). Inspection of conformation (106a) showed that the cyclic ether oxygen was eclipsed by the acetoxy group on C(8); in conformation (106b), however, the steric hindrance was much less, since the cyclic ether oxygen was now eclipsed by the hydrogen on C(8) and C(9) was eclipsed by the proton on C(6).

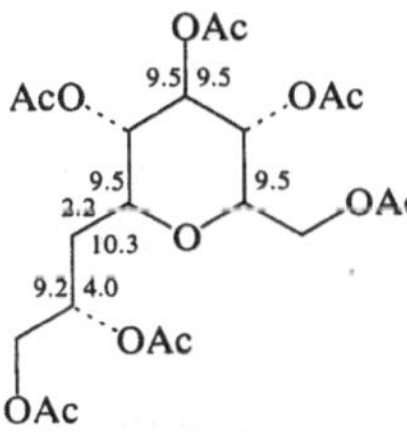

(105)

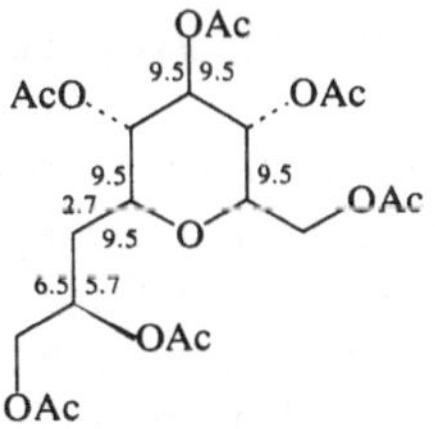

(106)

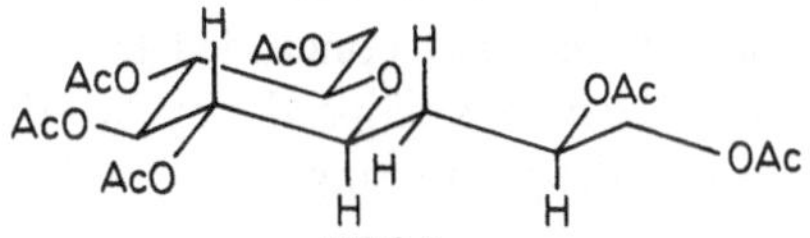

(105a)

(106a)

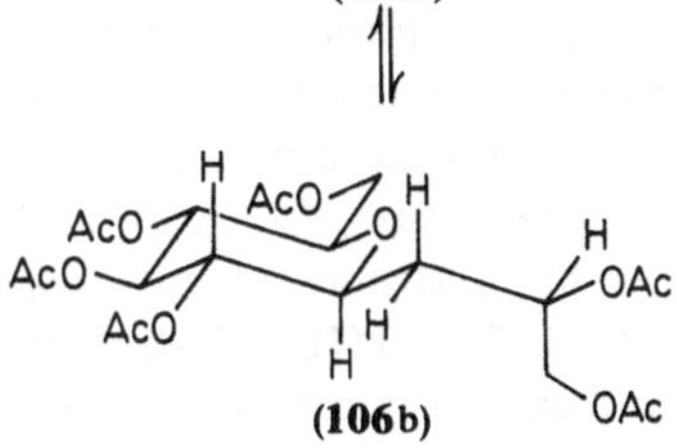

(106b)

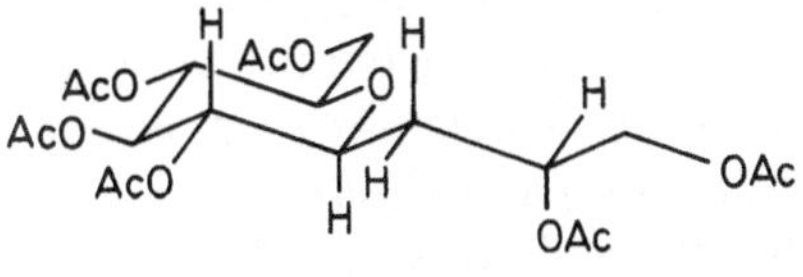

The vicinal coupling constants associated with the methylene protons on C(71), C(74), C(80), and C(110) in (15) and (18) indicated that the relative stereochemistries for the acetoxy groups on C(72), C(73), C(81), and C(109) were the same as in model compound (105), i.e. these acetoxy groups were all *trans,gauche* to cyclic ether oxygens. In (18) the relative stereochemistry of the acetoxy groups on C(108)−C(109) had to be *syn*, since the vicinal and geminal coupling constants associated with the protons on C(107) in (18) were comparable with those associated with the protons on C(53) in (55) and on C(6) in model compound (96) and not with those associated with the protons on C(6) in (95). In (15) the relative stereochemistry of the acetoxy groups on C(72)−C(73) could not be decided from the medium-sized value (5.1 Hz) for $J_{72,73}$ observed at 25° C. The MOORE group had originally proposed *syn* stereochemistry for the diol on C(72)−C(73) in palytoxin, since the diol group was readily cleaved by periodate and was very susceptible to borate complexation. Unfortunately the chemical evidence was ignored and the assignment of stereochemistry for C(72)−C(73) was changed to *anti* on the basis of a low temperature nmr experiment (28). At −60° C, $J_{72,73}$ in (15) had changed to 8.8 Hz, which was erroneously interpreted to mean that at this temperature the C(71)−C(74) segment was fully extended in the preferred conformation and that therefore the protons on C(72) and C(73) were *anti* to each other. This of course was not true, as was subsequently shown by KISHI's synthetic work (30) which unambiguously established *syn* stereochemistry for C(72)−C(73). In the preferred conformation of (15) at −60° C, the large value observed for $J_{72,73}$ actually suggested that C(71)−C(74) was not planar.

Relating the stereochemistries of C(17), C(24), C(25), C(64), C(65), and C(99) to neighboring tetrahydropyran rings by ^{1}H nmr analysis proved to be very difficult and in three cases could not be accomplished unambiguously from the coupling constants associated with the fully acetylated ozonolysis products (14), (15), and (17) and the periodate oxidation product (10). In assigning stereochemistry to these six chiral centers (28), the MOORE group had erroneously concluded that none of the substituents on these asymmetric carbons eclipsed a cyclic ether oxygen, an acetoxy group on C_β in a tetrahydropyran ring, or a carbon in a tetrahydropyran ring when the acyclic side chains were fully extended with C_α−C_β in the tetrahydropyran rings, since only large and small coupling constants were observed. Wrong assignments were therefore made at C(17), C(25) and C(64).

^{1}H nmr studies of model compounds (107) and (108) in collaboration with the MASAMUNE group showed that it was not possible to distinguish between the two stereochemistries at C(8) from coupling constants (38). For both (107) and (108), H-6 showed large and small couplings to H-7 and H-7′ and each of these protons in turn showed small and large couplings, respectively, to H-8. The coupling constants for (107) indicated that the side chain was fully extended in the preferred conformation as shown in (107a); however, the coupling constants for (108) showed that the side chain was bent in the preferred conformation as shown in (108b). Medium sized coupling constants had shown that (106) existed in both fully extended and non-extended conformations, but large and small coupling constants indicated that (108) existed predominantly in the non-extended conformation (108b). In the fully extended conformation (108a), the acetoxy group on C(8) was eclipsed by the tetrahydropyran ether oxygen. This steric hindrance was eliminated in (108b).

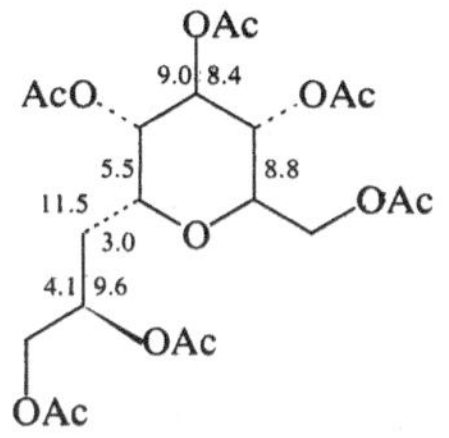

(107)

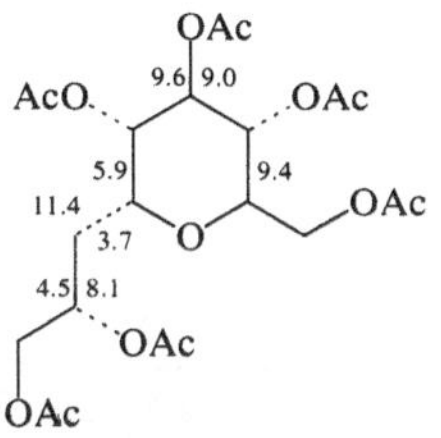

(108)

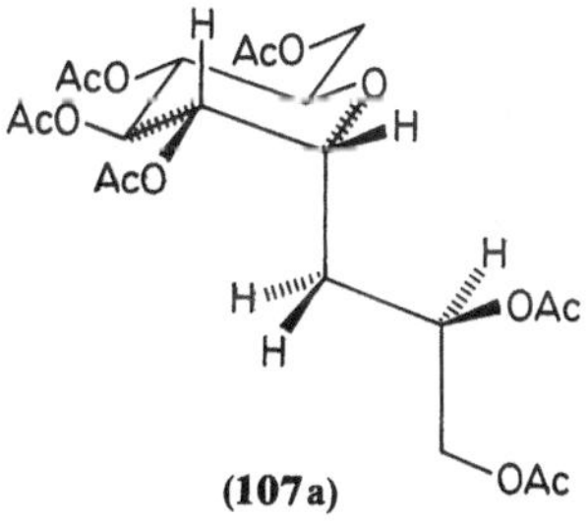

(107a)

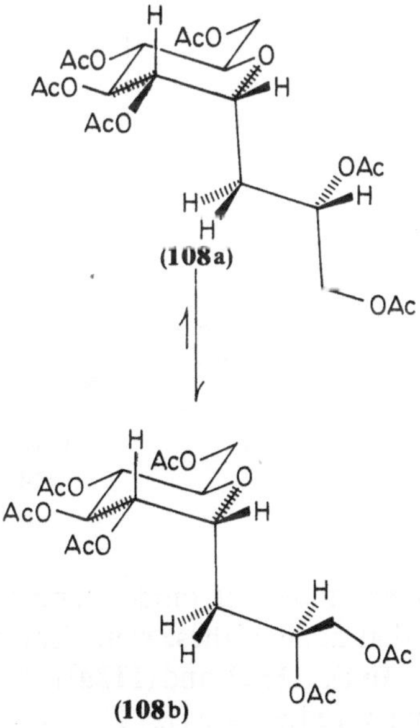

(108a)

(108b)

The model studies with (107) and (108) clearly showed that the stereochemistries of C(17) in (14) and of C(99) in (17) and (10) could not be determined from coupling constants. It was also not possible to decide between *syn* or *anti* stereochemistry for $C(16) - C(17)$ from the value of $J_{16,17}$ shown in **(14d)**.

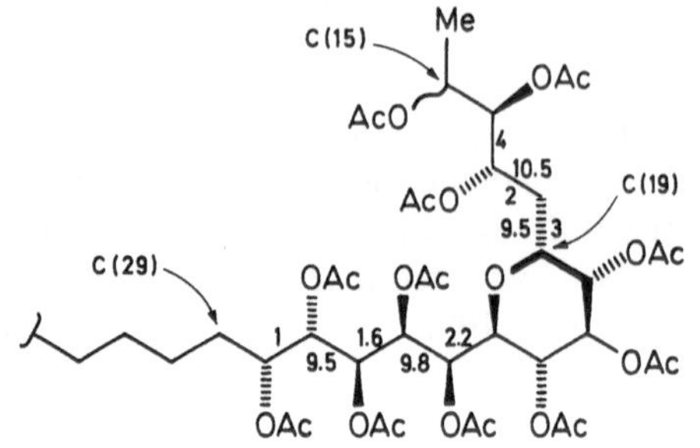

(14d)

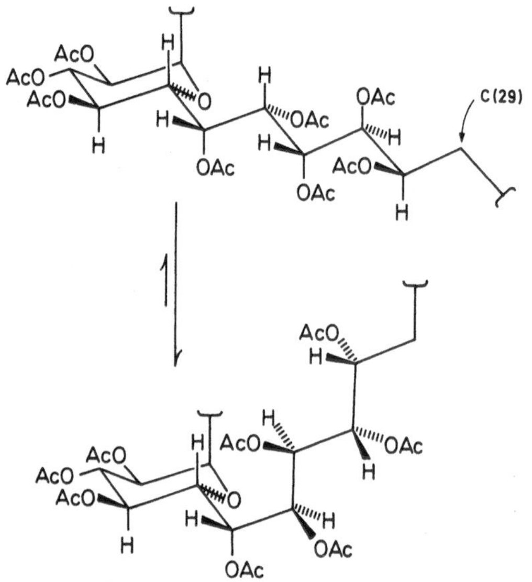

(14e)

[1]H nmr studies of model compounds **(109)**, **(110)**, **(111)**, and **(112)** showed that none of these stereoisomers had the conformations shown in **(109a)**, **(110a)**, **(111a)**, and **(112a)** (*38*). All of the vicinal coupling constants in the ring were $1.9 - 4.6$ Hz! This meant that the preferred conformation of the tetrahydropyran ring in these compounds was the chair conformation in which the three carbon side chain on $C(6)$ was equatorial and all of the other substituents on the ring were axial as shown in **(109b)**, **(110b)**, **(111b)**, and **(112b)**. In stereoisomers **(109)** and **(111)** the acetoxy group on $C(7)$ was

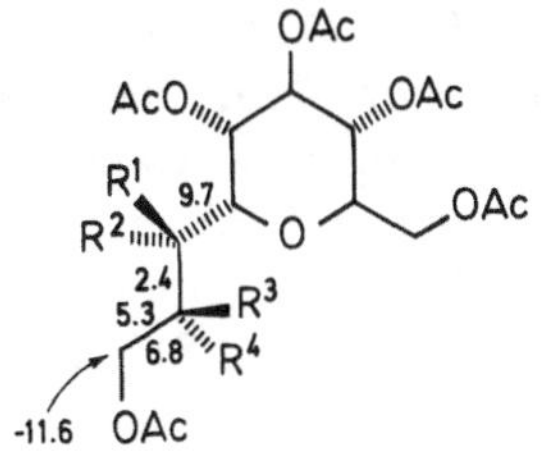

(109) R¹=R³=OAc; R²=R⁴=H
(110) R²=R⁴=OAc; R¹=R³=H
(111) R¹=R⁴=OAc; R²=R³=H
(112) R²=R³=OAc; R¹=R⁴=H

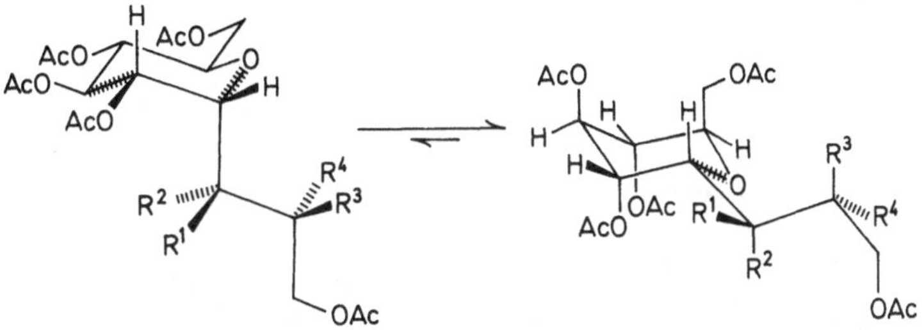

(109a)	R¹=R³=OAc; R²=R⁴=H	(109b)
(110a)	R²=R⁴=OAc; R¹=R³=H	(110b)
(111a)	R¹=R⁴=OAc; R²=R³=H	(111b)
(112a)	R²=R³=OAc; R¹=R⁴=H	(112b)

eclipsed by an equatorial acetoxy group on C(5) in conformations (109a) and (111a) and in stereoisomers (110) and (112) the acetoxy group on C(7) was eclipsed by both C(2) and C(4) in the tetrahydropyran ring in conformations (110a) and (112a). Steric hindrance was apparently greatly reduced when C(7) was attached equatorially to C(6). Surprisingly the five coupling constants associated with the three carbon side chain were found to be exactly the same for all four compounds. This strongly suggested that the preferred conformations of the four stereoisomers were (109b), (110c), (111c), and (112c). When the side chain in (109) was fully extended with C(6)−C(5)−C(4) as shown in (109b) the acetoxy groups on C(7) and C(8) were eclipsed by hydrogens on C(5) and C(6), respectively. When the side chains in (110), (111), and (112), however, were fully extended with C(6)−C(5)−C(4) as shown in (110b), (111b), and (112b), one or both of the acetoxy groups on C(7) and C(8) were eclipsed by either the axial acetoxy group on C(5) or the ether oxygen on C(6), respectively, or both. To eliminate the steric hindrance in (110b), (111b), and (112b), the non-extended conformations (**110c**), (**111c**), and (**112c**) had to be the preferred conformations. In conformation (**110c**) the axial acetoxy group on C(5) was now eclipsed by the hydrogen on C(7), the acetoxy group on C(8) eclipsed by the axial hydrogen on C(6), and carbon C(8) eclipsed by the equatorial proton on C(5). In conformation (**111c**), the acetoxy group on C(5), the acetoxy group on C(7), and C(8) were now eclipsed by hydrogens on C(7), C(5), and C(6), respectively. Finally in conformation (**112c**), the acetoxy group on C(5) and carbons C(8) and C(9) were now eclipsed by hydrogens on C(7), C(5), and C(6), respectively.

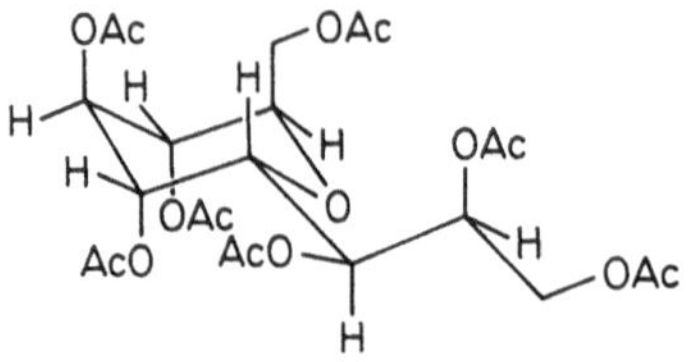

(**109b**)

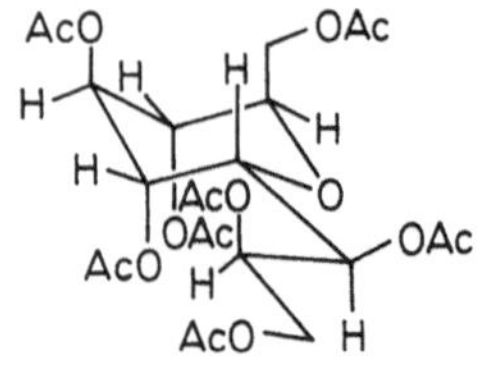

(**110c**)

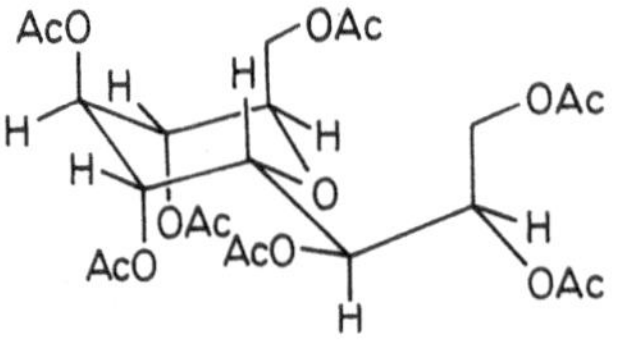

(**111c**)

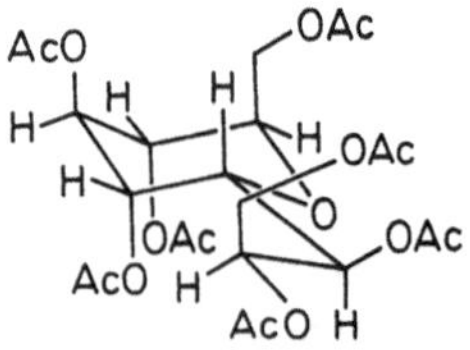

(**112c**)

The stereochemistry of C(65) in ozonolysis product (**15**) had to be as shown in (**15c**). Since C(65) was attached axially to the tetrahydropyran ring, the acetoxy group on C(65) had to be eclipsed by the equatorial proton on C(67) and not by carbons C(68) and C(70) in the ring. Opposite stereochemistry at C(65) would have surely resulted in inversion of the tetrahydropyran ring with concomitant equatorial orientation of C(65) and axial disposition of the other substituents on the ring. The model studies with (**109**), (**110**), (**111**), and (**112**), however, showed that it was not possible to distinguish between *syn* and *anti* stereochemistry for C(64) − C(65) on the basis of $J_{64,65}$ (*28*); synthesis, however, showed that the stereochemistry was *anti* (*30*).

¹H nmr studies were also carried out with model compounds (**113**) and (**114**). Both compounds showed similar coupling constants for the side chain which suggested that stereochemistry at C(8) in (**113**) and (**114**) could not be determined from coupling constants. This meant that it was not possible to assign stereochemistry at C(25) in (**14d**) from coupling constants (*28*).

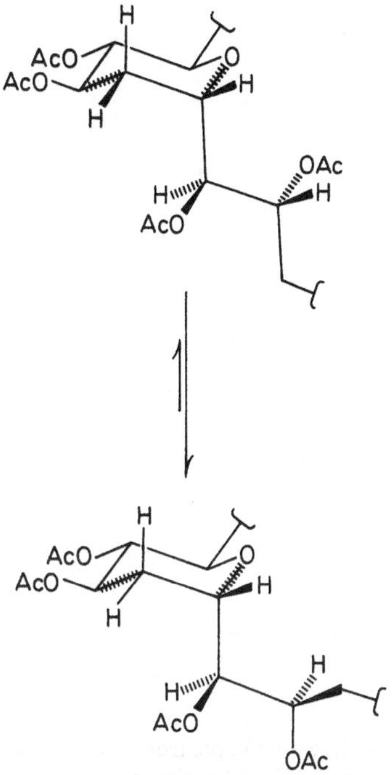

(15c)

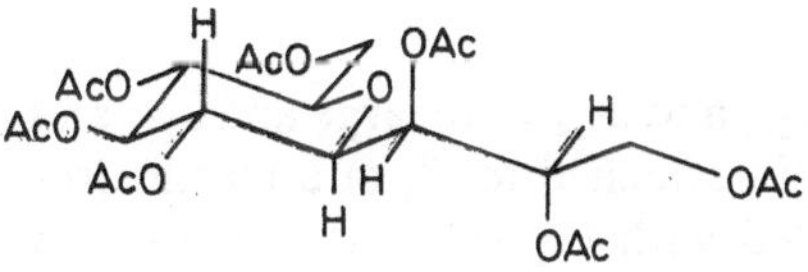

(113)

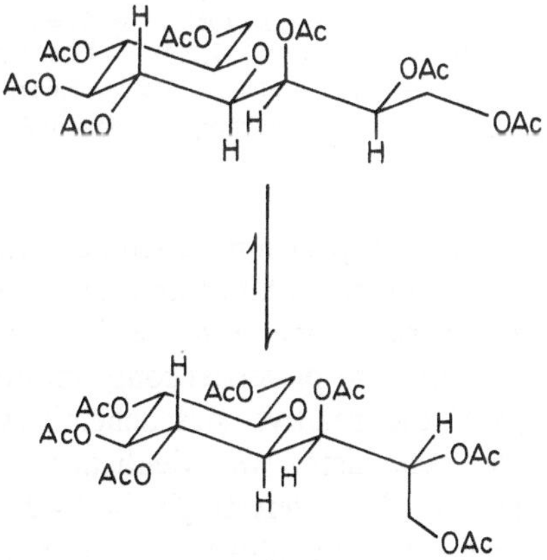

(114)

The MOORE group was able to relate the stereochemistries of the tetrahydrofuran and 2,6-dioxabicyclo[3.2.1]octane rings in the periodate oxidation product (11) by detailed ^{1}H nmr analysis. Spectral simulation (Fig. 9) indicated that the six protons on the C(116)−C(119) segment in (11) were arranged on the carbon chain as shown in (11g). The observed and

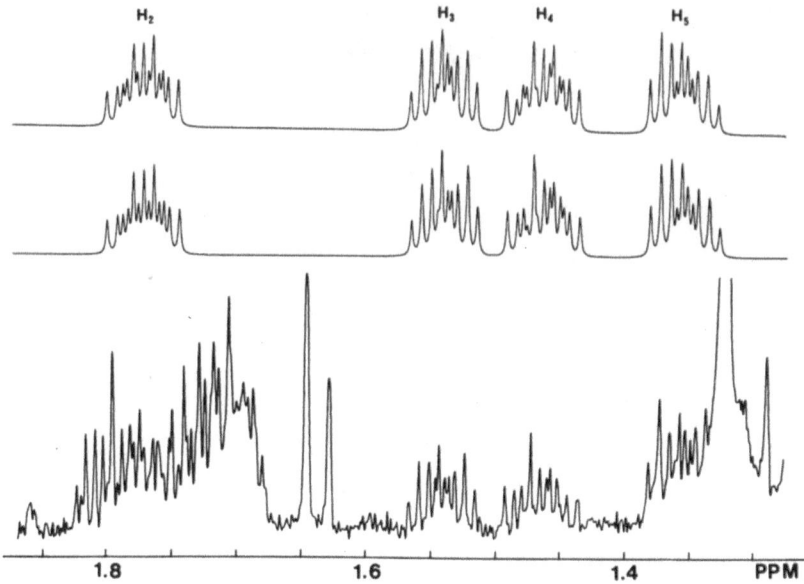

Fig. 9. The high field region, 1.26 – 1.87 ppm from TMS, of the 600 MHz ^{1}H nmr spectrum of compound (11) in 20% benzene-d_6 in chloroform-d. Lower trace is actual spectrum; part of peak indicated by X is due to a spinning side band of H_2O peak at 1.32 ppm. Middle trace is simulated spectrum of H_2, H_3, H_4, and H_5 resonances using the following parameters: $W(1) = 2394.6$, $W(2) = 1062.0$, $W(3) = 921.6$, $W(4) = 876.0$, $W(5) = 817.2$, $W(6) = 2514.6$, $J(1,2) = 8.2$, $J(1,3) = 5.1$, $J(2,3) = -13.6$, $J(2,4) = 5.3$, $J(2,5) = 10.4$, $J(3,4) = 10.8$, $J(3,5) = 5.5$, $J(4,5) = -13.8$, $J(4,6) = 8.4$, $J(5,6) = 5.6$ Hz. Upper trace is simulated spectrum using parameters above except $J(2,4) = 10.4$, $J(2,5) = 5.3$, $J(3,4) = 5.5$, $J(3,5) = 10.8$ Hz

calculated spectra matched best when $J_{2,5}$ and $J_{3,4}$ were large and $J_{2,4}$ and $J_{3,5}$ were small. Since an alkyl side chain attached to C_α of a tetrahydropyran ring appears to be fully extended with $C_\alpha - C_\beta$ rather than with $C_\alpha - O$ in the preferred conformation from X-ray crystallographic evidence (33), the relative stereochemistry shown in (11) was deduced. This stereochemistry was identical with that determined by X-ray crystallography (22). Interestingly the X-ray data indicated that the C(116) – C(119) segment had conformation (11h) and was fully extended with C(119) – O rather than with C(119) – C(120) in the solid state. The ^{1}H nmr data, on the other hand, showed that C(116) – C(119) was fully extended with carbons in both ring systems in the solution state.

To interrelate the stereochemistries of the tetrahydropyran ring in the C(92) – C(106) segment and the two ring systems in the C(107) – C(123) segment, the ozonolysis product (83) was hydroxylated with osmium tetroxide to diol (115). ^{1}H nmr analysis of the fully acetylated material (116)

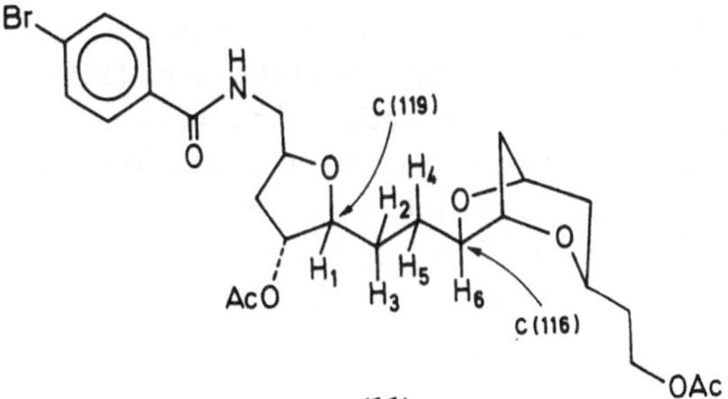

(11)

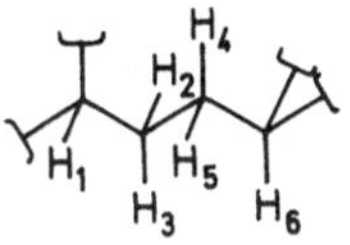

(11g)

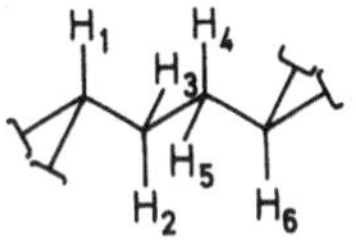

(11h)

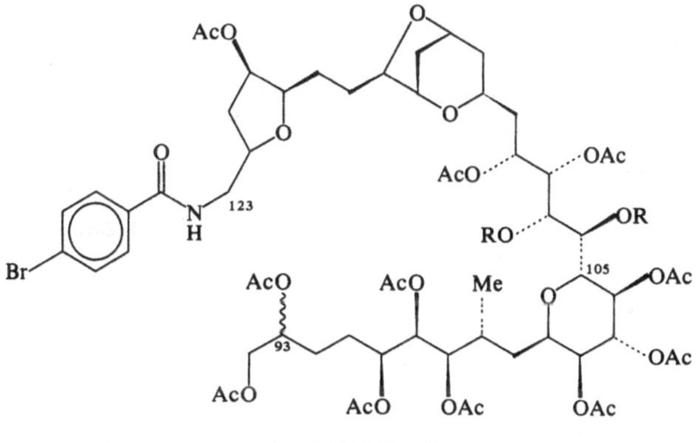

(115) R = H
(116) R = Ac

showed that hydroxylation had been stereospecific and that a single *anti*-1,2-diol had been produced. The relative stereochemistry shown in **(116a)** was assigned to the C(101) – C(109) segment of **(116)** since $J_{105,106}$ and $J_{106,107}$ were both 6.7 Hz and $J_{107,108}$ was 3.3 Hz (28). The proposed relative stereochemistry at C(105) – C(106) agreed with the finding of KISHI's group that the relative stereochemistry between the preexisting alkoxyl group and the adjacent hydroxyl group newly introduced by osmium tetraoxide oxidation is in all cases *erythro* in the major product (39). In the fully extended conformation **(116b)** the acetoxy group on C(106) was eclipsed by the acetoxy group on C(104) and the acetoxy group on C(107) was eclipsed by the ether oxygen on C(105). Steric relief, however, was not possible in non-extended conformations such as (116c) and (116d). In (116c) the acetoxy group on C(106) was now eclipsed by the acetoxy group on C(108) and in (116d) the acetoxy group on C(107) was now eclipsed by C(104). Other non-extended conformations exhibited even worse steric hindrance. In (116e), for example, C(108) was eclipsed by the ether oxygen on C(105). The preferred conformation of the C(103)−C(108) segment in (116) therefore appeared to be (116b).

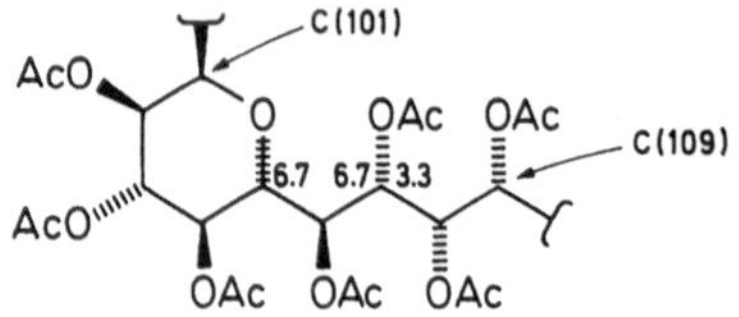

(116a)

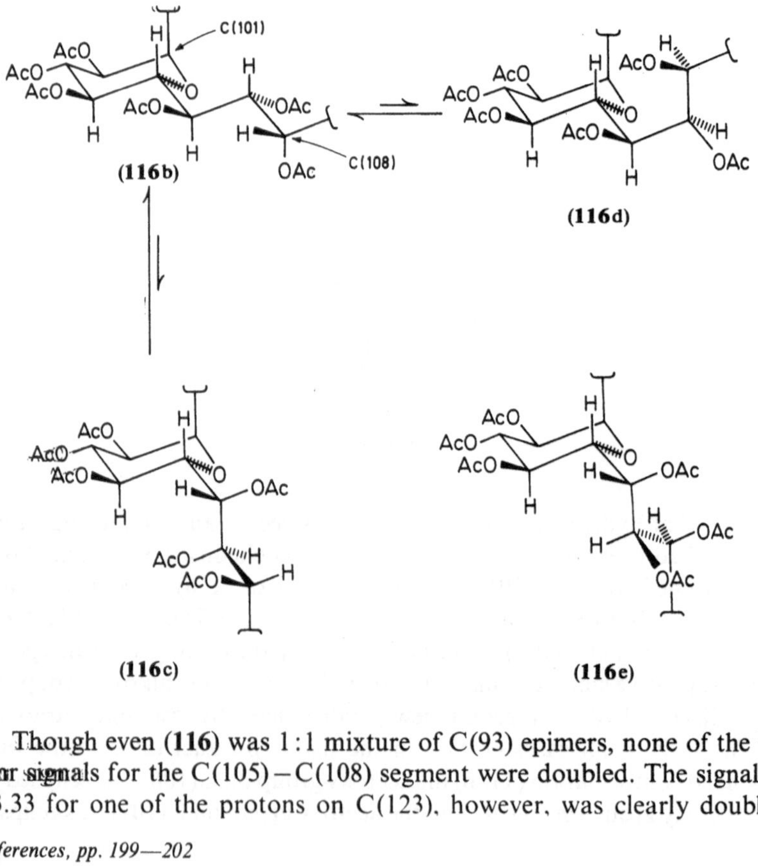

(116b)   (116d)

(116c)   (116e)

Though even (116) was 1:1 mixture of C(93) epimers, none of the ^{1}H nmr signals for the C(105)−C(108) segment were doubled. The signal at δ 3.33 for one of the protons on C(123), however, was clearly doubled

(Fig. 10), as were signals for protons in the vicinity of C(93). The doubling of the C(123) proton signal suggested that C(123) was spatially close to C(93) in the preferred conformation. Examination of a Dreiding model of (116) showed that it was possible for the two ends of the molecule to come in contact when the acyclic carbon segments, i.e. C(92)−C(100), C(106)−C(110), and C(117)−C(118), were all fully extended with carbons in the ether rings as shown in (116) (28).

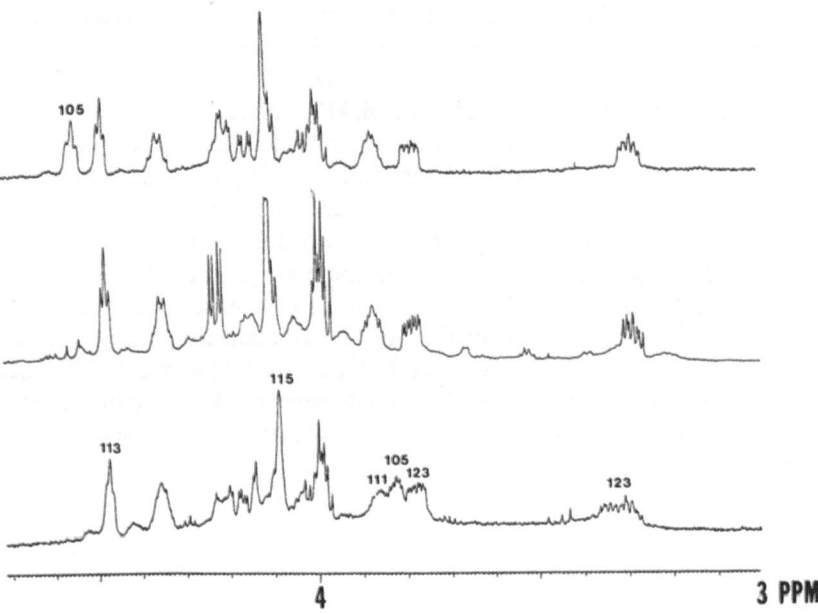

Fig. 10. Comparison of the low field region, 3.0−4.7 ppm from TMS, of the 600 MHz ¹H nmr spectra of (18) (center), (83) (top), and (116) (bottom) in chloroform-d

The ¹H nmr studies of (116) strongly suggested that C(24) in (14) had the stereochemistry indicated in (14d) since $J_{23,24}$ was 2.2 Hz. If C(24) had possessed the opposite stereochemistry, then $J_{23,24}$ should have been large. For the same reasons that C(105)−C(106) could not be bent in the preferred conformation of (116), C(23)−C(24) could not be bent in the preferred conformation of (14), no matter what the stereochemistry at C(24) was.

Since the stereochemistry of the C(24)−C(28) segment has been firmly established by synthesis, the preferred conformation of C(19)−C(29) is the non-extended one shown in (14e).

c) Periodate Oxidation

Acyclic *syn*-1,2-diols (*threo*-diols) are cleaved by periodate much faster than acyclic *anti*-1,2-diols (*erythro*-diols) and terminal glycols. For example, the relative rates of diol cleavage in sorbitol by periodate are in the following order: 3,4 > 2,3 ≫ 4,5 > 5,6 > 1,2 (*37*). The stereochemistry of many of the acyclic vicinal diol groups in palytoxin, therefore, could be predicted from relative rates of oxidation by periodate.

Compounds (**8**) and (**10**) (*19*), for example, were generally obtained in a 1:1 ratio and in relatively high yield on short-term periodate oxidation of palytoxin and its N-acylated derivatives. This strongly suggested that the diols at C(72) − C(73), C(87) − C(88), C(97) − C(98), and C(108) − C(109) were *syn*. Compounds (**7**) (*19*) and (**27, 28, 117, 118**) (*20, 34*), however, were formed in very low yield on short-term periodate oxidation of palytoxin, indicating that one or both of the diol functionalities flanking these tetrahydropyran rings in palytoxin were *anti*. Since the diol at C(72) − C(73) had already been assigned a *syn* stereochemistry, the diol at C(64) − C(65) had to be *anti* to explain the poor yield of (**7**). The diol at C(27) − C(28) was probably *syn*, since none of the periodate oxidation products possessing the 6,8-dioxabucyclo[3.2.1]octane ring, e.g. (**6**), had an *exo* substituent on C(36) longer than the C(28) − C(35) segment (*20*). Since one of the vicinal diols in the C(24) − C(28) segment of palytoxin was *syn*, the diol on C(16) − C(17) had to be anti to explain the low yields of (**27**), (**28**), (**117**), and (**118**).

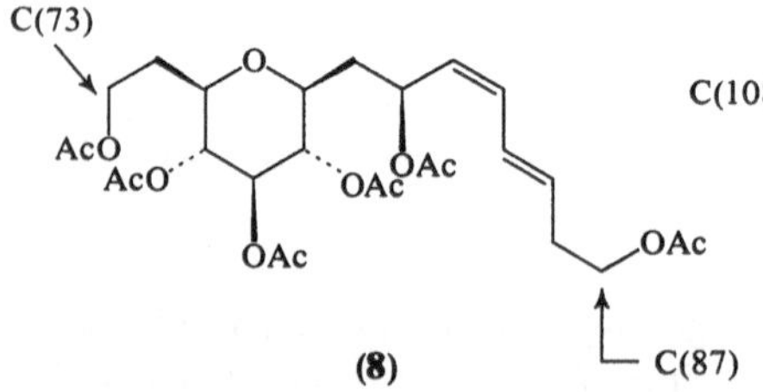

(8) 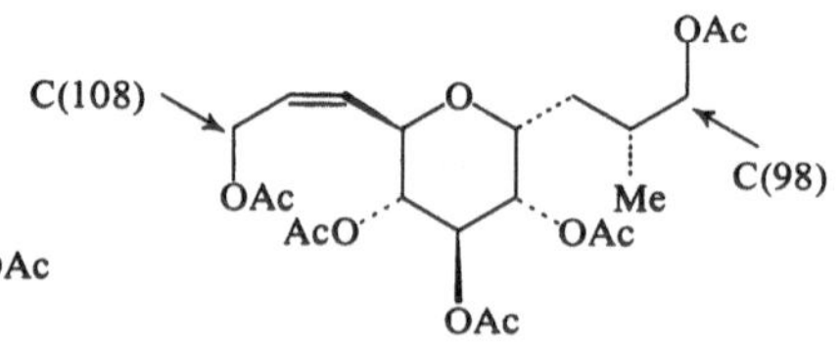   (10)

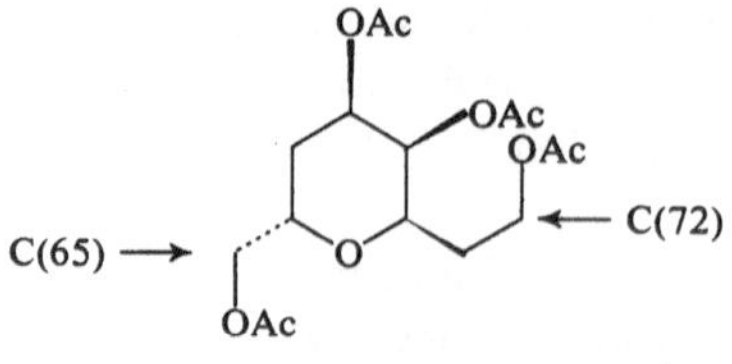

(7) 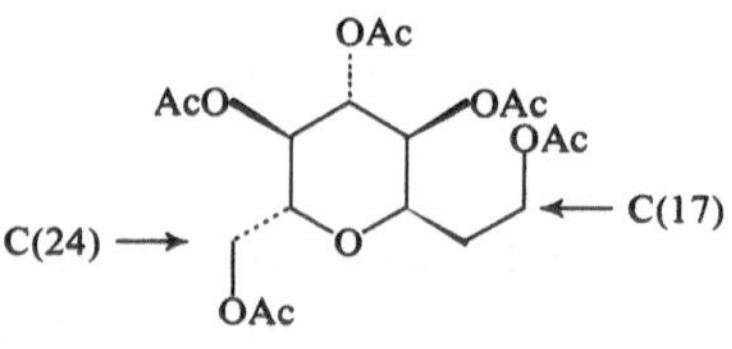   (27)

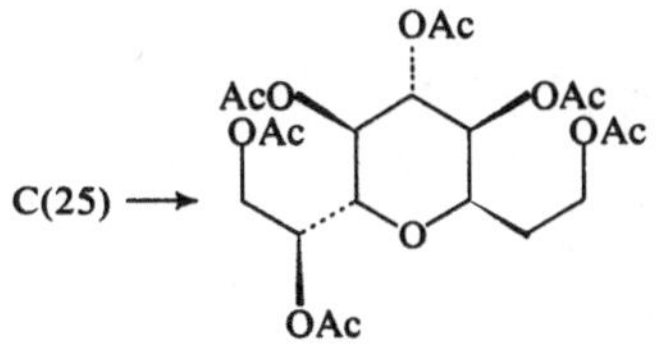

(117)

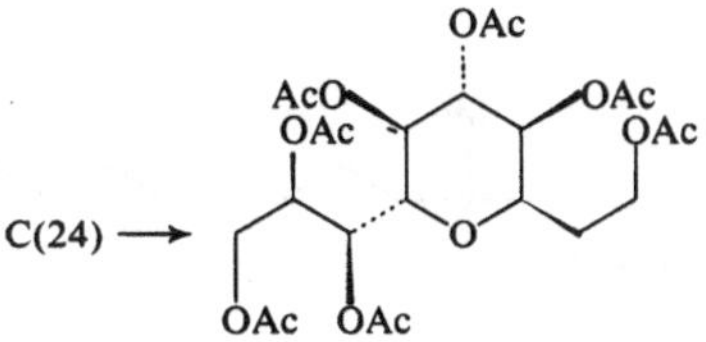

(28)

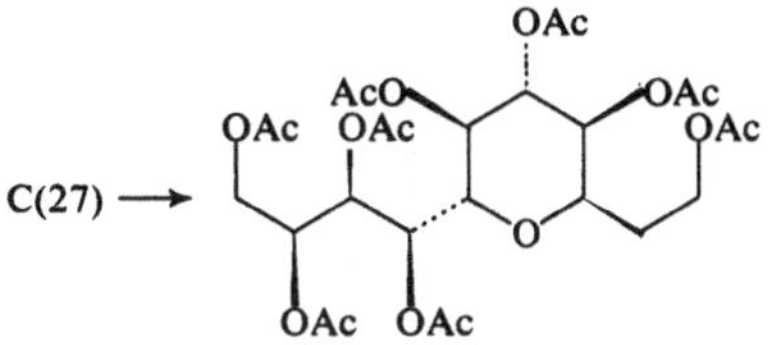

(118)

The periodate oxidation of ozonolysis product (80) to hexaol (81) (24) in good yield provided further proof that the diol at C(72) – C(73) was *syn* and that the diol at C(64) – C(65) was *anti*.

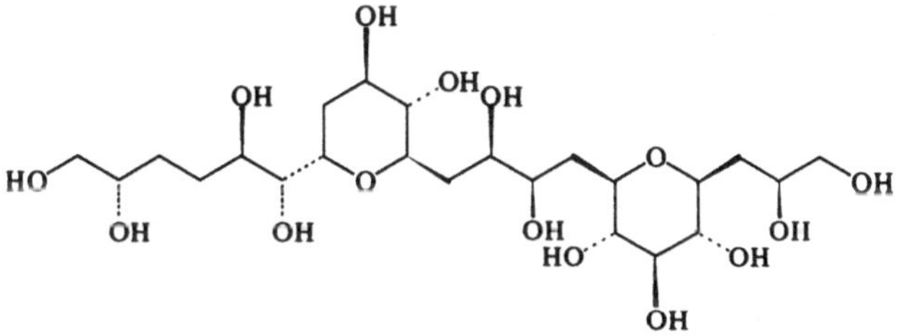

(80)

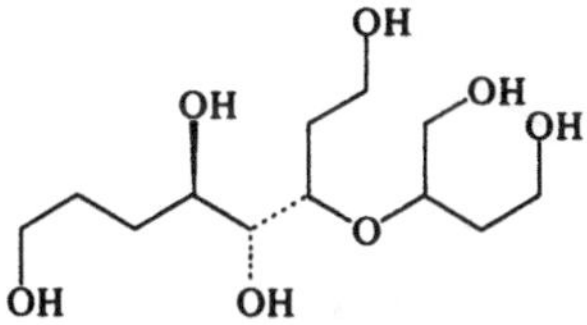

(81)

Careful oxidation of (75) with periodate followed by borohydride reduction and acetylation resulted in·(119) (31), strongly suggesting that the diol at C(25) – C(26) was *syn*.

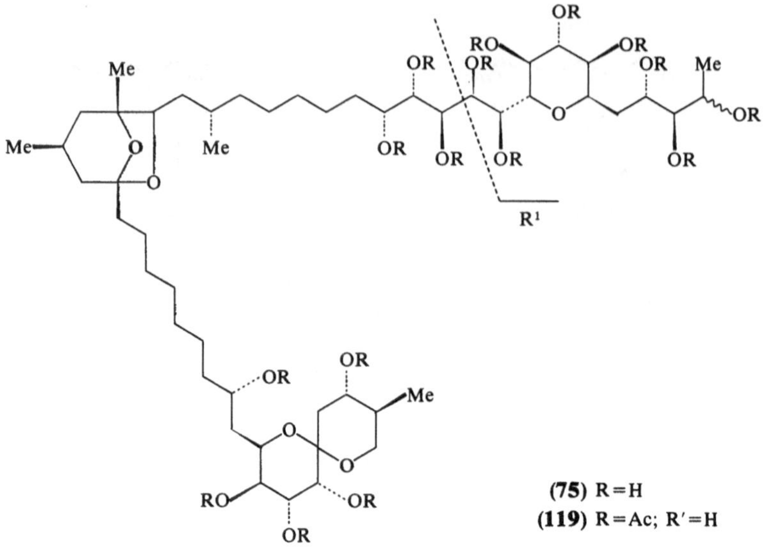

(75) R=H
(119) R=Ac; R'=H

In the hemiketal ring of palytoxin, the diol at C(54) – C(55) had to be *cis* to account for the facile formation of **(6)** (*20*) and **(120)** (*34*).

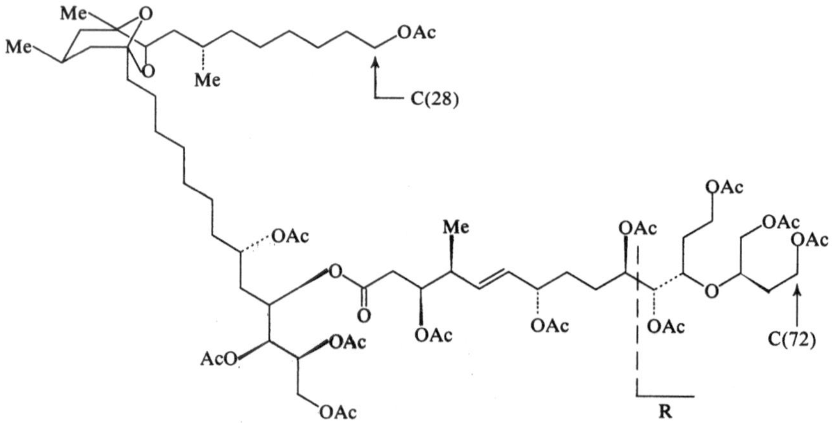

(120)
(6) R=H

Relative stereochemistry implied from the behavior of some vicinal diol groups towards oxidation is shown in Scheme XIII. Dashed lines indicate bonds which are readily cleaved by periodate.

References, pp. 199—202

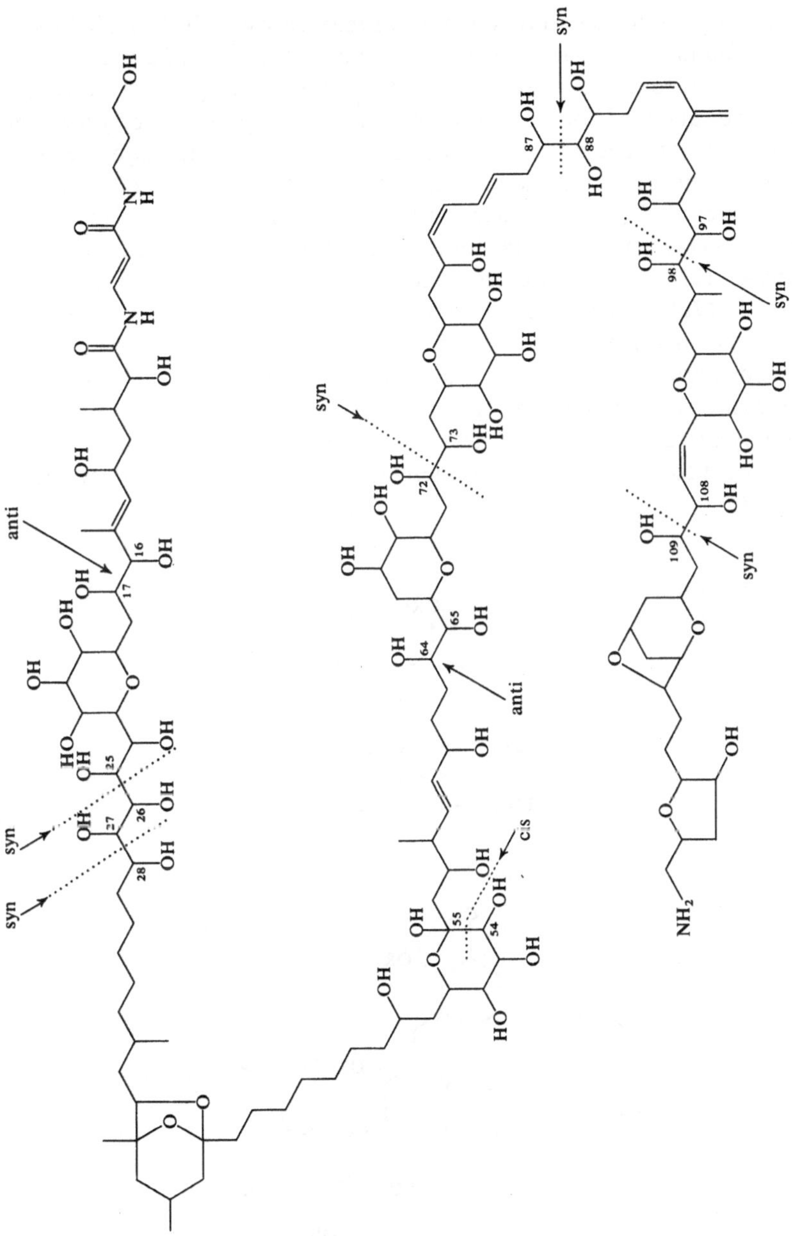

Scheme XIII

d) Borate Complexation

During ozonolysis studies of N-(p-bromobenzoyl)palytoxin the Moore group found that reductive workup of the reaction mixture with sodium borohydride followed by acetylation of the freeze-dried product with acetic anhydride and pyridine led to a complex mixture of partially acetylated and fully acetylated degradation products (36). Three of the major partially acetylated products were identified as diols (121), (122), and (104) (36). A small amount of (123) was also formed, but other triacetates of the parent pentaol (124) could not be found. Curiously other undecaacetates of tridecaol (125) could not be detected. Another heptaacetate of the parent nonaol (126), which was not fully characterized, was also found and tentatively identified as the 1,3-diol (127). Compound (127) was unreactive towards periodate oxidation and appeared to be fully acetylated in the tetrahydropyran ring from ^{1}H nmr analysis. Diols (104) and (127) were actually mixtures of C(93) epimers. Many partially acetylated derivatives of (128) were formed, but none of these compounds have been identified to date (34).

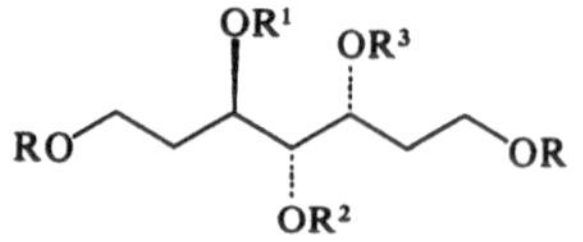

(121) $R=R^1=Ac$; $R^2=R^3=H$
(123) $R=R^2=Ac$; $R^1=R^3=H$
(124) $R=R^1=R^2=R^3=H$

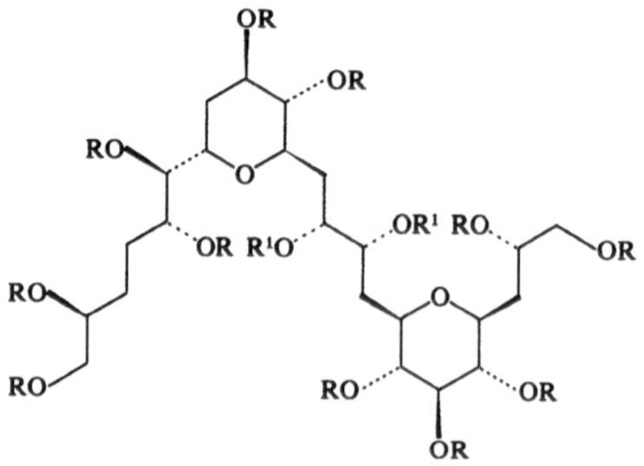

(122) $R=Ac$; $R^1=H$
(125) $R=R^1=H$

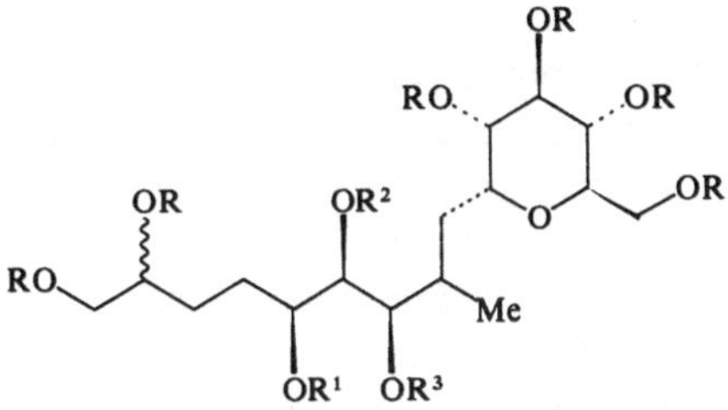

(17) R = R¹ = R² = R³ = Ac
(104) R = R¹ = Ac; R² = R³ = H
(126) R = R¹ = R² = R³ = H
(127) R = R² = Ac; R¹ = R³ = H

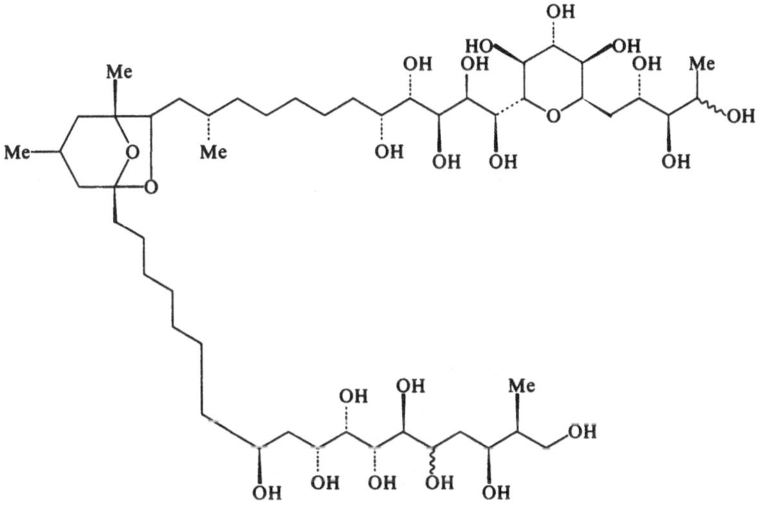

(128)

All of the partially acetylated compounds were formed as a result of borate complexation during the reductive workup procedure. Exposure of the borohydride-reduced ozonolysis products to the boric acid in the medium had produced certain borate ester complexes which had survived acetylation with acetic anhydride and pyridine. The acetylated borate complexes, however, had decomposed to diols during subsequent chromatography on silica. Unfortunately the implications of this borate complexation were not recognized at the time the first assignments of stereochemistry were made (28). It is now clear, however, that these diols pinpointed the positions of *syn*-1,2 and *syn*-1,3 diol groups in the acyclic segments of the palytoxin molecule.

In a follow-up study, the scope of this borate complexation reaction was investigated with several simpler carbohydrates of known stereochemistry (*36*). In the experimental procedure (Scheme XIV), the acyclic polyhydroxylated compound was first exposed to sodium borohydride or to boric acid alone if no reducible functionalities were present. The reaction mixture was then acidified with acetic acid and evaporated to dryness. During the evaporation step the borate complexes formed. The borate complexes were then acetylated and the acetylated borate complexes decomposed to diols by repeated treatment with methanol. The resulting diols were finally separated by HPLC on silica.

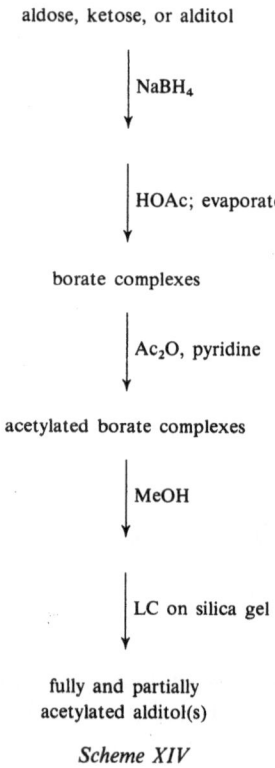

aldose, ketose, or alditol

NaBH$_4$

HOAc; evaporate

borate complexes

Ac$_2$O, pyridine

acetylated borate complexes

MeOH

LC on silica gel

fully and partially
acetylated alditol(s)

Scheme XIV

The total yield of diol generally varied from $20-50\%$. Results obtained with sixteen aldoses, ketoses, and alditols (*36*) indicated that *syn*-1,2-diols and *syn*-1,3-diols almost always predominated. *Anti*-1,3-diols, however, were sometimes formed in appreciable amounts. Interestingly, *anti*-1,2-diols were formed only to small extent, if at all, and terminal 1,2- and 1,3-diols could not be detected. As examples, the compositions of the diol

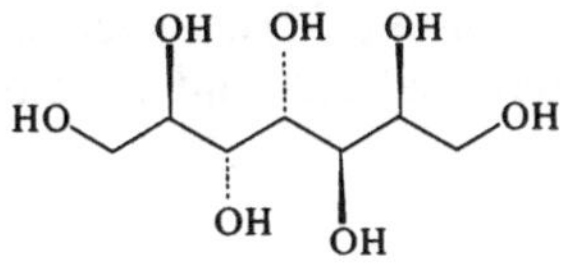

OH   OH   OH

HO⟍⟋⟍⟋⟍⟋⟍OH

OH   OH

(129)

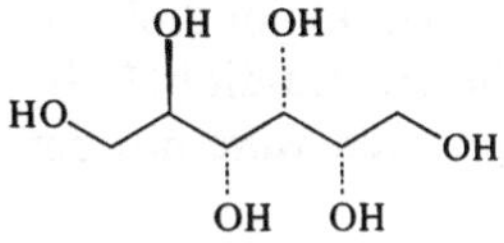

OH   OH

HO⟍⟋⟍⟋⟍OH

OH   OH

(130)

OH

HO⟍ ⟋O⟍ ⟋OH

HO⟍ ⟋OH

OH

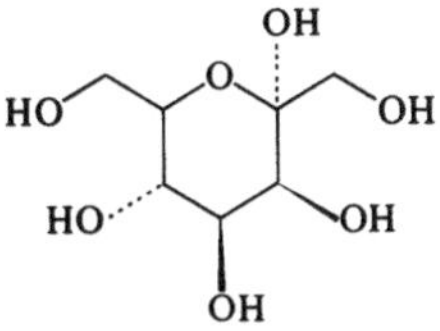

(131)

complements from perseitol **(129)**, sorbitol **(130)**, and D-mannoheptulose **(131)** are shown in Schemes XV, XVI, and XVII, respectively. Also shown in these schemes are vicinal coupling constants for the fully acetylated alditols and the various diols obtained from the borate complexation.

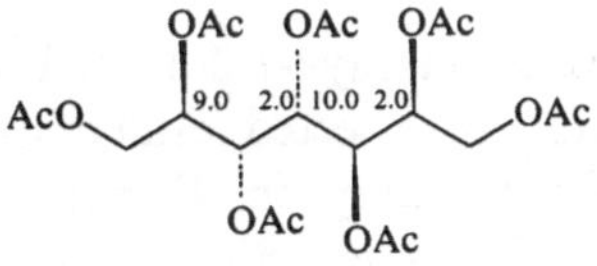

OAc   OAc   OAc

AcO⟍9.0 2.0⟋10.0 2.0⟍OAc

OAc   OAc

(94)

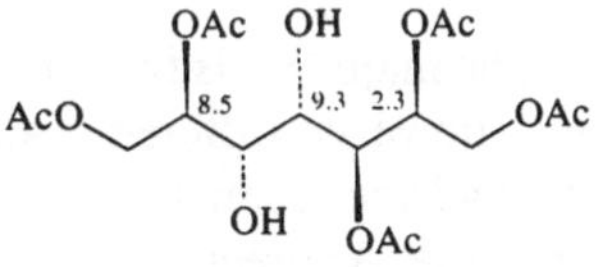

OAc   OH   OAc

AcO⟍8.5⟋9.3 2.3⟍OAc

OH   OAc

(132)   65%

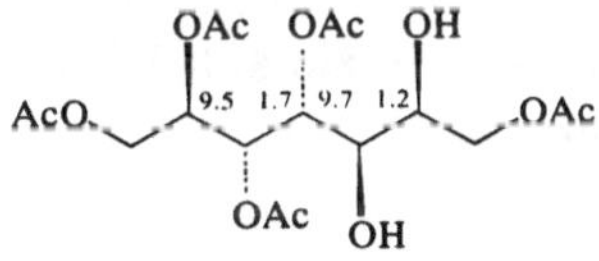

OAc   OAc   OH

AcO⟍9.5 1.7⟋9.7 1.2⟍OAc

OAc   OH

(133)   22%

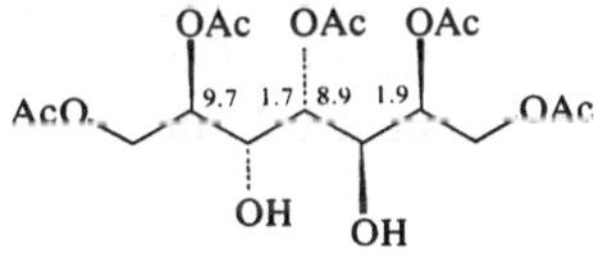

OAc   OAc   OAc

AcO⟍9.7 1.7⟋8.9 1.9⟍OAc

OH   OH

(134)   6%

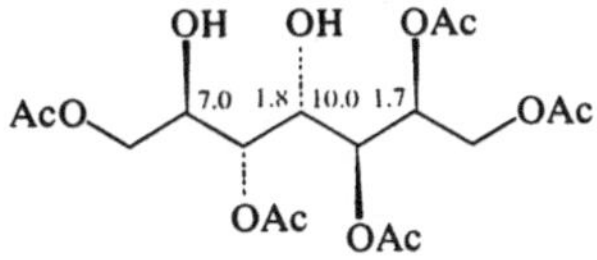

OH   OH   OAc

AcO⟍7.0 1.8⟋10.0 1.7⟍OAc

OAc   OAc

(135)   3%

OAc   OH   OH

AcO⟍7.9 1.6⟋9.4 2.0⟍OAc

OAc   OAc

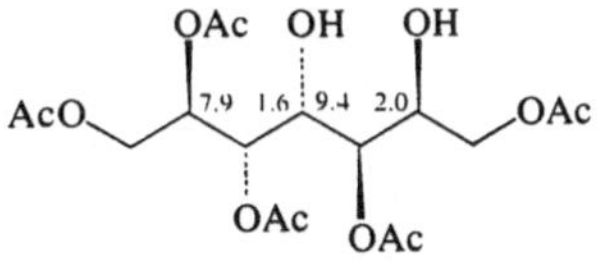

(136)   2%

Scheme XV

The major diols from perseitol were the *syn*-1,2-diols **(132)** and **(133)**. Minor amounts of the anti-1,3-diols **(134)**, **(135)**, and **(136)** were also formed. *Anti*-1,2-diols and terminal 1,2- and 1,3-diols were not detected.

The major diols from sorbitol were the *syn*-1,2-diols (**137**) and (**139**) and the *syn*-1,3-diol (**138**). *Anti*-1,2, *anti*-1,3, and terminal 1,2- and 1,3-diols, however, were not detected.

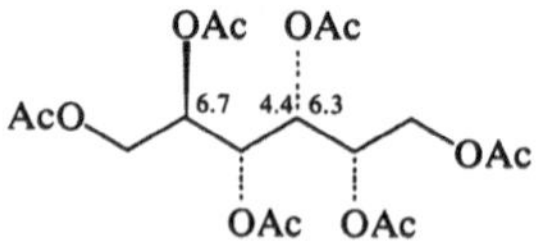

(**101**) 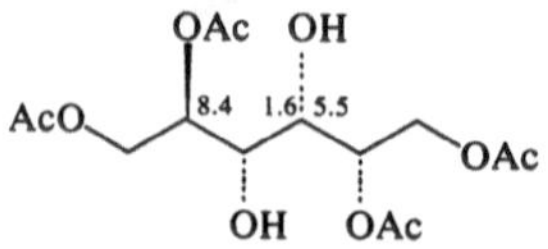   (**137**)   57%

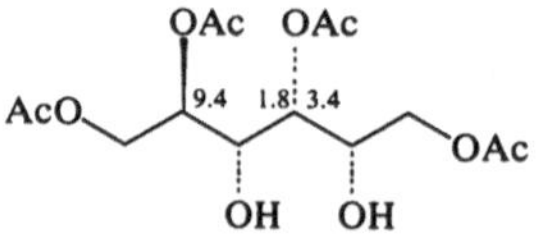

(**138**)   30% 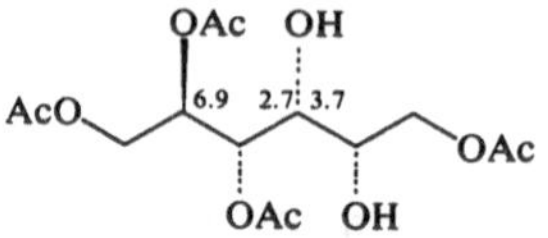   (**139**)   13%

Scheme XVI

The major diols from D-mannoheptulose were again *syn*-1,2-diols. The diols, however, differed in stereochemistry at C(2), an expected result of the borohydride reduction of the hemiketal functionality in D-manno-heptulose. Appreciable stereoselectivity occurred in the reduction. Diols of **D**-glycero-**D**-*manno*-heptitol (**140**) were formed in a 5:1 ratio over diols of perseitol (**129**). The major diol from (**140**) was the *syn*-1,2-diol (**141**). Only a minor amount of the *syn*-1,3-diol (**142**), however, was formed along with minor amounts of the *anti*-1,3-diols (**143**) and (**144**). Again anti-1,2-diols and terminal 1,2- and 1,3-diols were not found.

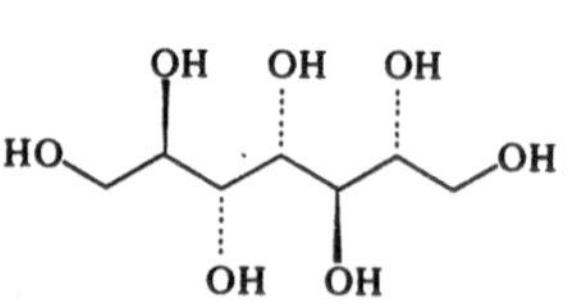

(**140**)

Syn-1,2- and *syn*-1,3-diol groups nearer the center of the alditol carbon chain appeared to complex more readily than *syn*-1,2- and *syn*-1,3-diol groups adjacent to the terminal carbons. Diols (**132**), (**137**), and (**141**), for example, were formed in higher than diols (**133**), (**138**), (**139**), and (**142**). Interestingly the susceptibility of a 1,2-diol to borate complexation appeared to be comparable to the diol's reactivity to periodate cleavage.

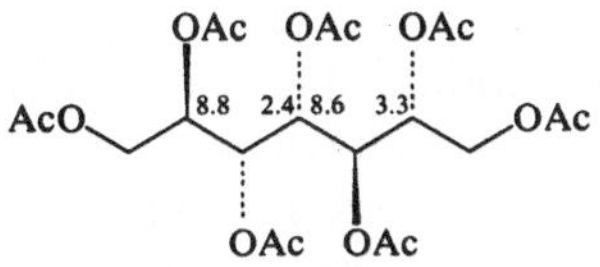

OAc OAc OAc
AcO   8.8 2.4 8.6 3.3   OAc

OAc OAc

(103)

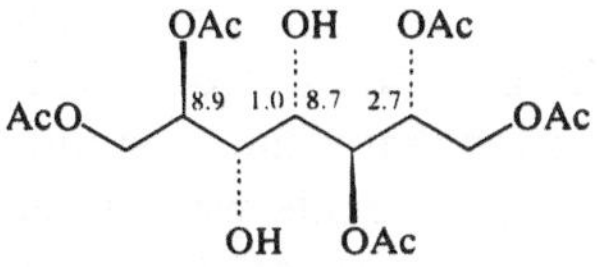

OAc OH OAc
AcO   8.9 1.0 8.7 2.7   OAc

OH OAc

(141) 66%

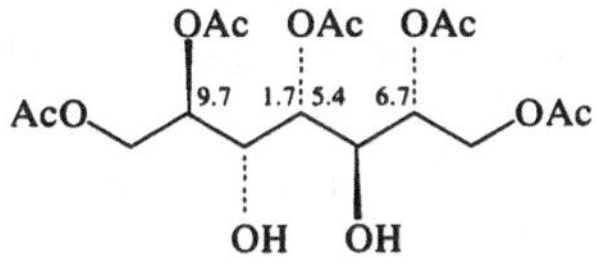

OAc OAc OAc
AcO   9.7 1.7 5.4 6.7   OAc

OH OH

(143) 8%

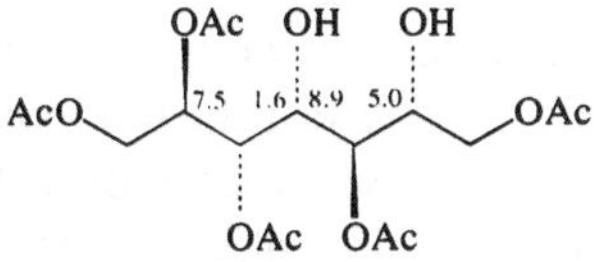

OAc OH OH
AcO   7.5 1.6 8.9 5.0   OAc

OAc OAc

(142) 4%

OH OH OAc
AcO   7.2 1.7 9.7 2.6   OAc

OAc OAc

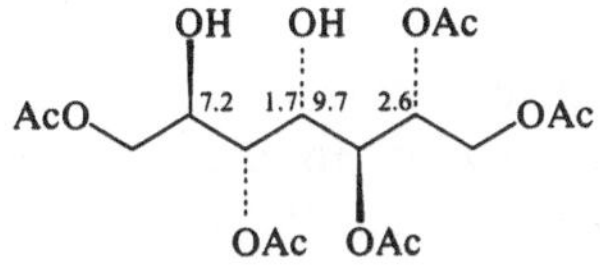

(144) 4%

Scheme XVII

Perseitol heptaacetate (**94**) and **D**-glycero-D-*manno*-heptitol heptaacetate (**103**) exhibited small values for $J_{2,3}$, 2.0 and 3.3 Hz, respectively,
indicating that compounds such as (**94**) and (**103**) can not be differentiated
by nmr analysis alone. It was possible to distinguish perseitol (**129**) from
D-*glycero*-D-*manno*-heptitol (**140**) by borate complexation as (**129**) led to
syn-1,2-diol (**133**), but (**140**) did not produce diol (**145**) as the stereochemistry at C(2) – C(3) was *anti*. Diols (**133**), (**134**), and (**136**) exhibited the same
small coupling constant $J_{2,3}$ (1.2 – 2.0 Hz) as perseitol heptaacetate (**94**).
Since the coupling constants were small in both the fully and partially
acetylated derivatives of perseitol the substituents on C(2) – C(3) had to be
syn. Diols (**142**) and (**143**) from borate complexation of **D**-*glycero*-D-
manno-heptitol (**140**), on the other hand, showed substantially larger values
for $J_{2,3}$, 5.0 and 6.7 Hz, respectively, than the fully acetylated alditol (**103**),
indicating that the stereochemistry at C(2) – C(3) was *anti*.

OAc OAc OH
AcO   OAc

OAc OH

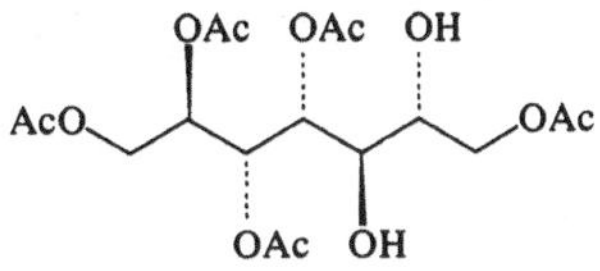

(145)

A comparison of coupling constants for several other fully acetylated alditols and diols obtained from borate complexation of the alditols (*36*) showed that vicinal coupling constants remained essentially the same if *syn*-1,3 substituents were absent. If *syn*-1,3 substituents were present in the alditol, however, the vicinal coupling constants became larger in the partially acetylated compounds if small to medium-sized coupling constants were observed in the fully acetylated alditol for *anti* stereochemistry. Conversely the coupling constants became smaller in the partially acetylated alditols if large to medium-sized coupling constants were observed in the fully acetylated alditol for *syn* stereochemistry. Sorbitol hexaacetate (**101**), for example, exhibited medium-sized values for $J_{2,3}$, $J_{3,4}$, and $J_{4,5}$ but diols (**137**), (**138**), and (**139**) generally showed smaller coupling values for $J_{2,3}$ and $J_{3,4}$ and larger values for $J_{4,5}$ (Scheme XVI), indicating that the relative stereochemistry was *syn,syn,anti*.

Borate complexation indicated that the relative stereochemistry of the C(96)−C(99) segment in palytoxin was *syn,syn,anti*. In the partially acetylated ozonolysis product (**104**), the values of $J_{96,97}$ (6.5 Hz), $J_{97,98}$ (1.5 Hz), and $J_{98,99}$ (8.5 Hz) were comparable to $J_{2,3}$, $J_{3,4}$, and $J_{4,5}$ observed in sorbitol 1,2,5,6-tetraacetate (**137**) (5.5, 1.6, and 8.4 Hz, respectively). $J_{97,98}$ was much smaller for (**104**) than for the fully acetylated compound (**17**) (5.5−6.4 Hz); also $J_{98,99}$ was much larger in (**104**) than in (**17**) (4.3−5.5 Hz). $J_{96,97}$ was slightly larger in (**104**) than in the fully acetylated compound; however, $J_{96,97}$ was expected to be much smaller in the *syn*-1,3-diol (**127**).

One feature of the borate complexation with ozonolysis products (**125**) and (**126**) was the total absence of cyclic 1,2-diols. Model studies with known disaccharides gave similar results and further showed that cyclic *cis*-1,2-diols are far less susceptible to borate complexation than acyclic *syn*-1,2-, *syn*-1,3-, and *anti*-1,3-diols. Borate complexation of 6-O-β-D-galactopyranosyl-D-glucitol (**146**) (the NaBH$_4$ reduction product of lactose), for example, led to *syn*-1,2-diol (**147**) and *anti*-1,3-diol (**148**); the *cis*-1,2-diol (**149**), however, was not formed.

(**146**)   (**147**)

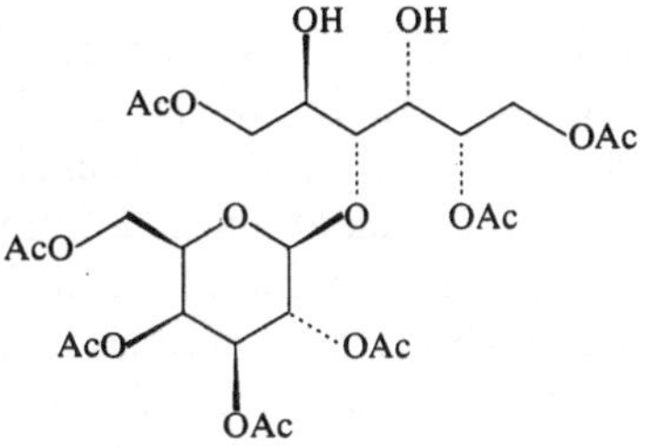

(148) 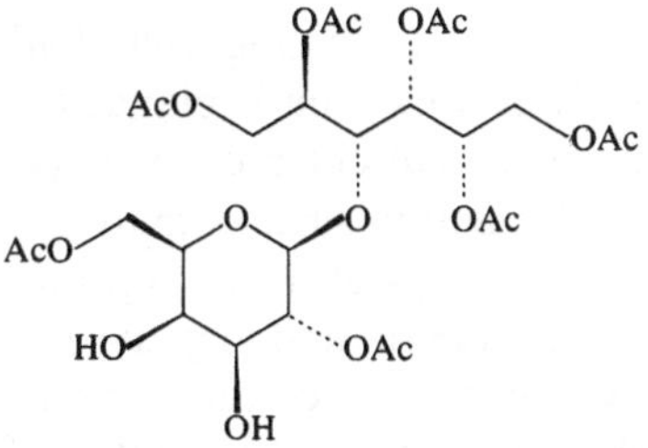   (149)

Although borate complexation of the various ozonolysis products from palytoxin has not yet been studied in detail, the preliminary data strongly suggest that this method could have been used to solve the relative stereochemistry of most, if not all, of the acyclic segments in the palytoxin molecule.

V.2 Methods Used for Absolute Stereochemistry

a) X-Ray Crystallography

The absolute stereochemistries of acetals (59) and (67) have been determined by X-ray crystallography (24). Thirteen asymmetric carbons in palytoxin are therefore 34S, 36R, 37R, 39S, 41S, 49S, 111S, 113R, 115S, 116R, 119R, 120R, and 122S.

b) Circular Dichroic Spectroscopy

The absolute configurations of C(10), C(11), and C(13) have been determined by a circular dichroism (CD) study of δ-lactone (92) (28). In ethanol the CD spectrum in ethanol of this degradation product revealed a single positive peak at 224 nm, [θ] +2300. The position of the CD peak as well as the coupling constants associated with the ring protons indicated that the δ-lactone ring was a half-chair while the positive sign for the CD peak required that the absolute configuration of (92) had to be as shown. The absolute stereochemistry in palytoxin was therefore 10S, 11R, 13R.

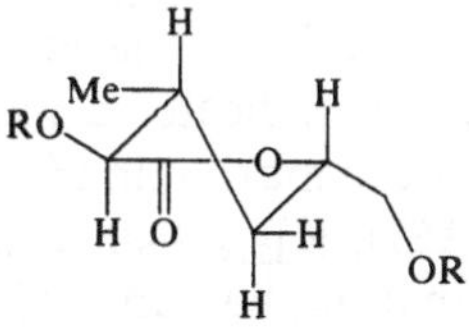

(68) R = Ac
(92) R = H

X-Ray crystallography had shown that the $C-CO-O-C$ portion of the δ-lactone ring is planar (40) and that the conformation of the ring is either a half-chair or boat (41). δ-Lactones that exist in the half-chair conformation such as malyngolide (42) show a CD peak at about 222 nm (43), the sign of which is positive if the absolute configuration of the half-chair is (150a) and negative if it is (150b) (Scheme XVIII). δ-Lactones that exist in the boat conformation, however, show a CD peak at about 215 nm, the sign of which is negative if the absolute configuration of the boat is (150c) and positive if it is (150d). (+)-5-Decanolide exists entirely in the half-chair conformation (150b) at −185° and its spectrum exhibits a negative peak at 222 nm (43). At room temperature, however, (+)-5-decanolide exists in both the boat and half-chair conformations and the CD spectrum displays two maxima, a positive one at 210 nm for the boat conformation (150d) and a negative one at 240 nm for the half-chair conformation. The presence of a positive peak at about 215 nm for the boat conformation produces a red shift in the negative peak for the half-chair conformation.

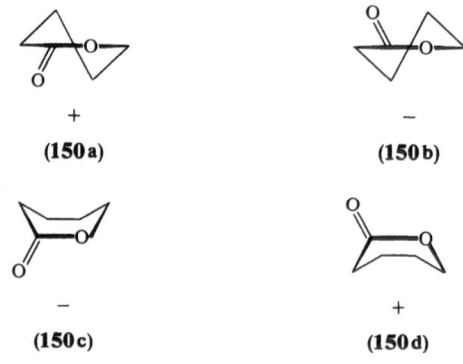

+   −

(150a)   (150b)

−   +

(150c)   (150d)

Scheme XVIII

TIUS and THURKAUF (44) synthesized (151), the enantiomer of (92) and the stereoisomers (152), (153), and (154) from (+)-pulegone and carried out a CD study in ethanol. Diols (151), $[\theta]_{224}$ −2300, and (152), $[\theta]_{223}$ +1040, exhibited CD maxima that were consistent with half-chair conformations, whereas (153), $[\theta]_{217}$ +730, and (154), $[\theta]_{217}$ −2440, showed CD maxima characteristic of boat conformations. Diacetate (155), $[\theta]_{222}$ −3100, also exhibited a CD maximum that indicated a half-chair conformation, but (156), $[\theta]_{215}$ −830, showed a CD maximum that was consistent with a boat conformation. Diacetate (157), $[\theta]_{214}$ +4000, displayed a CD maximum that favored a boat conformation, but (158), $[\theta]_{222}$ +1500, exhibited a CD maximum characteristic of a half-chair conformation. The conformations indicated by CD spectroscopy for the four diols and the diacetates (155),

(156), and (157) were supported by ^{1}H nmr data. The ^{1}H nmr data for (158), however, favored a boat instead of a half-chair conformation. The sign of the CD peak (Cotton effect) appeared to be dependent mainly on the absolute configuration of the half-chair or boat and unaffected by hydroxy, but not acetoxy, groups on the δ-lactone ring.

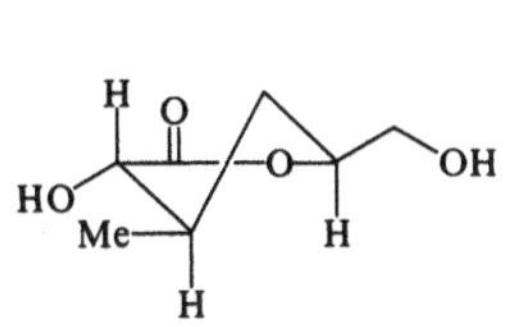

(151)

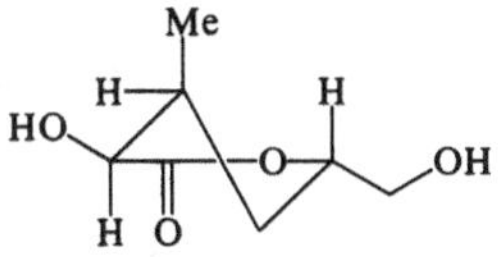

(152)

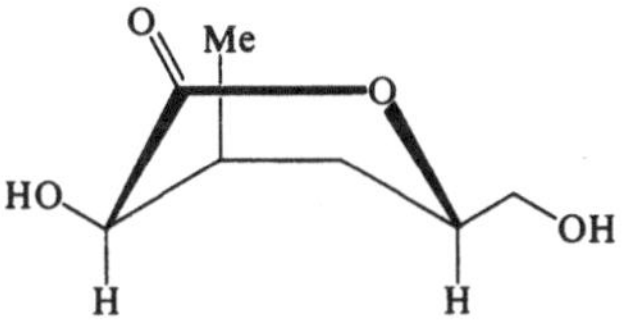

(153)

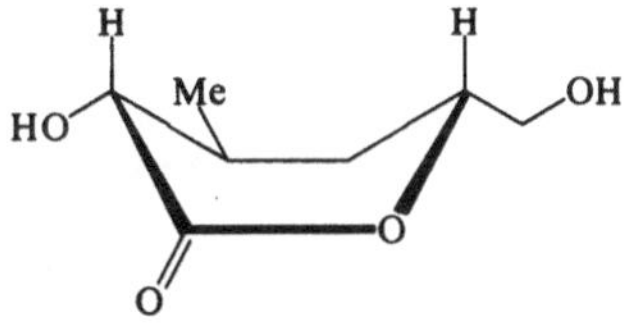

(154)

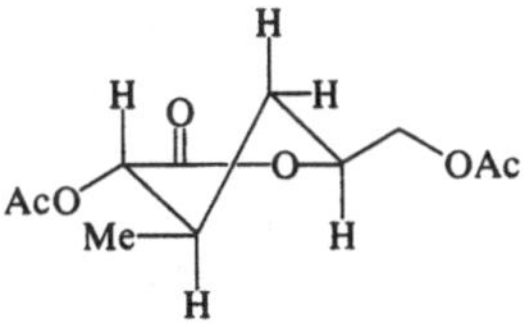

(155)

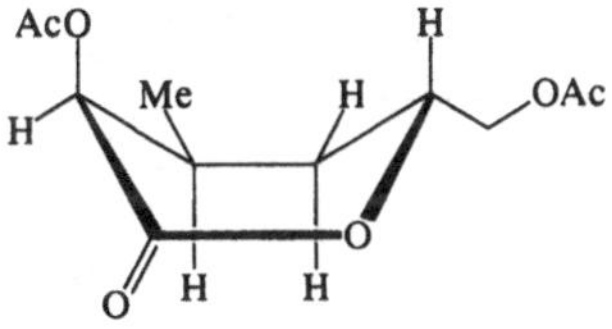

(156)

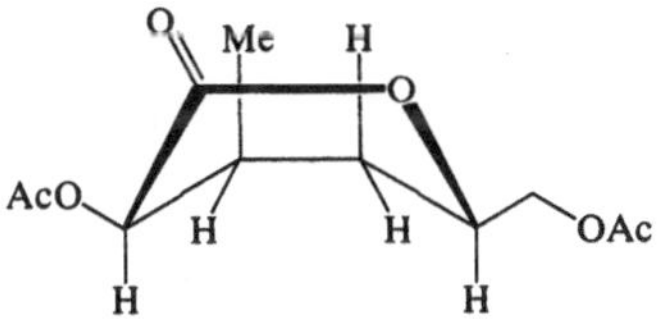

(157)

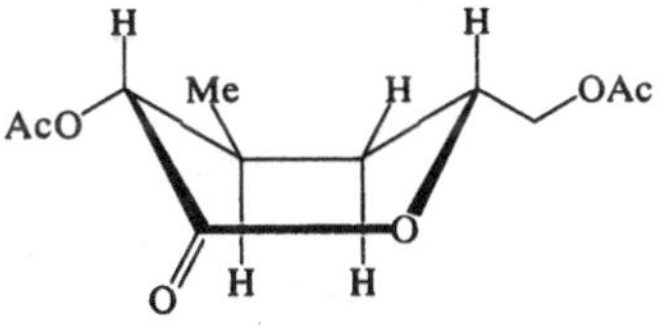

(158)

The absolute configurations of four other asymmetric carbons, viz. C(57), C(58), C(61), and C(99), could have been deduced by CD studies of δ-lactone type degradation products, but this was not done.

The absolute configurations of C(87) and C(89) in palytoxin were both deduced as R on comparing the CD spectrum of ozonolysis product (16) with the CD spectra of two closely related compounds (28). The CD spectrum of (16) in ethanol showed a maximum at 210 nm, [θ] +2000, comparable in sign, position, and intensity with those of (159), [θ]$_{212}$ +2500 in ethanol, and (160), $[\theta]_{212}$ +2900 in ethanol. Pentaacetates (159) and (160) were formed by borohydride reductions of 2-deoxy-D-glucose and 2-deoxy-D-galactose, respectively, followed by acetylation of the resulting 2-deoxyhexitols.

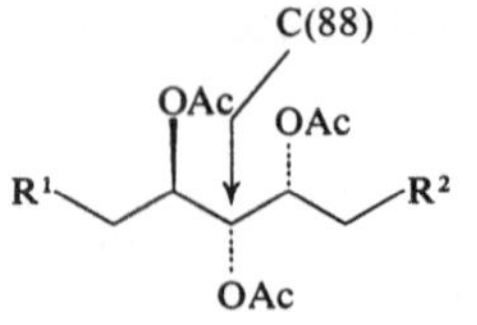

(16) $R^1 = R^2 = CH_2OAc$
(159) $R^1 = OAc$; $R^2 = CH_2OAc$
(160) $R^1 = CH_2OAc$; $R^2 = OAc$

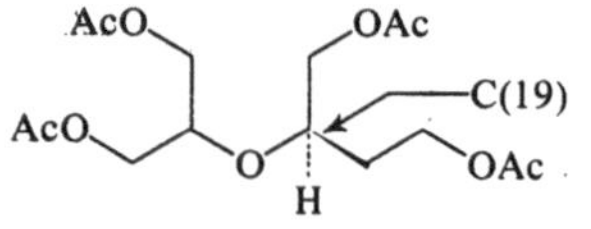

(30)

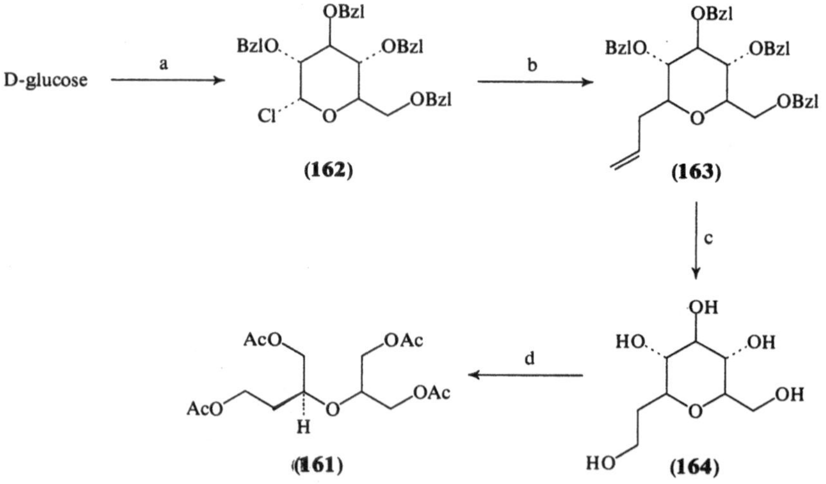

Reagents: a. 1. MeOH/Dowex 50 H$^+$ form. 2. C$_6$H$_5$CH$_2$Cl/KOH/dioxane/reflux. 3. H$^+$.
4. SOCl$_2$. b. CH$_2$=CH−CH$_2$MgBr/ether. c. 1. O$_3$/CH$_2$Cl$_2$/−78°C. 2. NaBH$_4$.
3. HCO$_2$H-MeOH/Pd-C. d. 1. NaIO$_4$/H$_2$O. 2. NaBH$_4$. 3. Ac$_2$O/py.

Scheme XIX

The absolute stereochemistry of the C(19)−C(23) tetrahydropyran ring in palytoxin was determined as 19R, 20S, 21R, 22S, 23S by comparison of the CD spectra of the periodate oxidation product (30) and the synthetic enantiomer (161). The degradation product (30) had a negative CD curve,

$[\theta]_{215}$ -100 in ethanol, which was opposite in sign to that of synthetic (161), $[\theta]_{215}$ $+100$ in methanol (28). Antipodal (161) was synthesized in a straightforward manner. D-Glucose was converted to the chloride (162) which was subjected to a Grignard reaction to give (163). Ozonolysis of (163), followed by borohydride reduction and debenzylation produced the tetrahydropyran (164). Periodate oxidation of (164) followed by boro-hydride reduction and acetylation resulted in (161) (Scheme XIX).

The absolute stereochemistry of the C(73) – C(81) segment of palytoxin was determined by a CD study of (165). Compound (165) was produced by ozonolysis of the periodate oxidation product (8), reductive workup with borohydride, and acetylation. Degradation product (165) exhibited a CD curve that was similar to that of synthetic (165), $[\theta]_{214}$ $+200$ in methanol. Compound (165) was constructed from (163) in a straightforward manner via the acetonide (166) and the nitrile (167) (Scheme XX) (45).

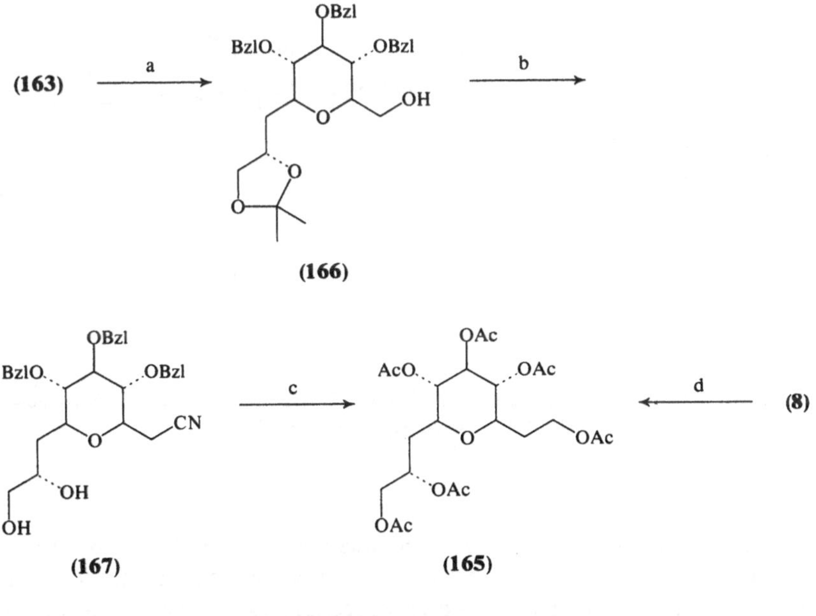

(166)

(167)   (165)

Reagents: a. 1. $OsO_4/NaClO_3/THF-H_2O$. 2. $NaBH_4$. 3. separate diols by chromatography. 4. $AcOH - Ac_2O(1:1)/3\%$ H_2SO_4. 5. $K_2CO_3/EtOH$. 6. acetone/p-TSA. b. 1. $TsCl/Et_3N/DMAP/0°C$. 2. $LiBr/THF$. 3. $KCN/18$-crown-6/MeCN/reflux. 4. 1N $HClO_4$-THF(1-1). c. 1. $DIBAL/CH_2Cl_2/0°C$. 2. $NaBH_4$. 3. HCO_2H-MeOH/Pd-C. 4. Ac_2O/py. d. 1. O_3. $NaBH_4$. 3. Ac_2O/py.

Scheme XX

c) Synthesis

The absolute configuration of all 64 asymmetric carbons in palytoxin have been unambiguously established by synthesis in the laboratory of KISHI (29, 30, 31, 46).

The absolute stereochemistries of the seven chiral centers at C(111), C(113), C(115), C(116), C(119), C(120), and C(122), which had been established by X-ray crystallography, were confirmed by synthesis of (168), produced from the periodate oxidation product (67) by hydrogenolysis (46). Phosphonium salt (169), synthesized from D-xylose in 18 steps, and aldehyde (170), prepared from 2-deoxy-D-ribose in 7 steps, were coupled in a Wittig reaction in the presence of lithium diisopropylamide (LDA) (Scheme XXI). Diimide reduction of the resulting olefin, debenzylation, and acetylation gave (168), $[\alpha]_D$ +56.0° (c 1.26, $CHCl_3$), which was identical in all respects with the degradation product.

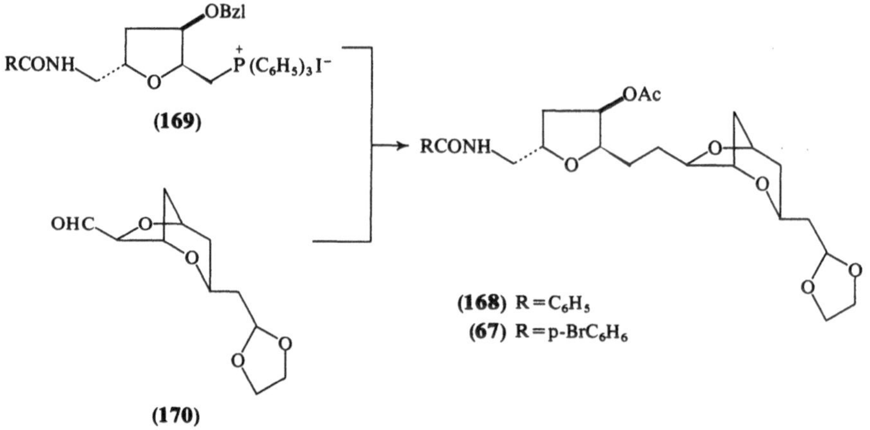

Reagents: 1. LDA/DMF-THF (2-1)/−78° C → RT. 2. NH=NH/dioxane/RT. 3. H_2/Pd-C/ AcOH-MeOH (1-20)/RT. 4. Ac_2O/Py-DMAP/RT.

Scheme XXI

The tetrahydropyran ring of (169) was constructed from D-xylose as shown in Scheme XXII. D-Xylose was first converted to the acetonide (171) in four steps according to a known procedure (47). Benzylation of the alcohol group in (171) followed by hydrolysis of the acetonide, silylation of the generated primary alcohol, and transformation of the dithiaacetal to an aldehyde resulted in the hemiacetal (172). A Wittig reaction then converted (172) into a 8:1 mixture of *trans* and *cis* (173), which on treatment with Triton B gave exclusively the saturated ester (174). Compound (174) was

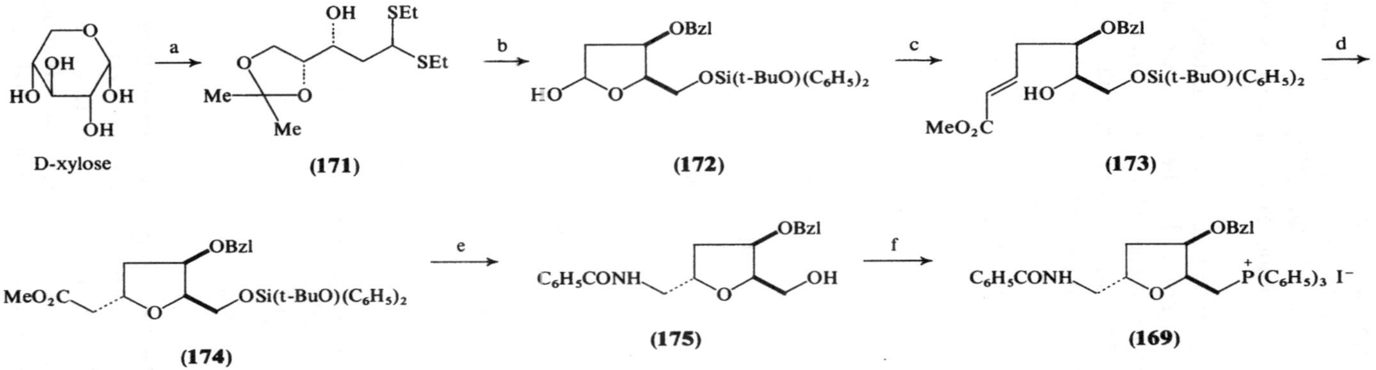

(173)

(172)

(169)

(175)

(171)

(174)

D-xylose

Reagents: a. 1. EtSH/HCl. 2. acetone/p-TSA. 3. t-BuOK. 4. LAH/THF. b. 1. C₆H₅CH₂Br/ NaH/THF-DMF (4-1)/RT. 2. 75% aq. AcOH/50°C. 3. (C₆H₅)₂(t-Bu)SiCl/imida- zole/DMF/RT. 4. HgO/HgCl₂/99% aq. acetone/60°C. c. (C₆H₅)₃P=CHCO₂Me/ C₆H₆/reflux. d. C₆H₅CH₂NMe₃OMe/MeOH/RT. e. 1. NH₂NH₂·H₂O/MeOH/ RT. 2. NOCl/CH₂Cl₂/−50°C. 3. C₆H₆/reflux. 4. 10% HCl/dioxane/70°C. 5. C₆H₅COCl/1N NaOH/Et₂O/0°C. f. 1. (CF₃CO)₂O/Py/−20°C ∼ −10°C. 2. (n-Bu)₄NI/CH₂Cl₂/reflux. 3. P(C₆H₅)₃/MeCN/120°C.

Scheme XXII

then converted to the acyl azide which rearranged to an isocyanate on heating. Hydrolysis, decarboxylation, and N-acylation produced (175) in an overall 60% yield from (171). Alcohol (175) was converted to the trifluoroacetate ester, then to the iodide, and finally to the phosphonium salt (169).

The 2,6-dioxabicyclo[3.2.1]octane ring of (170) was fabricated from 2-deoxy-D-ribose as shown in Scheme XXIII. The aldose was first subjected to a Wittig reaction to form the diene ester (176). After protection of the primary alcohol group, (176) was treated carefully with potassium t-butoxide to give (177) in an overall yield of 35% from 2-deoxy-D-ribose. Reduction of ester (177) with diisobutyl aluminium hydride (DIBAL), followed by protection of the resulting aldehyde as the ethylene ketal, desilylation, and Swern oxidation (48) of the liberated primary alcohol resulted in aldehyde (170).

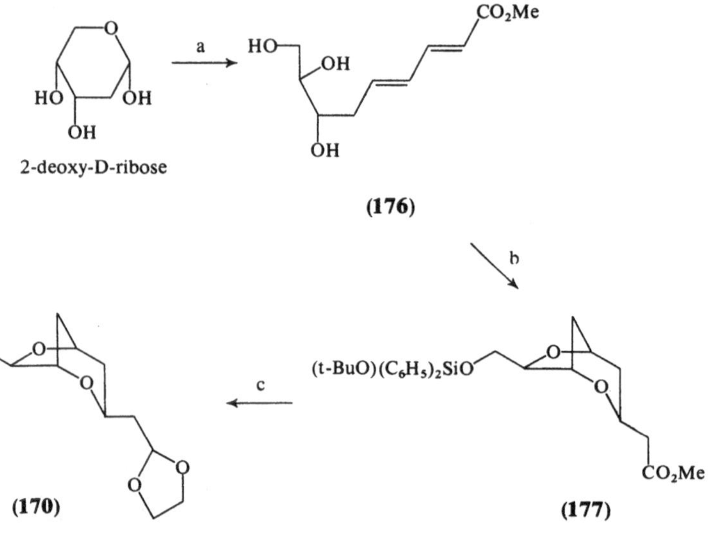

2-deoxy-D-ribose

(176)

(170)   (177)

Reagents: a. $(C_6H_5)_3P=CH-CH=CHCO_2Me/C_6H_6$/reflux. b. 1. $(C_6H_5)_2(t\text{-}Bu)SiCl$/imidazole/DMF/RT. 2. t-BuOK (0.3 eq.; 1 M in t-BuOH)/C_6H_6/RT. c. 1. DIBAL/ $CH_2Cl_2/-78°C$. 2. $(CH_2OH)_2/p$-TSA/C_6H_6/reflux. 3. $(n\text{-}Bu)_4NF$/THF/RT. 4. $(COCl)_2$/DMSO/$CH_2Cl_2/-60°C \rightarrow$ RT.

Scheme XXIII

With (168) in hand, the KISHI group next turned their attention to C(108) and C(109) in the degradation product (178), prepared by hydrogenolysis of the ozonolysis product (18) (29). In order to determine the absolute configurations of these two centers unambiguously, all four stereoisomers of (178) or its equivalent were synthesized. Aldehyde (179),

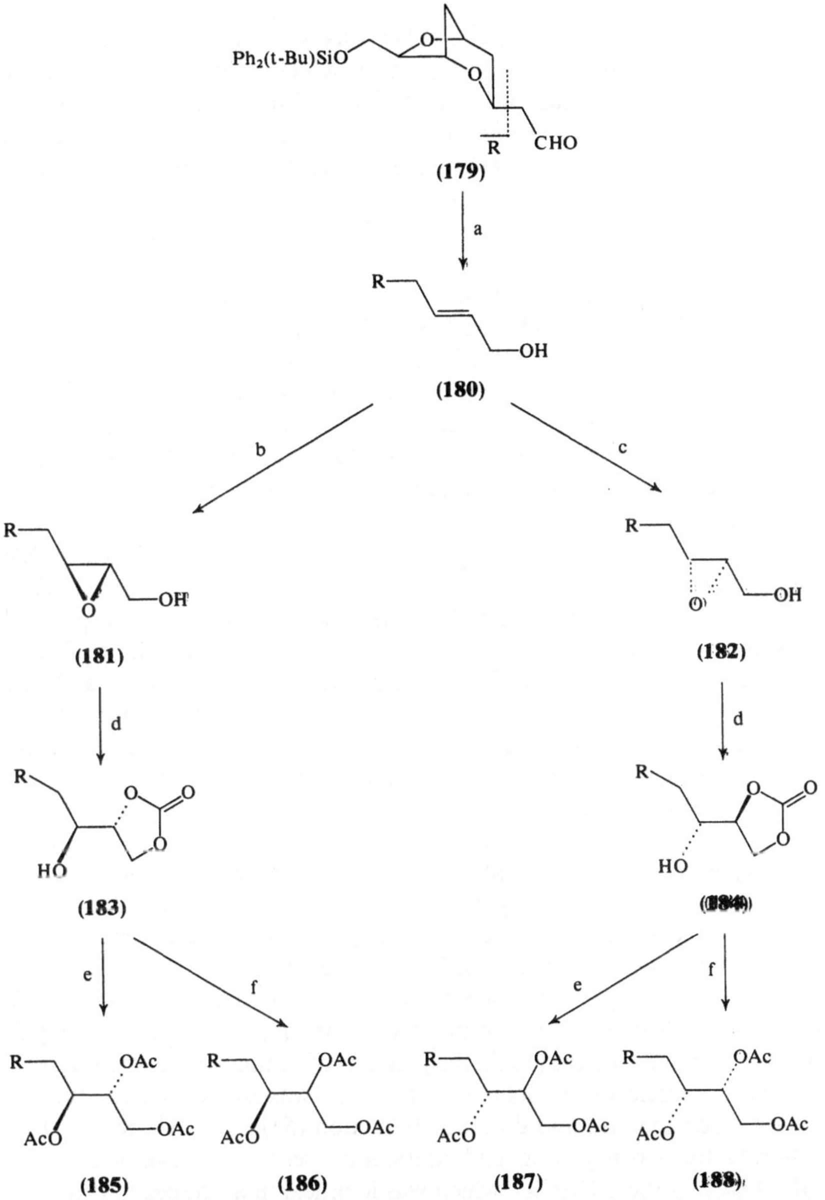

Reagents: a. 1. $(i\text{-PrO})_2P(O)CH_2CO_2Et/t\text{-BuOK}/THF/-78°C$. 2. DIBAL/hexane/$-78°C$.
b. $t\text{-BuO}_2H/(+)$-diethyl tartrate/$Ti(i\text{-PrO})_4/CH_2Cl_2/-23°C/2$ days. c. $t\text{-BuO}_2H/$
$(-)$-diethyl tartrate/$Ti(i\text{-PrO})_4/CH_2Cl_2/-23°C/2$ days. d. 1. $C_6H_5CH_2OCOCl/$
py/THF/$-23°C$. 2. $AlCl_3/Et_2O/-23°C$. e. 1. $NaOH/MeOH-H_2O$. 2. Ac_2O/py.
f. 1. $(COCl)_2/DMSO/CH_2Cl_2/-60°C \rightarrow$ RT. 2. $NaBH_4$. 3. separate alcohols
by chromatography. 4. $NaOH/MeOH-H_2O$. 5. Ac_2O/py.

Scheme XXIV

the DIBAL reduction product of (177), was converted to the *trans* allylic alcohol (180) in two steps (49) (Scheme XXIV). Sharpless asymmetric epoxidation (50) of (180) using (−)-diethyl D-tartrate yielded predominately epoxide (181) whereas (+)-diethyl tartrate resulted mainly in (182). Epoxides (181) and (182) were converted into the five-membered carbonate alcohols (183) and (184) in two steps. Mild base hydrolysis followed by acetylation gave the *erythro* isomers (185) and (186). Since the stereochemical outcome of asymmetric epoxidation of *cis* allylic alcohols is not always predictable (51), the *threo* isomers (187) and (188) were synthesized by Swern oxidation (48) of the carbonate alcohols (183) and (184), borohydride reduction of the resulting ketones, separation of the mixture of alcohols by chromatography, mild aqueous base hydrolysis to remove the carbonate protecting group, and acetylation. Comparison of the ^{1}H nmr spectra of (185), (186), (187), and (188) with that of the degradation product (178) showed that only (188) possessed the correct stereochemistry at C(108) and C(109). Tetraacetate (178), $[\alpha]_D$ +68.8° (c 0.52, CHCl$_3$), was synthesized from (169) and a derivative of (188) as shown in Scheme XXV and proved to be identical in all respects with degradation product (178). Triol (189), the precursor of (188), was first converted into the pivaloyl acetonide (190) in two steps. Desilylation and Swern oxidation of (190) yielded aldehyde (191) which was then coupled with (169) in a Wittig reaction. Reduction of the resulting olefin with diimide, debenzylation, removal of the pivaloyl and acetonide protecting groups with successive base and acid hydrolyses, and acetylation resulted in (178).

The absolute configurations of C(96), C(97), C(98), C(99), C(101), C(102), C(103), C(104), and C(105) were established by synthesis of ozonolysis product (17) (Scheme XXVI). The relative and absolute stereochemistry of the tetrahydropyran ring and C(99) was first determined by synthesis of (192), a more advanced degradation product prepared from (17) by aqueous base hydrolysis, periodate oxidation, borohydride reduction, and acetylation. 3,4,6-Tribenzyl-D-mannose 1,2-epoxide (193) (52) was reacted with the Grignard reagent of (S)-(+)-3-*tert*-butoxy-2-methyl-1-bromopropane (53) in the presence of lithiumcopper chloride (54) to give stereoselectively alcohol (194) which was isomerized to (195) by Swern oxidation (48) and diborane reduction (55). Debutylation of (195) followed by debenzylation and acetylation resulted in synthetic (192), $[\alpha]_D$ +64.7° (c 0.65, CHCl$_3$), which was identical in all respects with (192) from degradation of ozonolysis product (17). The C(99) diastereomer (196) was also synthesized, starting from (193) and the Grignard reagent of (R)-(−)-3-*tert*-butoxy-2-methyl-1-bromopropane (53), and compared with (192).

To complete the synthesis of (17), compound (195) was converted to the *trans* allylic alcohol (197). Using the Sharpless asymmetric epoxidation

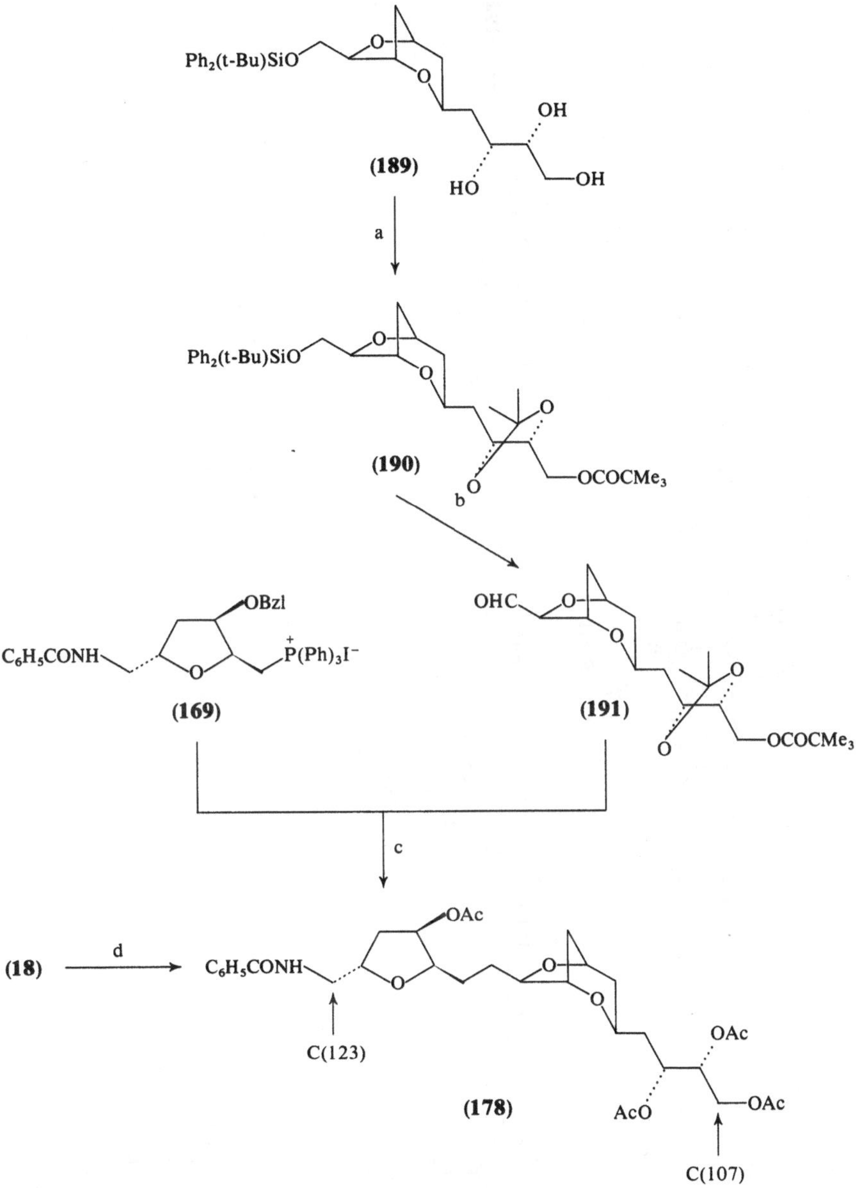

Reagents: a. 1. (Me)₃CCOCl/py. 2. Me₂C(OMe)₂/acetone/p-TSA. b. 1. (n-Bu)₄NF/THF.
2. (COCl)₂/DMSO/CH₂Cl₂/–60°C → RT. c. 1. LDA/DMF-THF. 2. NH=NH.
3. H₂/Pd-C/MeOH. 4. NaOH/H₂O. 5. AcOH/H₂O. 6. Ac₂O/py. d. H₂/Pd-C/
MeOH-AcOH.

Scheme XXV

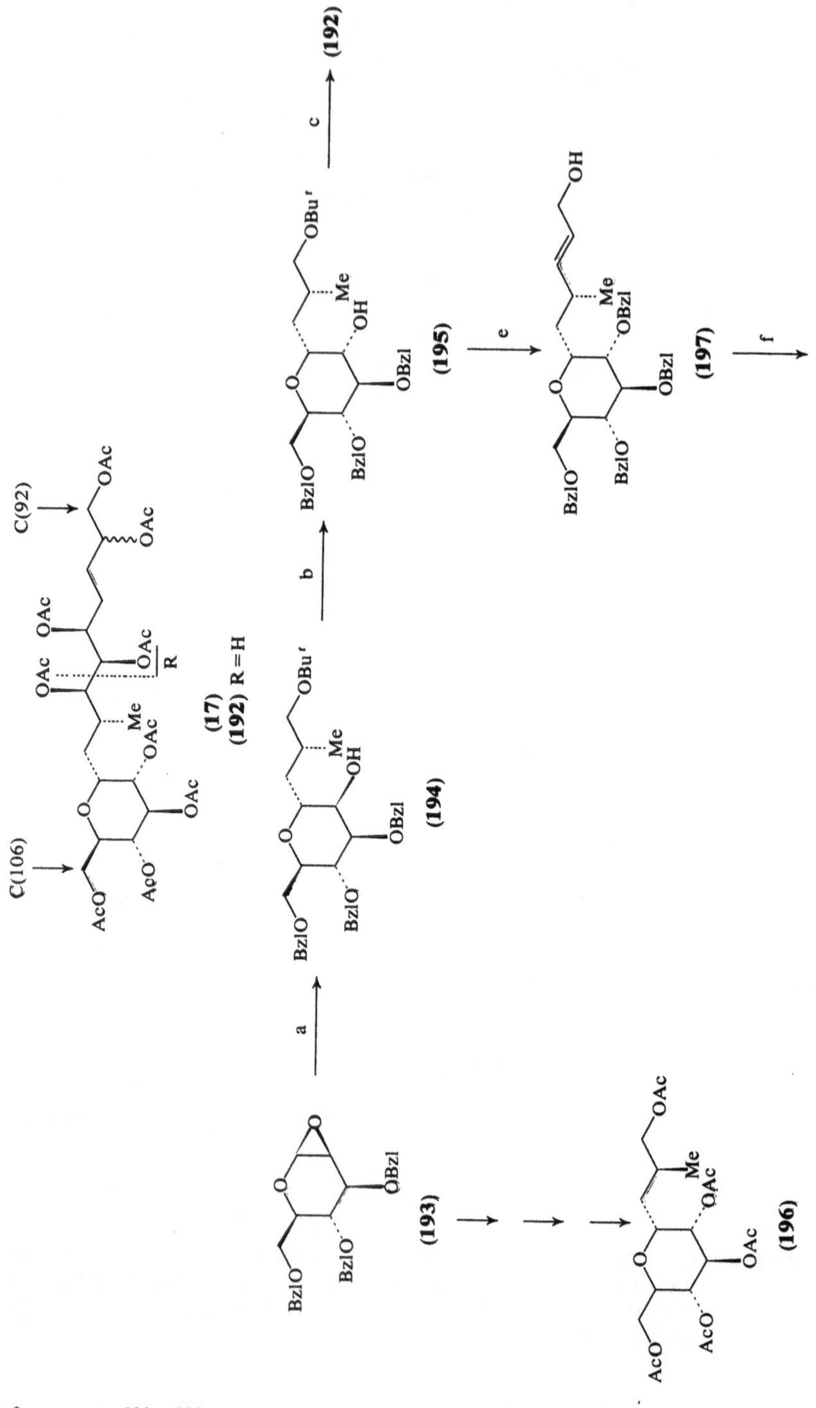

R. E. MOORE

C(92)

C(106)

(17)
(192) R = H

(192)

c

b

a

(194)

(195)

e

(197)

f

(193)

(196)

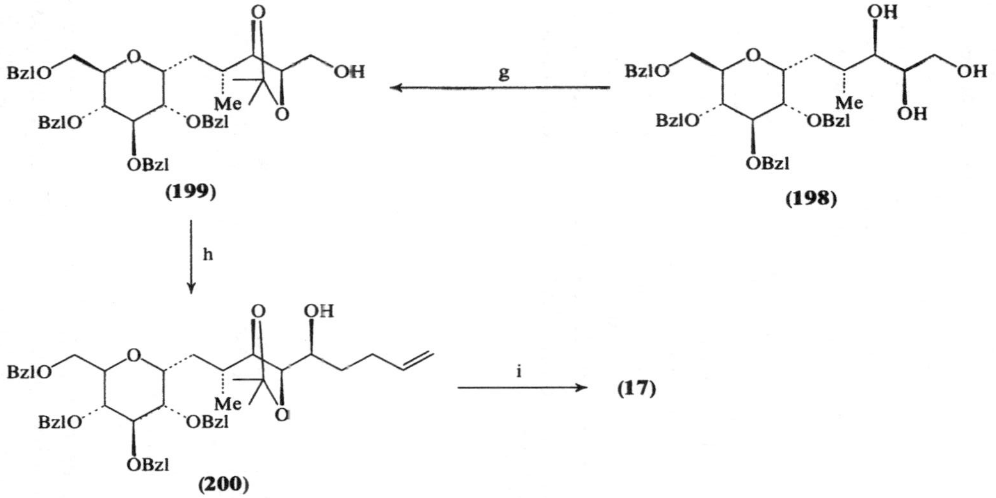

(198)

(199)

(200)

(17)

Reagents: a. (S)-(+)-3-tert-butoxy-2-methyl-1-magnesium bromide/Li₂CuCl₄. b. 1. (COCl)₂/ DMSO/ CH₂Cl₂/−60°C → RT. 2. diborane/THF/RT. c. 1. debutylation. 2. H₂/Pd-C/MeOH. 3. Ac₂O/py. d. (R)-(−)-3-tert-butoxy-2-methyl-1-magnesium bromide/Li₂CuCl₄. e. 1. benzylation. 2. debutylation. 3. (COCl)₂/DMSO/ CH₂Cl₂/−60°C → RT. 4. (i-PrO)₂P(O)CH₂CO₂Et/t-BuOK/THF/−78°C. 5. DIBAL/hexane/−78°C. f. 1. t-BuO₂H/(−)-diethyl tartrate/Ti(i-PrO)₄/CH₂Cl₂/ −23°C/2 days. 2. C₆H₅CH₂OCOCl/py/THF/−23°C. 3. AlCl₃/Et₂O/−23°C. 4. (COCl)₂/DMSO/CH₂Cl₂/−60°C → RT. 5. NaBH₄. 6. separate alcohols by chromatography. 7. NaOH/MeOH-H₂O. g. 1. Me₃CCOCl/py. 2. Me₂C(OMe)₂/ acetone/p-TSA. 3. LiAlH₄/THF. h. 1. (COCl)₂/DMSO/CH₂Cl₂/−60°C → RT. 2. CH₂=CHCH₂CH₂MgBr. i. 1. OsO₄. 2. aqueous AcOH. 3. H₂/Pd-C/MeOH. 4. Ac₂O/py.

Scheme XXVI

reaction, (197) was transformed into the *threo* triol (198) and then into acetonide (199). Swern oxidation of (199) gave an aldehyde which was reacted with the Grignard reagent of 3-butenyl bromide to yield a 2:1 mixture of alcohols. Osmium tetroxide oxidation of the minor alcohol (200) followed by acid hydrolysis, debenzylation, and acetylation produced nonaacetate (17). The two C(93) epimers (17A), $[\alpha]_D$ +36.2° (c 0.41, CHCl$_3$), and (17B), $[\alpha]_D$ +31.1° (c 0.13, CHCl$_3$), in (17) were separated by preparative silica gel thin layer chromatography and were found to be identical with those from the ozonolysis product (17). Other nonaacetates were prepared from (197) *via* the two *erythro* triols and the other *threo* triol corresponding to (198), but none were found to be identical with the degradation product (17).

The absolute stereochemistry at C(96) in synthetic (17) was determined in the following manner. The major alcohol accompanying (200) from addition of 3-butenylmagnesium bromide to the Swern oxidation product of (199) was subjected to benzylation, aqueous acid hydrolysis, periodate oxidation, and sodium borohydride reduction to afford (*R*)-2-(benzyloxy)hex-5-en-1-ol (201), $[\alpha]_D$ −9.4°, which was identical with synthetic (201), $[\alpha]_D$ −11.7°, prepared from D-glyceraldehyde acetonide in eight steps (29) (Scheme XXVII). Therefore, C(96) had the *S* configuration in the minor alcohol (200) and (17).

Reagents: 1. (C$_6$H$_5$)$_3$P=CHCO$_2$Et/C$_6$H$_6$/RT. 2. H$_2$/5% Rh-Al$_2$O$_3$/hexane. 3. DIBAL/CH$_2$Cl$_2$/−78°C. 4. (C$_6$H$_5$)$_3$P=CH$_2$/THF/0°C. 5. AcOH-H$_2$O(1-1)/Et$_2$O-THF(1-1)/55°C. 6. TrCl/py/80°C. 7. C$_6$H$_5$CH$_2$Br/NaH/DMF-THF(1-4)/RT. 8. AcOH-H$_2$O(4-1)/80°C.

Scheme XXVII

The absolute configurations of C(87), C(88), and C(89) were determined by synthesis of ozonolysis product (16) and compound (203) (Scheme XXIX), a more advanced degradation product of the partial ozonolysis product (85). Pentaacetate (16) was synthesized from 2-deoxy-D-glucose in 13 steps (Scheme XXVIII). The optical rotations of synthetic (16) and (16) from degradation of palytoxin were found to be identical, $[\alpha]_D$ +53.4° (c 0.46, CHCl$_3$). To determine how the skeleton of (16) was connected to the C(92)−C(123) segment of palytoxin, (85) was further ozonized and treated with acetic anhydride and pyridine to give the *trans-*

(16)

(85)

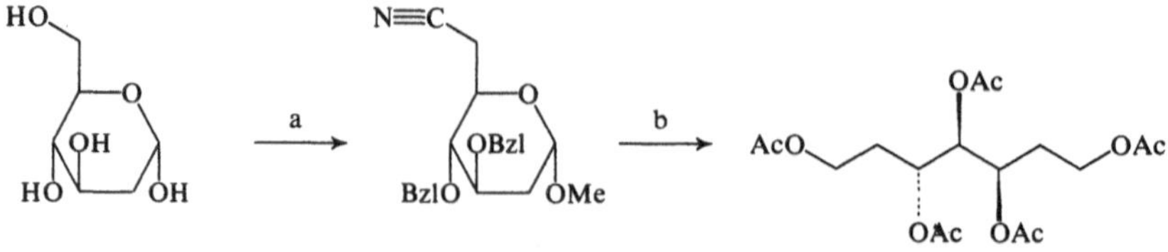

2-deoxy-D-glucose   (202)   (16)

Reagents: a. 1. Dowex-50X8-400/MeOH/RT. 2. TrCl/py/90°C. 3. C₆H₅CH₂Br/NaH/THF-
          DMF(4-1)/RT. 4. HCl/CH₂Cl₂/−78°C. 5. MsCl/Et₃N/CH₂Cl₂/0°C. 6. NaCN/
          DMSO/90°C. b. 1. DIBAL/toluene/−78°C. 2. NaBH₄/EtOH/0°C. 3. 1N H₂SO₄/
          dioxane/70°C. 4. NaBH₄/EtOH/0°C. 5. Ac₂O/DMAP-py/RT. 6. H₂/Pd-C/
          AcOH-MeOH(1-20)/RT. 7. Ac₂O/DMAP-py/RT.

Scheme XXVIII

α,β-unsaturated aldehyde (202). Synthesis of (203) and the C(88) diastereomer (204) was accomplished from nitrile (202) in seven steps
(Scheme XXIX). Synthetic (203) and (203) from degradation of palytoxin
were found to have identical ¹H nmr spectra, establishing the absolute
stereochemistry at C(87), C(88), and C(89) as 87R, 88S, and 89R.

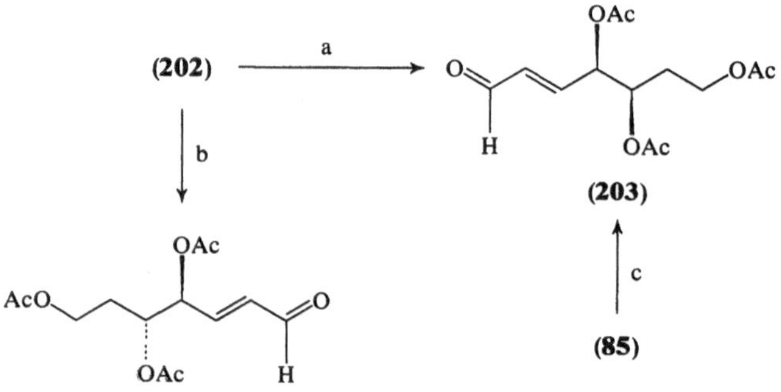

Reagents: a. 1. TFA-H₂O(4-1)/RT. 2. NaBH₄/EtOH/0°C. 3. DIBAL/toluene/−78°C→RT.
          4. Dowex-50X8-400/MeOH/RT. 5. H₂/Pd-C/AcOH-MeOH(1-20)/RT. 6. Ac₂O/
          DMAP-py/RT. 7. 1N HCl/AcOH/0°C. b. conditions not reported. c. 1. O₃.
          2. Ac₂O/py.

Scheme XXIX

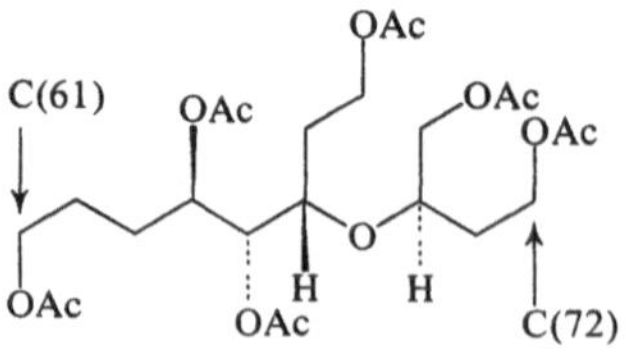

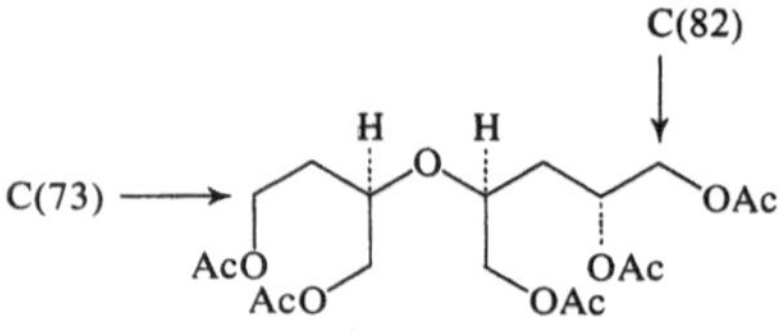

(82)   (205)

References, pp. 199—202

Degradation products (82) and (205) were useful in determining the absolute configurations of seven asymmetric carbons in palytoxin, *viz.* C(64), C(65), C(66), C(70), C(75), C(79), and C(81). The synthesis of (82) was accomplished as shown in Scheme XXX. *Trans*-allylic alcohol (206), prepared from 2-deoxy-D-glucose in seven steps, was subjected to Sharpless

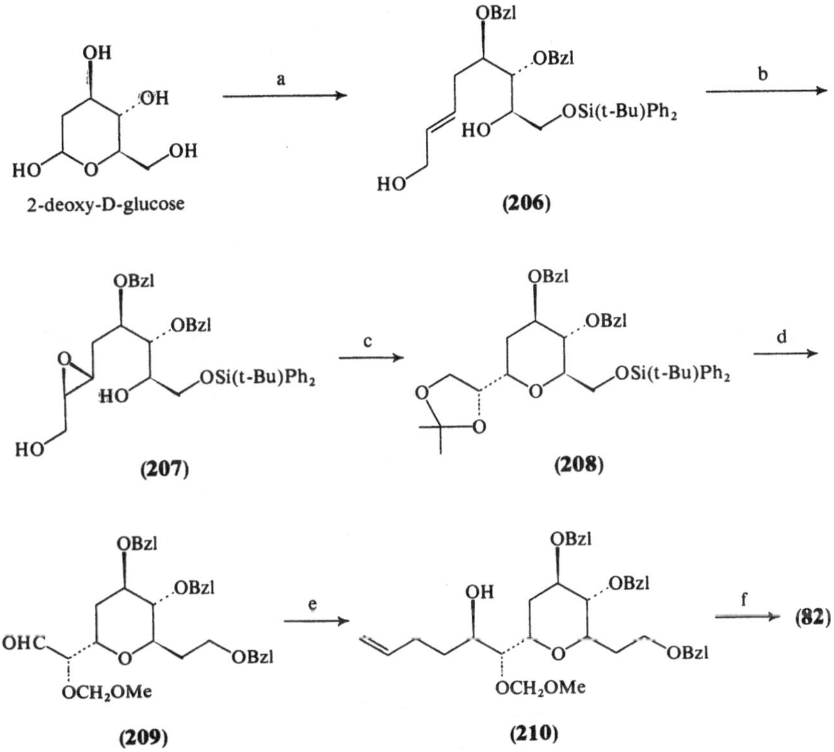

Reagents: a. 1. Dowex-50X8-400/MeOH/RT. 2. TrCl/py/90°C. 3. C$_6$H$_5$CH$_2$Br/NaH/THF-DMF(4-1)/RT. 4. 1N H$_2$SO$_4$/dioxane/80°C. 5. (Ph)$_2$(*t*-Bu)SiCl/imidazole/DMF/−60°C → −35°C. 6. Ph$_3$P=CHCO$_2$Et/C$_6$H$_6$/90°C. 7. DIBAL/C$_6$H$_6$-CH$_2$Cl$_2$(4-1)/0°C. b. *t*-BuO$_2$H/(−)-diethyl tartrate/Ti(*i*-PrO)$_4$/CH$_2$Cl$_2$/−23°C/ 2 days. c. camphorsulfonic acid/acetone/RT. d. 1. (*n*-Bu)$_4$NF. 2. MsCl/Et$_3$N/ CH$_2$Cl$_2$/0°C. 3. NaCN/DMSO/90°C. 4. DIBAL/toluene/−78°C → RT. 5. NaBH$_4$/EtOH/0°C. 6. PhCH$_2$Br/NaH/THF-DMF(4-1)/RT. 7. AcOH-H$_2$O(4-1)/50°C. 8. Ph$_3$(*t*-Bu)SiCl/imidazole/DMF/RT. 9. MeOCH$_2$Br/KH/ THF-DMF(4-1)/RT. 10. (*n*-Bu)$_4$NF/THF/RT. 11. (COCl)$_2$/DMSO/CH$_2$Cl$_2$/ −60°C → RT. e. 1. 3-butenyl magnesium bromide/ether. 2. separation of alcohols by chromatography. f. 1. Dowex-50X8-400/MeOH/50°C. 2. O$_3$/MeOH/ −78°C, followed by NaBH$_4$/MeOH/0°C workup. 3. Ac$_2$O/DMAP-py/RT. 4. H$_2$/ Pd-C/AcOH-MeOH(1-20)/RT. 5. NaIO$_4$/MeOH-H$_2$O(5-1)/0°C. 6. NaBH$_4$/ MeOH/0°C. 7. Ac$_2$O/DMAP-py/RT.

Scheme XXX

asymmetric epoxidation using (D-(−)-diethyl tartrate to give the expected epoxide (**207**). Treatment of (**207**) with camphorsulfonic acid yielded the acetonide (**208**) as the sole product. Compound (**208**) was converted into aldehyde (**209**) in eleven steps and reacted with 3-butenylmagnesium bromide to give a 6:1 mixture of C(64)-alcohols. The minor alcohol (**210**) was transformed into hexaacetate (**82**) in seven steps. Upon comparison of

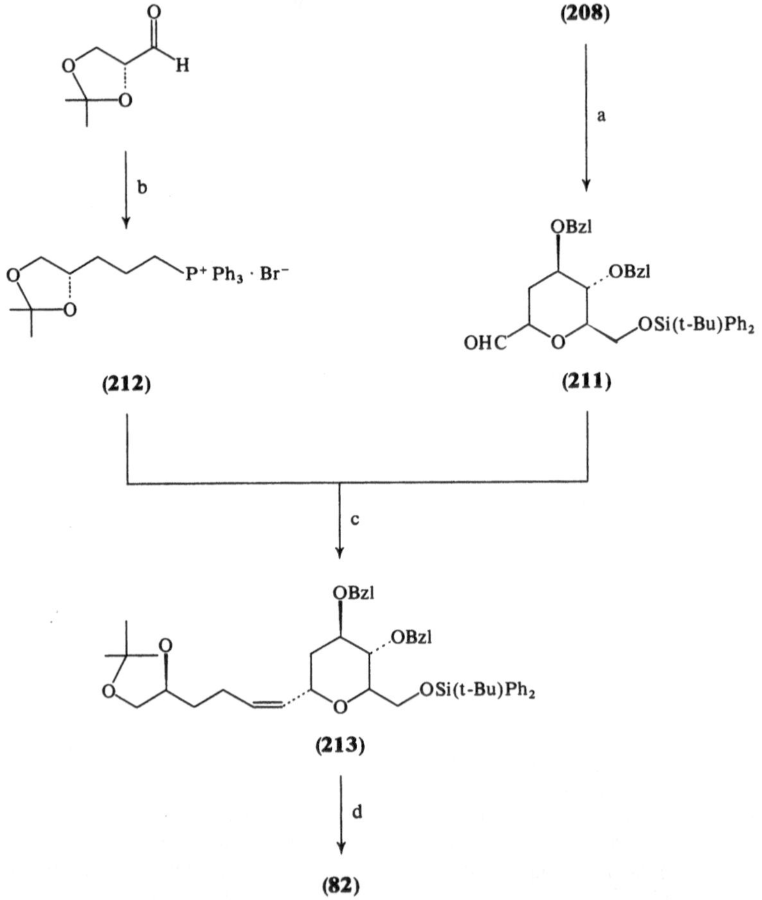

Reagents: a. 1. $AcOH-H_2O(4-1)/50°C$. 2. $NaIO_4/MeOH-H_2O(5-1)/0°C$. b. 1. $(C_6H_5)_3P=CHCO_2Et/C_6H_6/RT$. 2. $H_2/5\%$ $Rh-Al_2O_3/hexane/RT$. 3. $LiAlH_4/Et_2O/40°C$. 4. $NBS/Ph_3P/CH_2Cl_2/RT$. 5. $Ph_3P/sulfolane/110°C$. c. Wittig reaction. d. 1. $(n-Bu)_4NF/THF/RT$. 2. $MsCl/(Et)_3N/CH_2Cl_2/0°C$. 3. $NaCN/DMSO/90°C$. 4. $DIBAL/toluene/−78°C$. 5. $NaBH_4/MeOH/0°C$. 6. $PhCH_2Br/NaH/THF-DMF(4-1)/RT$. 7. $OsO_4/NMO/acetone-H_2O(5-1)/RT$. 8. $Ac_2O/DMAP-py/RT$. 9. $H_2/Pd-C/AcOH-MeOH(5-1)/RT$. 10. $AcOH-H_2O(4-1)/RT$. 11. $NaIO_4/MeOH-H_2O(5-1)/0°C$, followed by $NaBH_4$ workup. 12. $Ac_2O/DMAP-py/RT$.

Scheme XXXI

spectroscopic and optical data, synthetic (82), $[\alpha]_D$ +7.8° (c 0.16, CHCl$_3$), was found to be identical with the degradation product (82). The relative stereochemistry between C(64) and C(65) was shown to be *erythro* by the synthesis shown in Scheme XXXI. Wittig reaction of aldehyde (211), prepared in two steps from (208), with phosphonium salt (212), synthesized from D-glyceraldehyde acetonide in five steps, produced the *cis*-olefin (213). After addition of C(72), osmium tetroxide oxidation furnished a 7:1 mixture of the two possible *erythro* diols. Only the major *erythro* diol could be converted to hexaacetate (82).

The synthesis of (217), the antipode of (205), was accomplished as shown in Scheme XXXII. Degradation product (205) had been prepared by ozonolysis of (42) followed by reductive workup with borohydride and acetylation. *Trans*-allylic alcohol (215) was first synthesized from 4,6-O-ethylidene-D-glucopyranose (214). Asymmetric epoxidation using L-(+)-diethyl tartrate yielded the expected epoxide (216). Regioselective reduction of the epoxide ring in (216) followed by acetylation, debenzylation, periodate oxidation, borohydride reduction, and acetylation resulted in (217). The C(81) isomer of (211) was also synthesized by the same synthetic route except that D-(−)-diethyl tartrate was used in the asymmetric epoxidation. Comparison of spectroscopic and optical data showed that synthetic (211), $[\alpha]_D$ +9.8° (c 0.05, CHCl$_3$), was the antipode of degradation product (205), $[\alpha]_D$ −8.9° (c 0.09, CHCl$_3$).

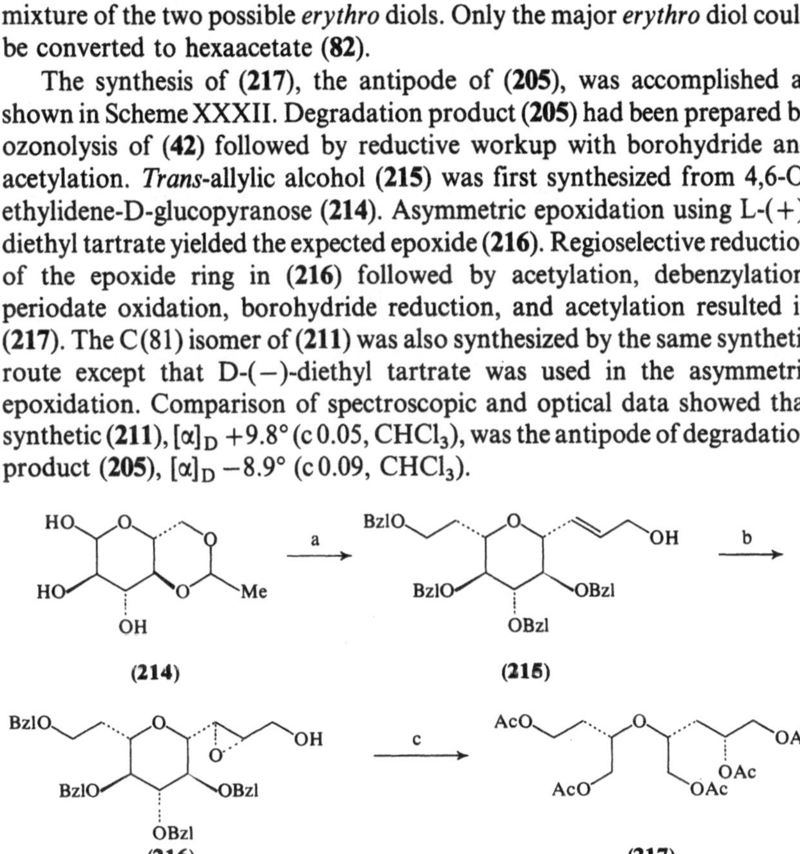

(214)   (215)

(216)   (217)

Reagents: a. 1. Ph$_3$P=CHCO$_2$Me/MeCN/reflux/60 h. 2. 0.005 M KOH/MeOH. 3. separate methyl esters by chromatography (56). 4. LiAlH$_4$/THF/0°C. 5. C$_6$H$_5$CH$_2$Br/NaH/THF-DMF(4-1)/RT. 6. 1N H$_2$SO$_4$/dioxane/80°C. 7. TrCl/py/90°C. 8. C$_6$H$_5$CH$_2$Br/NaH/THF-DMF(4-1)/RT. 9. AcOH-H$_2$O(4-1)/70°C. 10. (COCl)$_2$/DMSO/CH$_2$Cl$_2$/−60°C → RT. 11. Ph$_3$P=CHCO$_2$Et/C$_6$H$_6$/80°C. 12. DIBAL/C$_6$H$_6$-CH$_2$Cl$_2$(4-1)/0°C. b. Sharpless' asymmetric epoxidation using (+)L-diethyl tartrate. c. 1. regioselective reductive epoxide opening. 2. acetylation. 3. debenzylation. 4. periodate oxidation. 5. borohydride reduction. 6. acetylation.

Scheme XXXII

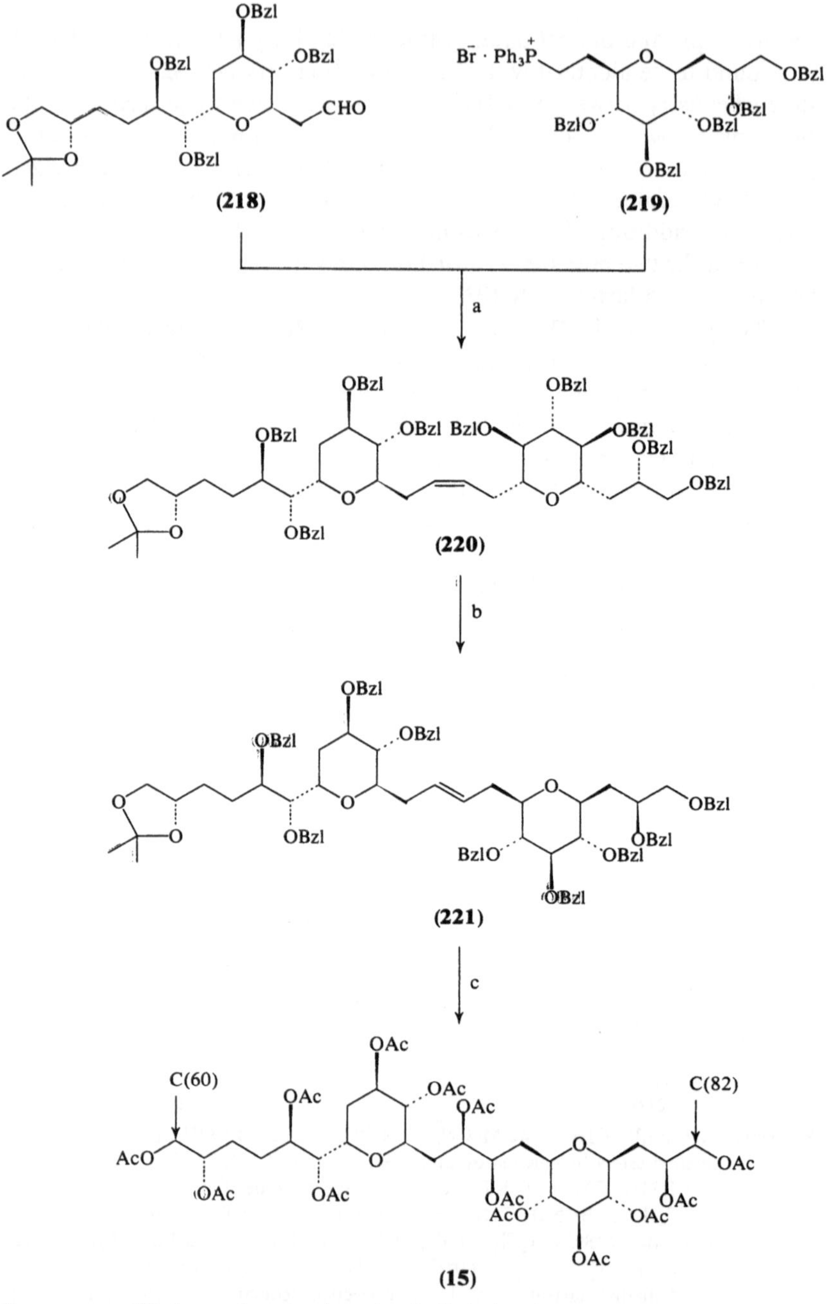

Reagents: a. Wittig reaction. b. photochemically-induced isomerization. c. 1. OsO₄. 2. Ac₂O/py.

Scheme XXXIII

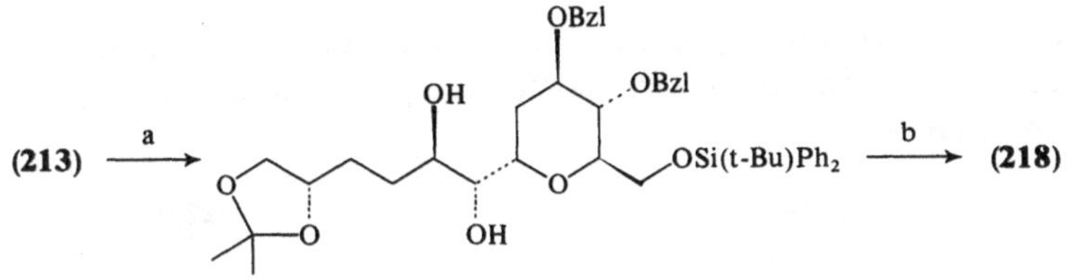

(222)

Reagents: a. OsO$_4$/N-methylmorpholine N-oxide (NMO)/acetone-H$_2$O(5-1)/RT.   b. 1. C$_6$H$_5$CH$_2$Br/NaH/THF-DMF(4-1)/RT.   2. (n-Bu)$_4$NF/THF/RT.   3. MsCl/ Et$_3$N/CH$_2$Cl$_2$/0°C.   4. NaCN/DMSO/90°C.   5. DIBAL/toluene/−78°C → RT.

Scheme XXXIV

The absolute configurations of C(61), C(68), C(69), C(72), C(73), C(76), C(77), and C(78) were determined by synthesis of the ozonolysis product (15) as outlined in Scheme XXXIII. Wittig reaction of aldehyde (218), prepared from (213) in six steps (Scheme XXXIV), with phosphonium salt (219), synthesized from 2,3,4,6-tetrabenzyl-D-glucopyranose (30, 57) in 16 steps (Scheme XXXV), produced the *cis*-olefin (220). Osmium tetroxide oxidation of (220), followed by acetylation, aqueous acid hydrolysis, debenzylation, and acetylation led to a 1:1 mixture of C(72), C(73)-*erythro* tridecaacetates, neither of which was identical with (15). Photochemical isomerization of (220) gave the *trans*-olefin (221) which after hydroxylation with OsO$_4$, acetylation, aqueous acid hydrolysis, debenzy-

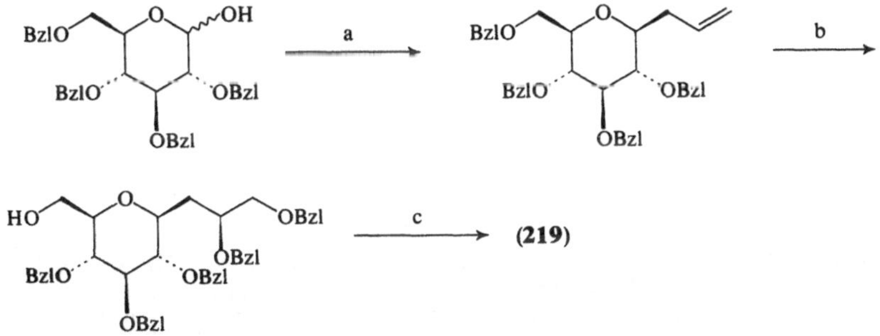

(223)

Reagents: a. 1. (COCl)$_2$/DMSO/CH$_2$Cl$_2$/−60°C → RT.   2. CH$_2$=CHCH$_2$MgBr/Et$_2$O/ −78°C.   3. Et$_3$SiH/BF$_3$·Et$_2$O/MeCN/−10°C.   b. 1. OsO$_4$/NMO/acetone-H$_2$O(5-1)/RT.   2. separation of diols by chromatography.   3. Ac$_2$O/DMAP-py/CH$_2$Cl$_2$/RT.   4. H$_2$/Pd-C/AcOH-MeOH(1-20)/RT.   5. TrCl/py/90°C.   6. MeONa/MeOH/RT.   7. C$_6$H$_5$CH$_2$Br/NaH/THF-DMF(4-1)/RT.   8. AcOH-H$_2$O(4-1)/70°C.   c. 1. MsCl/Et$_3$N/CH$_2$Cl$_2$/0°C.   2. NaCN/DMSO/90°C.   3. DIBAL/toluene/−78°C → RT.   4. NaBH$_4$/EtOH/0°C.   5. NBS/Ph$_3$P/CH$_2$Cl$_2$/ RT.   6. Ph$_3$P/sulfolane/110°C.

Scheme XXXV

lation, and acetylation yielded a 2:1 mixture of the C(72), C(73)-*threo* tridecaacetates. The major *threo* compound proved to be (15), $[\alpha]_D +18.4°$ (c 0.69, CHCl₃).

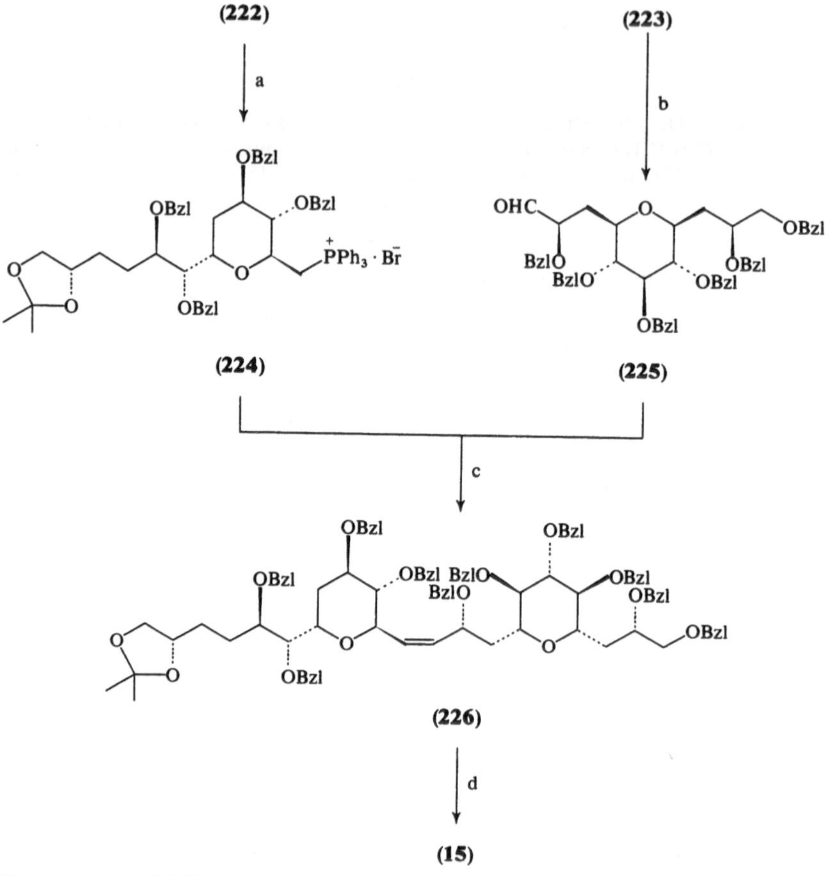

(222)   a →

(223)   b →

(224)   **(225)**

c

(226)

d

(15)

Reagents: a. 1. PhCH₂Br/NaH/THF-DMF(4-1)/RT. 2. (n-Bu)₄NF/THF/RT. 3. NBS/ Ph₃P/CH₂Cl₂/RT. 4. Ph₃P/110°C. b. 1. (COCl)₂/DMSO/CH₂Cl₂/−60° → RT. 2. Ph₃P=CHCO₂Et/C₆H₆/80°C. 3. DIBAL/C₆H₆-CH₂Cl₂(4-1)/0°C. 4. Sharpless epoxidation using L-(+)-diethyl tartrate. 5. DIBAL/Et₂O/RT. 6. TrCl/py/ 90°C. 7. PhCH₂Br/NaH/THF-DMF(4-1)/RT. 8. AcOH-H₂O(3-1)/70°C. 9. (COCl)₂/DMSO/CH₂Cl₂/−60°C → RT. c. Wittig reaction. d. 1. hydroboration. 2. aqueous acid hydrolysis. 3. debenzylation. 4. acetylation.

Scheme XXXVI

The absolute stereochemistry of C(73) was established by the following synthesis. Wittig reaction of phosphonium salt (224), prepared in four steps from (222), and aldehyde (225), synthesized in nine steps from (223), produced *cis*-olefin (226), which after sequential subjection to hydroboration, aqueous acid hydrolysis, debenzylation, and acetylation yielded a mixture of tridecaacetates (Scheme XXXVI). One of these tridecaacetates proved to be identical with the palytoxin degradation product (15). The assignment of absolute stereochemistry at C(73) was based on the Sharpless asymmetric epoxidation step used in the synthesis of (225) and on the fact that (225) could be converted to a meso heptaacetate (227) on borohydride reduction, debenzylation, and acetylation. Since the relative stereochemistry between C(72) and C(73) was *threo*, the absolute stereochemistry at C(72) was now rigorously established.

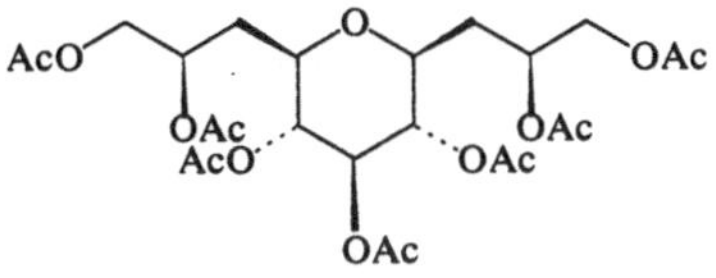

(227)

Additional proof of absolute stereochemistry for C(72) and C(73) was obtained by degradation of (15) and the major diol from osmium tetroxide oxidation of (221) to 3R,4R-hexa-1,3,4,6-tetraol-3,4-acetonide-1,6-diacetate (228), $[\alpha]_D$ +44° (c 0.02, CHCl$_3$) (Scheme XXXVII). The absolute configuration of the (228) from degradation was confirmed by comparison of the optical rotation with that of an authentic sample, $[\alpha]_D$ +(44.5° (c 0.44, CHCl$_3$), synthesized from (−)-diethyl tartrate in eight steps (Scheme XXXVIII) (30).

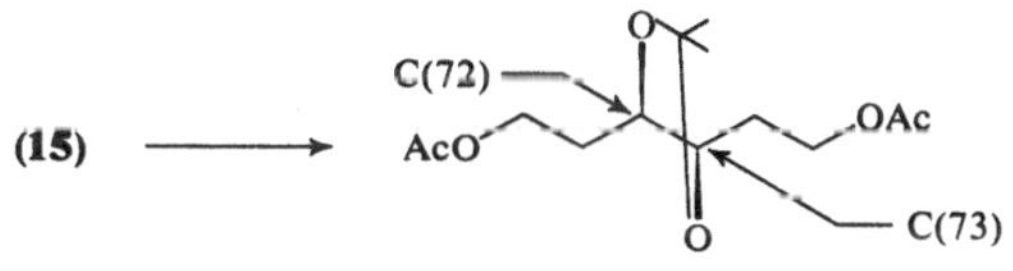

(228)

Reagents: 1. NaOMe/MeOH/RT. 2. MeC(OMe)$_2$Me/Dowex-50X8-400/RT. 3. Pb(OAc)$_4$/ C$_6$H$_6$-CH$_2$Cl$_2$/RT, followed by addition of MeMgI at room temperature. 4. (COCl)$_2$/DMSO/CH$_2$Cl$_2$/ −60°C → RT. 5. MCPBA/Na$_2$HPO$_4$/CH$_2$Cl$_2$/RT. 6. LiAlH$_4$/THF/RT. 7. Ac$_2$O/py.

Scheme XXXVII

The KISHI group next proceeded to establish the absolute configurations of C(57) and C(58) in triacetate (229) by synthesis. This more advanced degradation product had been prepared from the ozonolysis product (14) by base hydrolysis, periodate oxidation, borohydride reduction, and acetylation (30). Starting with S-(+)-3-hydroxy-2-methylpropionic acid (230), KISHI reported the synthesis of a 2-methyl-1,3,5-pentanetriol tri-

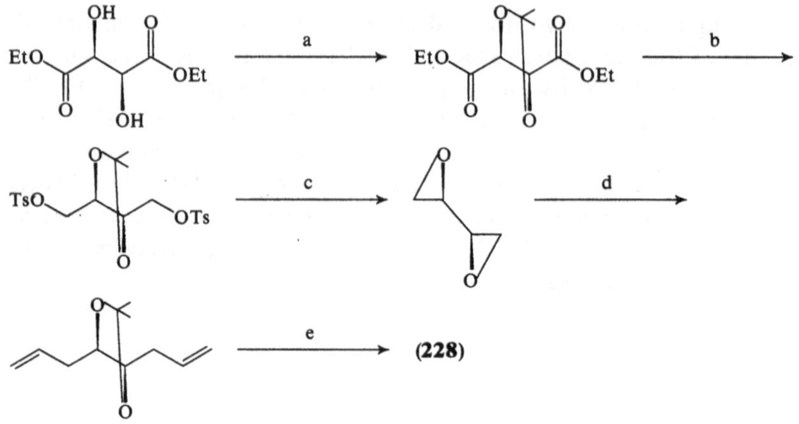

Reagents: a. MeC(OMe)$_2$Me/p-TSA/C$_6$H$_6$. b. 1. LiAlH$_4$/Et$_2$O. 2. TsCl/py. c. 2 N HCl/MeOH, followed by KOH workup. d. 1. CH$_2$=CHMgBr/CuI/Et$_2$O. 2. MeC(OMe)$_2$Me/p-TSA/acetone. e. O$_3$/MeOH, followed by NaBH$_4$ workup and acetylation.

Scheme XXXVIII

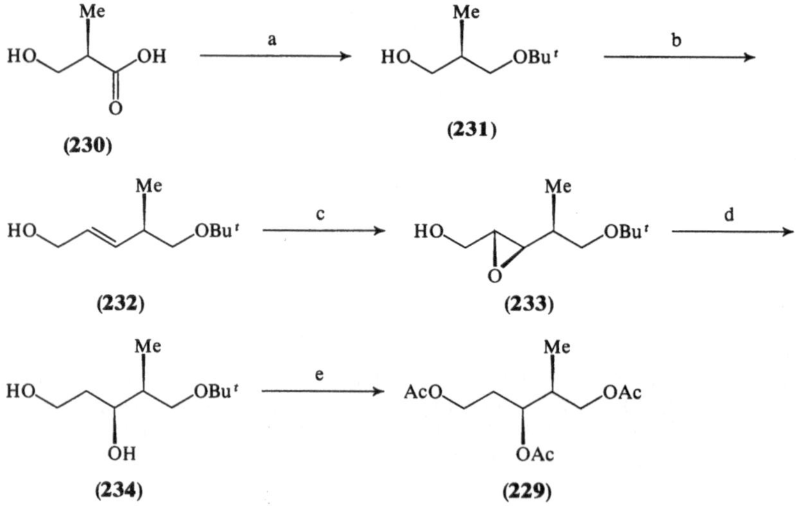

Reagents: a. 1. PhNCO/C$_6$H$_6$/RT. 2. BH$_3 \cdot$ Me$_2$S/THF/RT. 3. isobutylene/H$_3$PO$_4$-BF$_3 \cdot$ Et$_2$O/CH$_2$Cl$_2$/$-72°$ C $\rightarrow$ 5° C. 4. KOH/EtOH/reflux. b. 1. (COCl)$_2$/DMSO/CH$_2$Cl$_2$/$-60°$ C $\rightarrow$ RT. 2. (i-PrO)$_2$P(O)CH$_2$CO$_2$Et/t-BuOK/THF/$-78°$C$\rightarrow$0°C. 3. DIBAL/CH$_2$Cl$_2$-hexane(1-1)/$-78°$C. c. Sharpless asymmetric epoxidation using D-($-$)-diethyl tartrate. d. Red-Al/THF/0° C. e. 1. TFA/0° C, followed by 10% NaOH workup. 2. Ac$_2$O/py.

Scheme XXXIX

acetate (229), $[\alpha]_D -14.2°$ (c 0.73, CH_2Cl_2), which had a ^1H nmr spectrum and an optical rotation that were identical with those of the degradation product (30) (Scheme XXXIX). In the synthesis (230) was first converted into a *t*-butyl ether (231), a compound which had previously been described by COHEN (53). Sharpless asymmetric epoxidation of the *trans*-allylic alcohol (232), formed in three steps from (231), using D-(−)-diethyl tartrate led to epoxide (233) which was reduced to diol (234). After acid hydrolysis and acetylation, triacetate (229) was obtained.

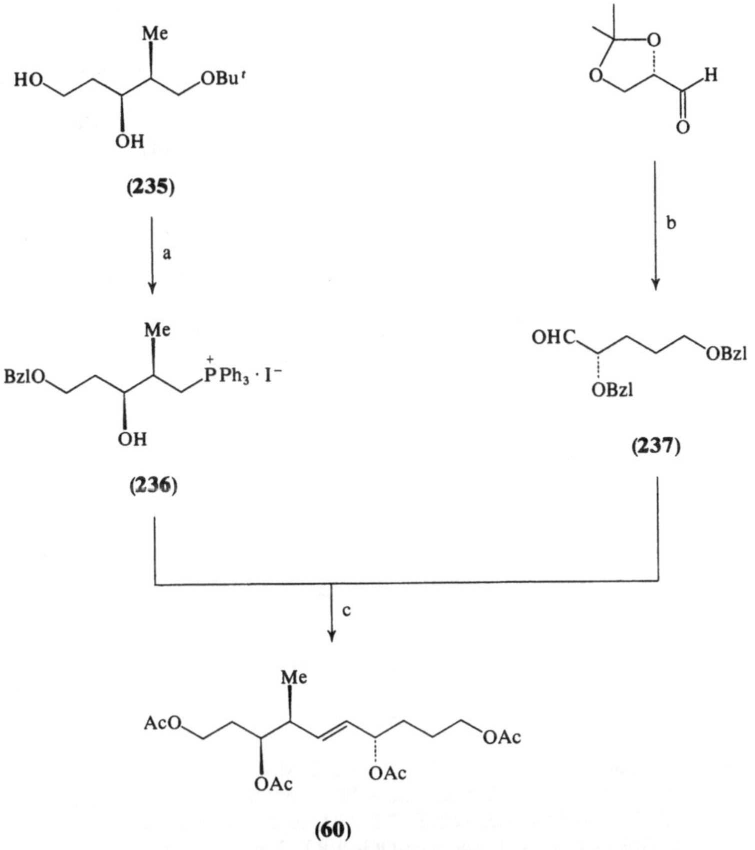

(235)

(236)

(237)

(60)

Reagents: a. 1. $PhCH_2Br/NaH/THF$-DMF(4-1)/RT. 2. TFA/0°C, followed by 10% NaOH workup. 3. $MsCl/Et_3N/CH_2Cl_2/0°C$. 4. $LiI/DMF/60°C$. 5. Ph_3P/C_6H_6/reflux.
b. 1. $Ph_3P=CHCO_2Et/EtOH/RT$. 2. H_2/Rh-Al_2O_3/hexane. 3. $LiAlH_4/Et_2O/$0°C. 4. $PhCH_2Br/NaH/THF$-DMF(4-1)/RT. 5. conc. HCl/EtOH/RT. 6. $PhCHO/CSA$/toluene/reflux. 7. $LiAlH_4/AlCl_3/Et_2O/RT$. 8. $(COCl)_2/DMSO/$$CH_2Cl_2/-60°C \rightarrow RT$. c. 1. Wittig reaction. 2. debenzylation. 3. acetylation.

Scheme XL

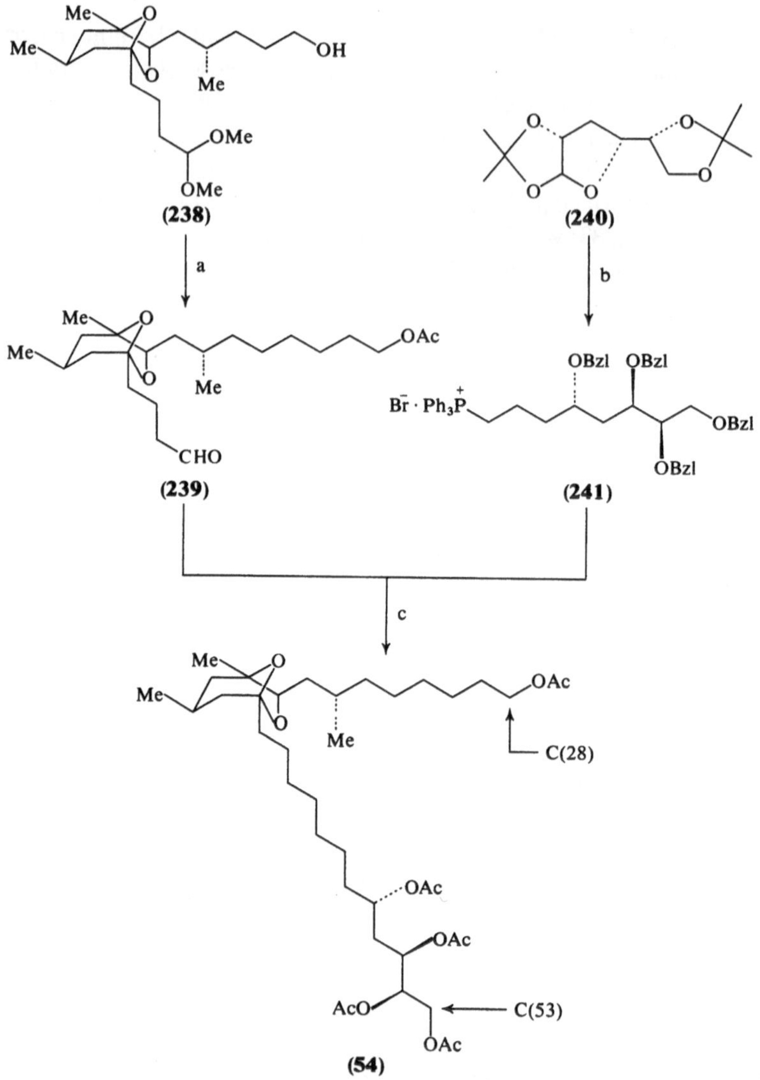

(238)

(240)

(239)

(241)

(54)

— C(28)

— C(53)

Reagents: a. 1. ethylene glycol/CSA/C_6H_6/reflux. 2. MsCl/Et_3N/CH_2Cl_2/0°C. 3. NaCN/ EtOH/75°C. 4. DIBAL/toluene/−78°C. 5. Ph_3P=$CHCO_2$Et/dichloroethane/ 95°C. 6. H_2/Pd-C/EtOAc/RT. 7. LiAlH$_4$/Et_2O/0°C. 8. Ac_2O/py/RT. 9. AcOH- H_2O(4-1)/50°C. b. 1. AcOH-H_2O(3-2)/RT. 2. PhCH$_2$Br/NaH/THF-DMF(4-1)/ RT. 3. 1N HCl/THF/50°C. 4. Ph_3P=$CHCO_2$Et/C_6H_6/RT. 5. H_2/Pd-C/ EtOAc/RT. 6. LiAlH$_4$/Et_2O/RT. 7. Ph_2 (t-Bu)SiCl/imidazole/DMF/0°C. 8. PhCH$_2$Br/NaH/THF-DMF(4-1)/RT. 9. (n-Bu)$_4$NF/THF/RT. 10. MsCl/Et_3N/ CH_2Cl_2/0°C. 11. LiBr/DMF/50°C. 12. Ph_3P/C_6H_6/100°C. c. 1. Wittig reaction. 2. hydrogenation-hydrogenolysis. 3. acetylation.

Scheme XLI

Tetraacetate (**60**) was now synthesized by a Wittig reaction of phosphonium salt (**236**) and aldehyde (**237**), followed by debenzylation and acetylation (Scheme XL). Starting with (**237**) and the antipode of (**236**), the C(61) diastereomer of (**60**) was also synthesized. Only the synthetic (**60**) showed a ^{1}H nmr spectrum that was identical with that of the degradation product. The stereochemistry of C(61) was now firmly established. Aldehyde (**237**) had been synthesized from D-glyceraldehyde acetonide in eight steps. Phosphonium salt (**236**) had been prepared from an intermediate, presumably having structure (**235**), used in the synthesis of (**229**).

The absolute configurations of C(34), C(36), C(37), C(39), C(41), and C(49), which had been established by X-ray crystallography, were confirmed by synthesis of periodate oxidation product (**54**) (Scheme XLI). The stereochemistry at C(51) and C(52) was also established by this synthesis. Using S-(−)-citronellal as the starting material, the KISHI group first prepared (**238**) (*58*) and converted this bicyclic acetal alcohol into the aldehyde acetate (**239**) (*31*). Wittig reaction of (**239**) with the phosphonium salt (**241**), prepared from 3-deoxy-D-galactose 1,2,5,6-diacetonide (**240**) (*59*) in twelve steps, produced an olefin which after hydrogenation-hydrogenolysis and acetylation yielded pentaacetate (**54**), $[\alpha]_D$ +54° (c 0.85, $CHCl_3$). The synthetic material and the degradation product were found to have the same ^{1}H nmr spectra and optical rotations.

For the synthesis of (**238**) S-(+)-citronellal (**242**) was first converted to the methyl ketone (**243**) (Scheme XLII) which was then coupled to the cyclopentenyl compound (**244**), prepared from S-(−)-3-hydroxy-2-methylpropionic acid in seven steps, by a Grignard reaction. Subsequent debenzylation of the resulting alcohol (**245**) followed by ozonolysis and ketalization gave the bicyclic ketal (**238**).

The absolute stereochemistry at C(51) and C(52) having been determined, the configurations at C(53) and C(54) were studied next. Proton nmr studies of spiro ketals (**76**) and (**246**), which had been formed in a 1:1 ratio by treatment of ozonolysis product (**73**) with dilute hydrochloric acid (*31*) followed by acetylation, strongly suggested that the stereochemistry was as shown. The assignment was ultimately confirmed by synthesis.

Cis-α,β-unsaturated ketone (**247**) was first formed in an eighteen step synthesis from 3-deoxy-D-galactose 1,2,5,6-diacetonide (**240**) (*59*) (Scheme XLIII). A mixture of two unsaturated spiro-6,6-ketals, (**249a**) and (**249b**) which differed in structure only at the spiro center was then produced by treatment of (**247**) with aqueous acetic acid (Scheme XLIV). The major isomer (**249a**) afforded a single tetraacetate (**250**) on osmium tetroxide oxidation and acetylation. Compound (**250**) was also produced by osmium tetroxide oxidation of (**247**) followed by spiroketalization and acetylation. The major *erythro* diol (**251**) from OsO_4 oxidation of (**247**) yielded a 1:2 mixture of spiroketal tetraacetates. The minor tetraacetate

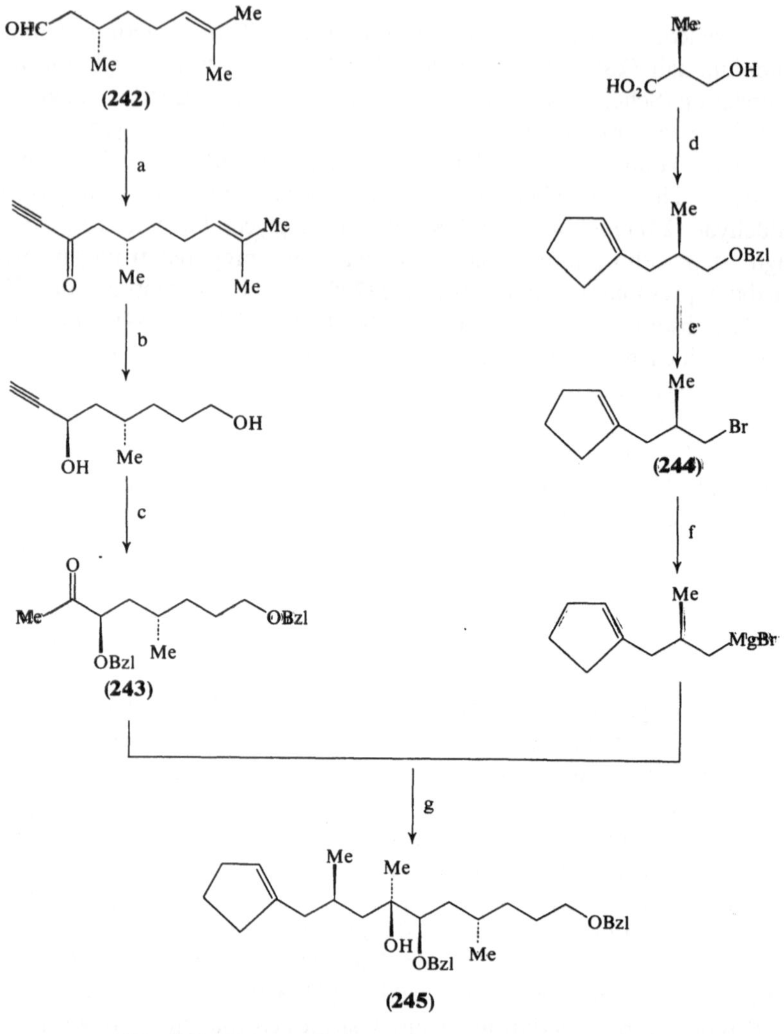

(242)

(243)

(244)

(245)

Reagents: a. LiC≡CH/THF/−78°C, followed by Jones oxidation. b. 1. O₃/1.5 eq.
MeOH/Sudan Red 7B/CH₂Cl₂/−78°C, followed by treatment with Me₂S/
−78°C → RT. 2. LiAlH₄/(+)-Darvon alcohol/Et₂O/−78°C. c. 1. PhCH₂Br/
NaH/THF-DMF (4-1)/0°C → RT. 2. HgCl₂/H₂O-MeOH(1-250)/60°C. d. 1.
PhCH₂OC(=NH)CCl₃/CF₃SO₂OH/c-C₆H₁₂-CH₂Cl₂(2-1)/RT. 2. LiAlH₄/Et₂O/

0°C → RT. 3. TsCl/py/0°C → RT. 4. 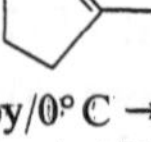 ⟨cyclopentenyl⟩—MgBr/Li₂CuCl₄/THF/RT/

2.5 days. e. 1. Li/liq. NH₃/THF. 2. MsCl/py/0°C → RT. 3. LiBr/DMF/60°C.
f. Mg/I₂/THF/reflux. g. 1. inverse addition/−78°C. 2. Li/liq. NH₃/THF.
3. O₃/MeOH/−78°C, followed by treatment with camphorsulfonic acid/Me₂S/
−78°C → RT.

Scheme XLII

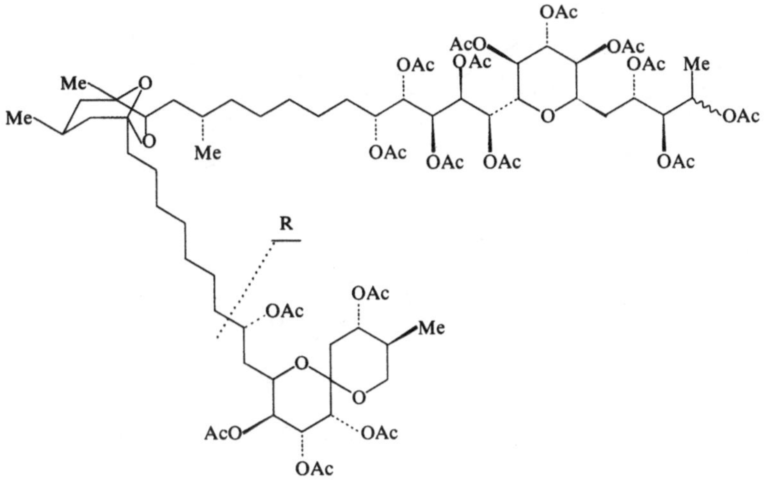

(76)

(246)   R=

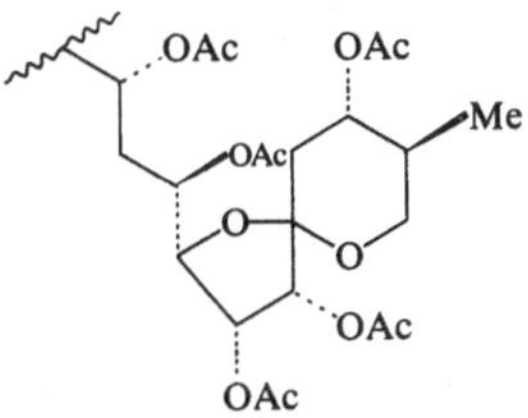

was identical with (250) whereas the major tetraacetate was identical with the one derived from OsO₄ oxidation and acetylation of (249b). These two tetraacetates differed only in stereochemistry at the spiro carbon.

Coupling of the spiroketal segment to the segment containing the 6,8-dioxabicyclo[3.2.1]octane ring was achieved as shown in Scheme XLV. Compound (238) was converted to the phosphonium salt (252) and tetraacetate (250) was transformed into aldehyde (253). Wittig reaction of (252) and (253) followed by hydrogenation-hydrogenolysis and Swern oxidation (48) gave aldehyde (254). Coupling of (254) with phosphonium salt (255), prepared from D-xylose, in a second Wittig reaction, followed by hydrogenation-hydrogenolysis, deacetonization, and acetylation (Scheme XLVI) yielded octaacetate (119), $[\alpha]_D$ +112° (c 0.18, CHCl₃), which proved to be identical in all respects with degradation product (119). A similar sequence of reactions, starting from (254) and the antipode of (255) gave the C(27), C(28) diastereomer of (119) which was not identical with the degradation product on comparison of spectral and optical properties. This synthesis not only confirmed the stereochemistry of C(53) and C(54), but established the absolute configurations of C(27) and C(28).

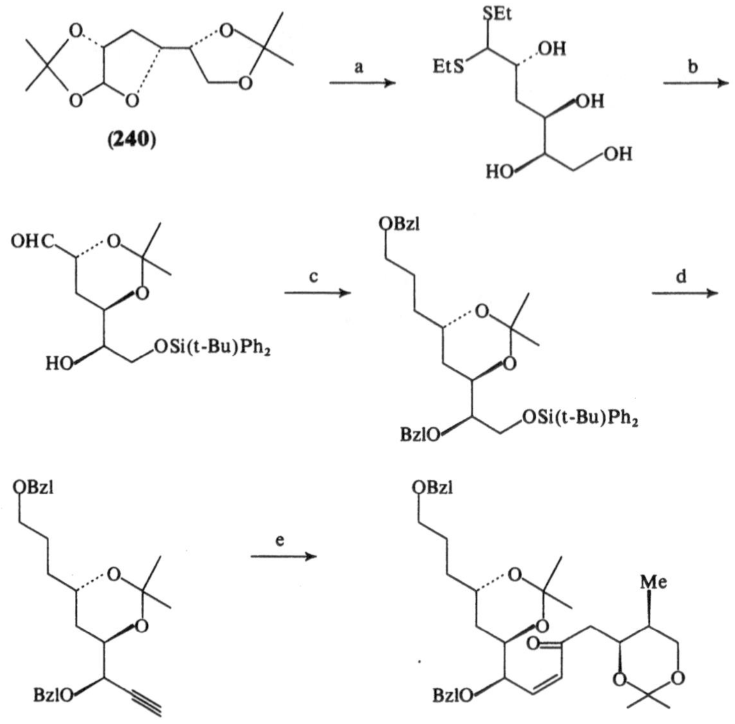

(240)

(247)

Reagents: a. 1. AcOH-H₂O(3-2)/RT. 2. Ph₂(t-Bu)SiCl/imidazole/DMF/0°C. 3. PhCH₂Br/
NaH/THF-DMF(4-1)/RT. 4. (n-Bu)₄NF/THF/RT. 5. EtSH/conc. HCl/0°C.
b. 1. Ph₂(t-Bu)SiCl/imidazole/DMF/0°C. 2. MeC(OMe)₂Me/CSA/RT. 3.
HgCl₂-HgO/acetone-H₂O(5-1)/0°C.   c. 1. (i-PrO)₂P(O)CH₂CO₂Et/t-BuOK/
THF/−78°C. 2. H₂/Pd-C/AcOEt/RT. 3. LiAlH₄/Et₂O/RT. 4. PhCH₂Br/NaH/
THF-DMF(4-1)/RT. d. 1. (n-Bu)₄NF/THF/RT. 2. Swern oxidation (48). 3.
PhHgCCl₂Br/Ph₃P/C₆H₆/reflux. e. 1. n-BuLi/THF/−78°C, followed by addition
of (248). 2. CrO₃·2Py/CH₂Cl₂/RT. 3. H₂/Lindlar catalyst/hexane/RT.

Scheme XLIII

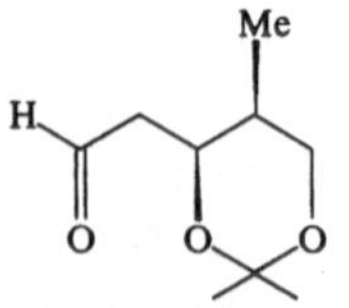

(248)

References, pp. 199—202

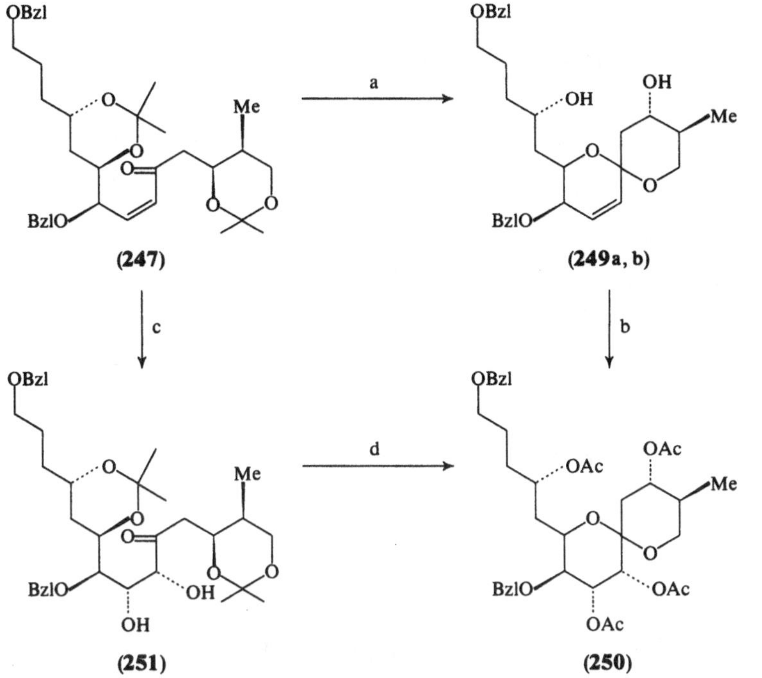

Reagents: a. AcOH-H$_2$O. b. 1. OsO$_4$. 2. Ac$_2$O/pyridine. c. OsO$_4$. d. 1. AcOH-H$_2$O.
2. acetylation.

Scheme XLIV

The absolute stereochemistry of the tetrahydropyran ring in degradation products (28) and (78) was determined by synthesis of the more advanced degradation product (30) (*31*), [α]$_D$ +34.0° (c 0.11, CHCl$_3$), and its enantiomer (161) (*28*). Synthesis of (30) was achieved in a straightforward manner from alcohol (257), which had been prepared from 2,3,4-tribenzyl-1,6-anhydro-D-glucopyranose (256) (*57*) (Scheme XLVII). Comparison of spectral and optical data confirmed the identity of synthetic (30) and the degradation product (30). The absolute configurations of C(19) and C(23) were therefore established.

Only four diastereomers had to be considered as structural possibilities for degradation product (28), a minor product of periodate oxidation of (73), since the stereochemistry of the tetrahydropyran ring was known from ^{1}H nmr data. Starting with (257) all four diastereomeric heptaacetates were

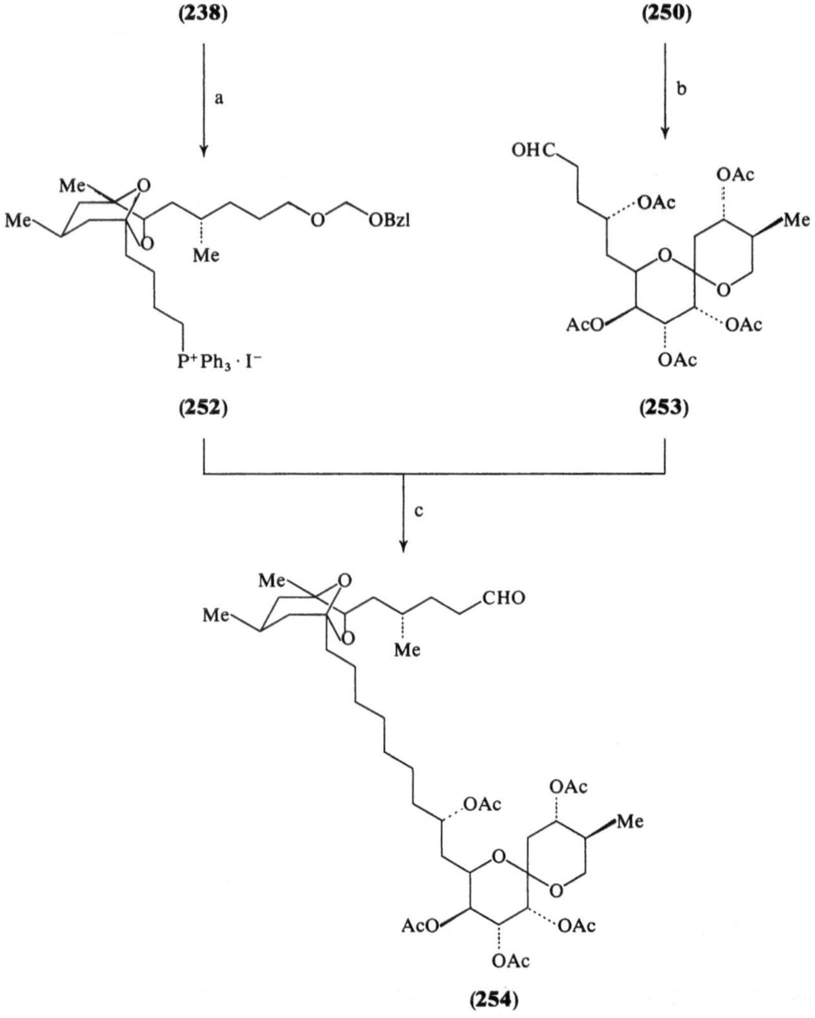

(238)   **(250)**

(252)   **(253)**

(254)

Reagents: a. 1. PhCH$_2$OCH$_2$Cl/(i-Pr)$_2$EtN/DMF/RT. 2. MeSH/BF$_3$·Et$_2$O/CH$_2$Cl$_2$/RT.
3. HgCl$_2$/CaCO$_3$/MeCN-H$_2$O(4-1)/RT. 4. NaBH$_4$/EtOH/0°C. 5. MsCl/py/0°C.
6. NaI/acetone/50°C.   7. Ph$_3$P/sulfolane/110°C.   b. 1. H$_2$/Pd-C/EtOH/RT.
2. Ph$_2$(t-Bu)SiCl/imidazole/DMF/0°C. 3. Ac$_2$O/py. 4. (n-Bu)$_4$/NF/THF/RT.
5. Swern oxidation (48). c. 1. Wittig reaction. 2. hydrogenation-hydrogenolysis.
3. Swern oxidation (48).

Scheme XLV

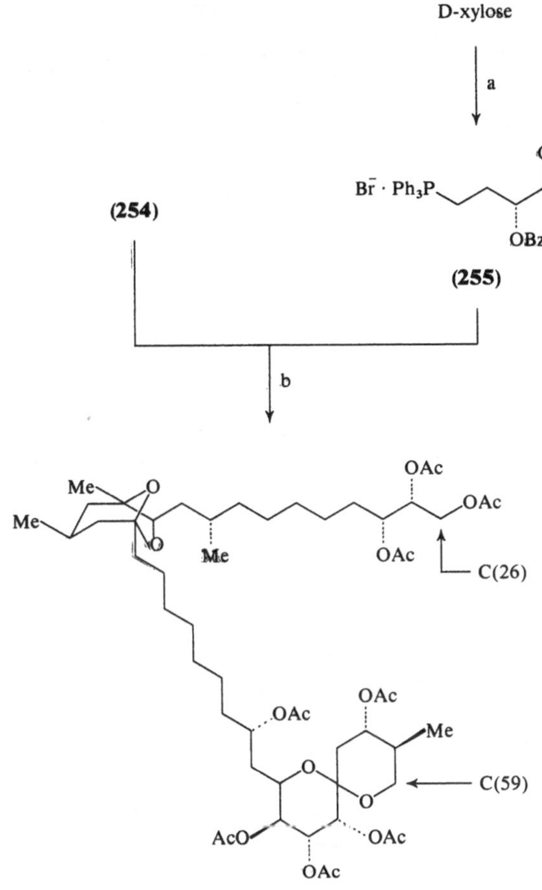

Scheme XLVI

Reagents: a. 1. EtSH/HCl. 2. acetone/H⁺. 3. t-BuOK. 4. LiAlH₄/THF. 5. PhCH₂Br/
NaH/THF-DMF(4-1)/RT. 6. HgCl₂-HgO/acetone-H₂O(5-1)/0°C. 7. LiAlH₄/
Et₂O/0°C. 8. NBS/Ph₃P/CH₂Cl₂/0°C. 9. Ph₃P/C₆H₆/reflux. b. 1. Wittig
reaction. 2. hydrogenation-hydrogenolysis. 3. deacetonization. 4. acetylation.

Scheme XLVI

prepared. One of the synthetic *threo* heptaacetates (28) was found to be
identical with the degradation product on comparison of ¹H nmr spectra.
No details on the synthesis of (28) were given in reference (*31*), but the
authors indicated that methods similar to those shown in Scheme XXIV
were used to introduce C(25) and C(26) and establish known stereochemistry at C(24) and C(25). Presumably the allyl side chain was degraded to a
two-carbon side chain before chain extension at C(24).

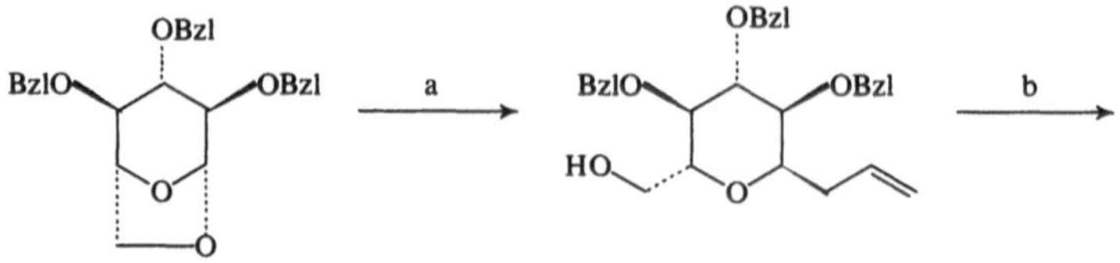

(256)   **(257)**

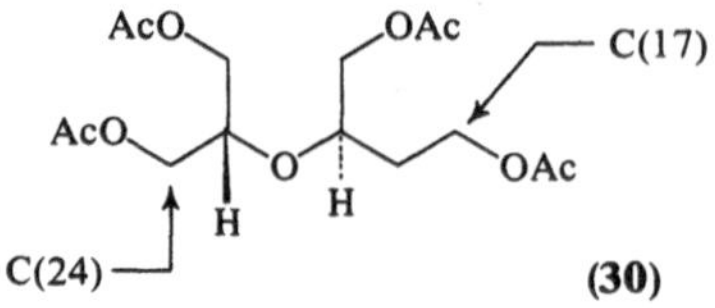

(30)

Reagents: a. $CH_2=CHCH_2TMS/BF_3 \cdot Et_2O/MeCN/0°C \rightarrow RT/20h$. b. 1. $O_3/MeOH/$
$-78°C$, followed by $NaBH_4$ workup. 2. $H_2/Pd-C/AcOH-MeOH/RT$. 3.
$NaIO_4/MeOH-H_2O(3-1)/0°C$, followed by $NaBH_4$ workup. 4. $Ac_2O/py/RT$.

Scheme XLVII

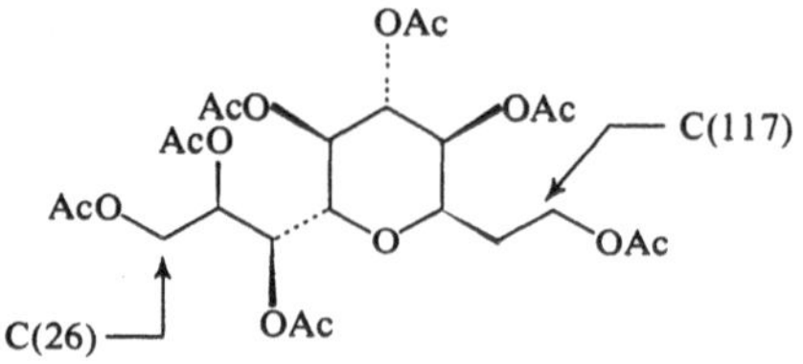

(28)

Using methods similar to those employed in the preparation of **(28)**, the
KISHI group synthesized alcohol **(258)** and the C(26) diastereomer. The
absolute configuration at C(26) was unambiguously established by employing L-glyceraldehyde in the synthesis. Compound **(258)** was converted to
cis-α,β-unsaturated ketone **(259)**. Osmium tetroxide oxidation led to two
erythro diols which were separated. One of the diols, after borohydride
reduction, deacetonization, debenzylation, and acetylation, gave a mixture
of two decaacetates **(78)** (Scheme XLVIII) which differed only in
stereochemistry at C(15) *(60)*. This synthetic mixture **(78)** proved to be
identical with the degradation product **(78)** which was also a mixture of
C(15) diastereomers. The two decaacetates could be separated by thin layer
chromatography. This synthesis, however, established only the relative
stereochemistry between C(16) and C(17) in **(78)**. Details of the synthesis
were not reported in reference *(31)*.

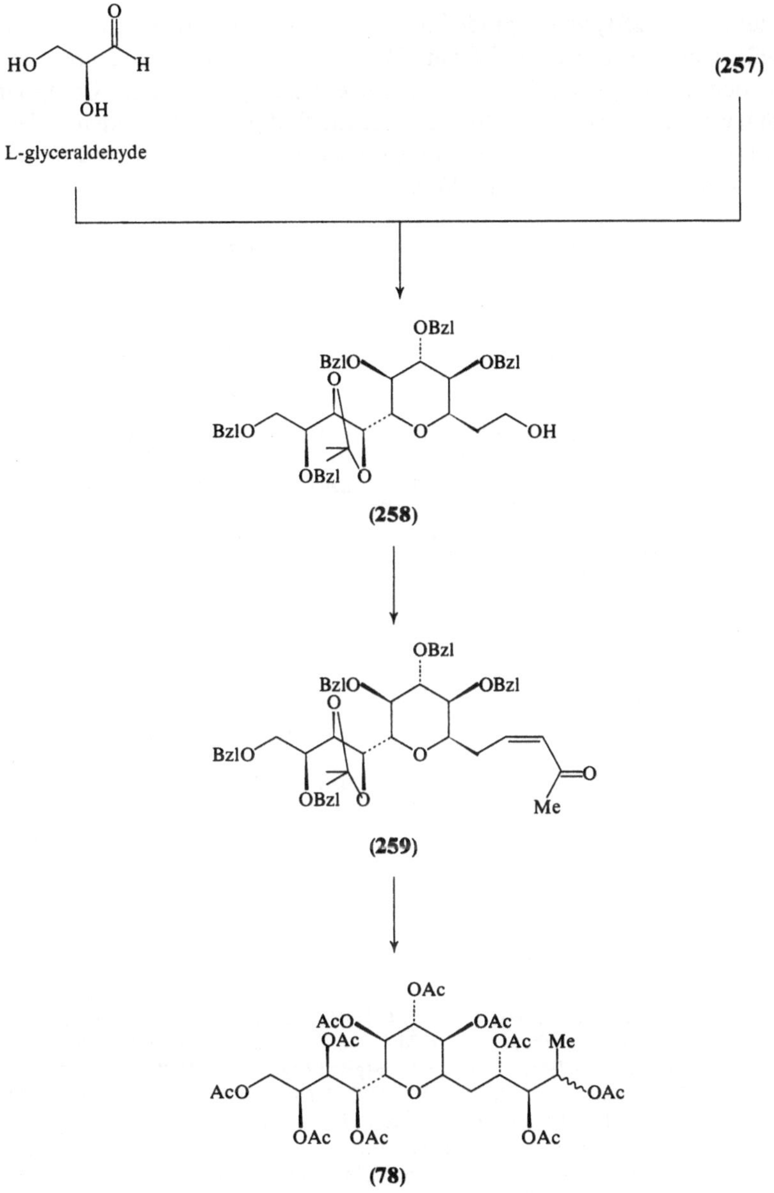

Scheme XLVIII

The absolute configurations at C(16) and C(17) were decided from the following experiments (Scheme XLIX). The diol (260) with the unnatural configurations at C(16) and C(17) obtained from osmium tetroxide oxidation of (259) was degraded to tetraacetate (261). *Trans*-allylic alcohol (262), which was prepared from (257) in seven steps, was converted to epoxide (263) by Sharpless' asymmetric epoxidation reaction. Compound (263) was then transformed to a tetraacetate that proved to be identical with (261), indicating that the absolute stereochemistry at C(16) and C(17) in (78) was opposite to that in (261).

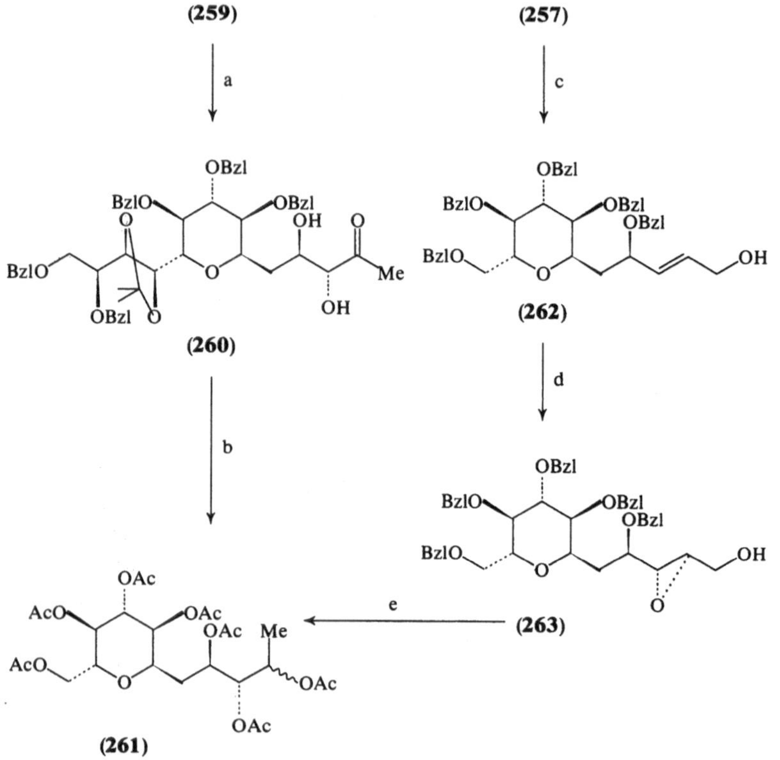

Reagents: a. OsO$_4$. b. 1. NaBH$_4$. 2. PhCH$_2$Br/NaH. 3. aqueous AcOH. 4. NaIO$_4$, followed by NaBH$_4$ workup. 5. H$_2$/Pd-C. 6. Ac$_2$O/py. c. 1. PhCH$_2$Br/NaH. 2. O$_3$/MeOH/−78°C. 3. CH$_2$=CHMgBr/Et$_2$O. 4. PhCH$_2$Br/NaH, followed by TLC separation. 5. O$_3$/MeOH/−78°C. 6. (i-Pr)$_2$P(O)CH$_2$CO$_2$Et/t-BuOK/THF. 7. DIBAL/CH$_2$Cl$_2$-C$_6$H$_6$. d. t-BuO$_2$H/(−)-diethyl tartrate/Ti(i-PrO)$_4$/CH$_2$Cl$_2$/ −23°C/2 days. e. 1. PhCH$_2$OCOCl/py. 2. AlCl$_3$. 3. MeOCH$_2$Br/(i-Pr)$_2$EtN. 4. aqueous NaOH. 5. NaIO$_4$. 6. MeMgI, followed by TLC separation. 7. conc. HCl/MeOH. 8. H$_2$/Pd-C. 9. Ac$_2$O/py.

Scheme XLIX

TIUS and THURKAUF (28, 44) established the absolute stereochemistry at C(10), C(11), and C(13) by synthesis of (151), the antipode of δ-lactone (92) (Scheme L). *R*-5-Methylcyclohex-2-en-1-one (264), prepared from (+)-

pulegone by the procedure of OPPOLZER and PETRZILKA (61), was reduced stereoselectively to *cis* alcohol (265) with diisobutylaluminium hydride. After ozonolysis and borohydride reduction, the resulting triol (266) was converted to an acetonide and oxidized to a carboxylic acid (267). Upon removal of the protecting group by mild acid hydrolysis, δ-lactone (268) was spontaneously formed during work-up. The stereochemistry at the carbon bearing the lactone oxygen was inverted by converting (268) to an epoxyester (62) which rearranged to (269) on treatment with boron trifluoride etherate. Protection of (269) as the 2-ethoxyethyl ether followed by oxidation with Vedejs' reagent (63) furnished a 8.5:1 mixture of alcohols (270). The major alcohol after removal of the protecting group gave (151).

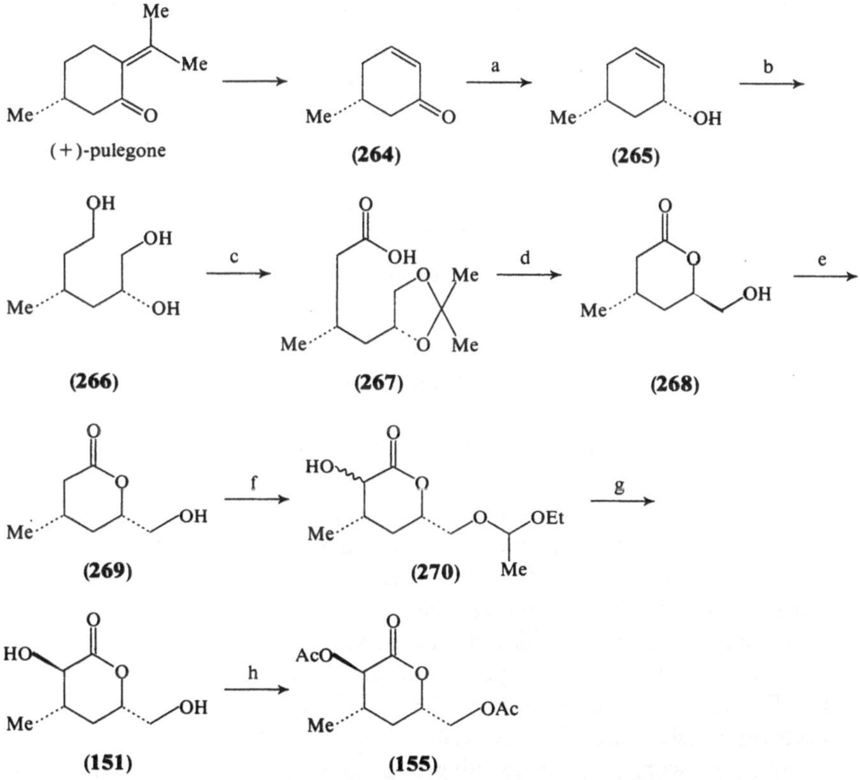

Reagents: a. DIBAL/CH$_2$Cl$_2$/0°C, followed by workup with NaF/H$_2$O. b. O$_3$/CH$_2$Cl$_2$/ −78°C, followed by NaBH$_4$ workup. c. 1. acetone/p-TSA/RT. 2. PDC/ DMF/RT. d. 1N HCl-THF(1-1)/RT. e. 1. TsCl/CH$_2$Cl$_2$/Et$_3$N/DMAP/RT. 2. PhCH$_2$ONa/THF/RT. 3. BF$_3$·Et$_2$O. f. 1. EtOCH=CH$_2$/pyridinium tosylate. 2. LDA/THF/RT, followed by addition of MoOPh. g. aqueous HCl. h. Ac$_2$O/py/RT.

Scheme L

Acetylation of (151) produced diacetate (155). Both (151) and (155) had ^{1}H nmr spectra that were identical with those of (92) and (68), respectively, from ozonolysis of palytoxin. The CD spectra of δ-lactones (155), $[\theta]_{222}$ −3100, and (68), $[\theta]_{224}$ +3900, both in ethanol, had opposite signs, indicating that the antipodes of (92) and (68) had been synthesized.

The KISHI group also determined the absolute configuration of C(10), C(11), and C(13) by synthesis of degradation product (271). Compound (271) had been produced from the periodate oxidation product (4) by ozonolysis, borohydride reduction, acetylation, lithium aluminium hydride reduction, and acetylation (30). Starting with S-(+)-3-hydroxy-2-methylpropionic acid, tetraacetate (271), $[\alpha]_D$ +17° (c 0.17, CHCl$_3$), was synthesized in twenty-three steps. The description of this synthesis which appears in the supplementary section of ref. (30) contains two errors.

(271)

VI. Conclusions

The palytoxins from Hawaiian *Palythoa toxica*, Hawaiian *P. tuberculosa*, Jamaican *P. mammilosa*, Puerto Rican *P. caribaeorum*, and an unidentified Tahitian *Palythoa* sp. have the same structure, since all exhibit identical ^{1}H and ^{13}C nmr spectra. Eventhough a direct comparison of the palytoxins from Hawaiian and Okinawan *P. tuberculosa* has not been made, both toxins appear to be identical since the same degradation products are produced on periodate oxidation and ozonolysis.

The complete structure of palytoxin has been elucidated and is shown in (1). The absolute configuration of the sixty-four asymmetric carbons have been rigorously established by KISHI's synthetic work.

In earlier work (16) the Hawaii group had suggested that the palytoxins from *P. toxica*, *P. mammilosa*, and a Tahitian *Palythoa* sp. might have slightly different structures since subtle differences could be detected in their ^{13}C nmr spectra. Furthermore the HIRATA group had obtained an unusual periodate oxidation product (60) (9) from the palytoxin in Okinawan *P. tuberculosa* which suggested that their palytoxin might be different from the palytoxins isolated by the MOORE group. Also MACFARLANE's report

(14) that the ^{252}Cf-plasma desorption mass spectrum of palytoxin from Okinawan *P. tuberculosa* exhibited ions for two components provided additional evidence that more than one palytoxin existed and that each palytoxin sample might be a mixture of two or more closely related compounds. When the MOORE group found that the palytoxins from Hawaiian *P. toxica* and *P. tuberculosa* and the Tahitian *Palythoa* sp. could be degraded to esters (6) and (56) by periodate oxidation, this prompted then to erroneously propose an anhydroketal structure for the C(51) − C(55) segment of palytoxin from the Tahitian *Palythoa* sp. and to propose pyranose and furanose hemiketal structures that were in equilibrium with each other for the C(51) − C(55) segments of the palytoxins from Hawaiian *P. toxica* and *P. tuberculosa* (20). It also led them to suggest (20) that the two components in the palytoxin from Okinawan *P. tuberculosa* might be a related C(54) ketal and hemiketal, as these structures explained the formation of the periodate product (60). Since all of the palytoxins underwent rapid periodate oxidation, MOORE had to propose that hydrolysis of the anhydroketal form, if present, to the corresponding hemiketal form would have had to be facile.

The ready deuterium exchange of the protons on C(54) and C(56) in D_2O, the facile bond cleavage of C(54) − C(55) by periodate at 0° C, and the ease of formation of a α,β-unsaturated ketone at C(55) − C(56) − C(57), however, were better accommodated by a hemiketal structure at C(55) rather than an anhydro structure for the C(51) − C(55) segment in the palytoxin from the Tahitian *Palythoa* sp. Unlike this palytoxin sample (20), 2,5-anhydrotagatose was recovered unchanged and showed no deuterium exchange in D_2O when heated to 55° C for 24 hours (15). Moreover the ^{13}C chemical shift of C(55) in the palytoxin from the Tahitian *Palythoa* sp. was 100.2 ppm in D_2O (16), essentially the same as that reported (100.3 ppm) for the chemical shift of C(55) in the palytoxin from Okinawan *P. tuberculosa* in D_2O (15). The chemical shift for C(55) was comparable with those for the hemiketal carobons in the various forms of D-tagatose, which ranged from 98.7 to 103.6 ppm in D_2O (64) (Scheme LI) and not with the chemical shift of the ketal carbon of 2,5-anhydrotagatose (272) (65), which was 109.6 ppm in D_2O (15). These data indicated that C(55) was not an anhydroketal carbon in palytoxin.

(272)

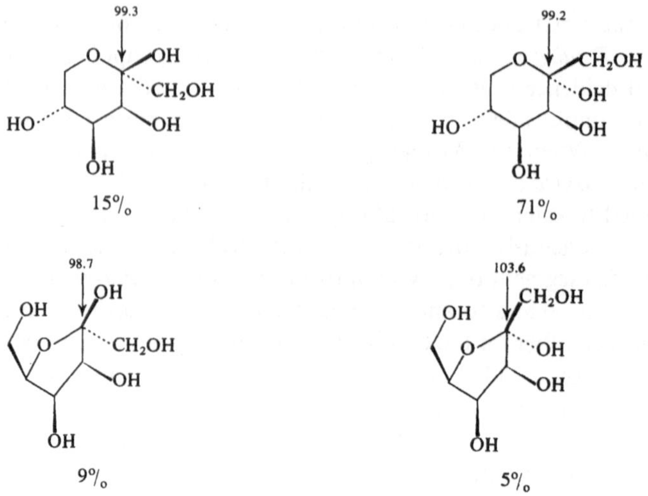

99.3

OH
··CH₂OH
OH

HO···

ÖH

15%

99.2

CH₂OH
··OH
OH

HO···

ÖH

71%

98.7

OH
OH
···CH₂OH
OH

ÖH

9%

103.6

CH₂OH
··OH
OH

ÖH

5%

Scheme LI

Recent work by the MOORE group (66) has now shown that palytoxin from Hawaiian *P. toxica* and *P. tuberculosa* exists in only one form. If ultrafiltration is used instead of reverse-phase chromatography in the desalting steps of the isolation procedure (1), a single component of structure (1) is always obtained. The ^{13}C nmr spectrum in dimethyl sulfoxide-d_6 shows a single peak at 99.8 ppm for the C(55) hemiketal carbon. Artifacts, however, are produced when palytoxin is absorbed onto polyethylene and other reverse-phase materials. As a result the signal for the C(55) hemiketal carbon is frequently missing and multiple carbon signals are seen for the methyl group on C(58) and the olefinic double bond at C(59) − C(60).

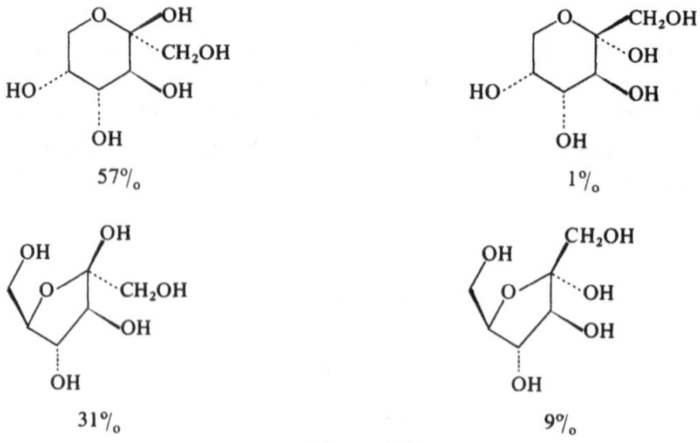

OH
··CH₂OH
OH

HO···

OH

57%

CH₂OH
··OH
OH

HO···

OH

1%

OH
OH
···CH₂OH
OH

ÖH

31%

CH₂OH
OH
··OH
OH

ÖH

9%

Scheme LII

In aqueous solution ketohexoses exist as equilibrium mixtures of α- and β-pyranose and furanose forms. Carbon-13 nmr studies of 3.7 M D-fructose in D_2O at pH 5 and 36°C, for example, show that β-pyranose and β-furanose forms predominate over α-pyranose and α-furanose forms (67) (Scheme LII). Carbon-13 nmr studies of tagatose, however, which has relative stereochemistry that is similar to that of the hemiketal system in palytoxin, indicate that the α-pyranose form predominates over the β-pyranose and the two furanose forms (64).

For palytoxin the six-membered hemiketal (1A) is favored over pyranose (1B) and the five-membered hemiketals (1C) and (1D). In (1A) both C(50) and C(56) are attached equatorially and the anomeric hydroxyl group is oriented axially. The hydroxyl groups on C(54) and C(55) are therefore *cis*, which agrees with the stereochemistry predicted by the facile periodate cleavage of C(54)−C(55) to give esters (6) and (56).

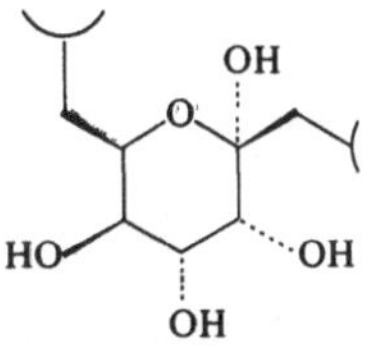

(1 A)

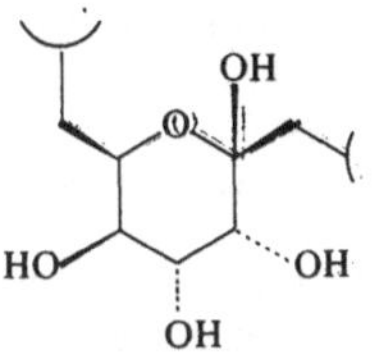

(1 B)

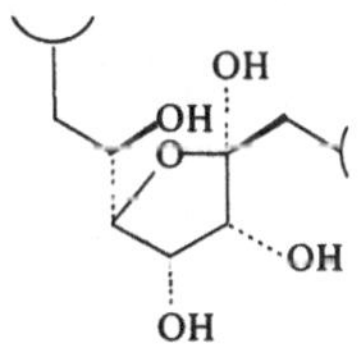

(1 C)

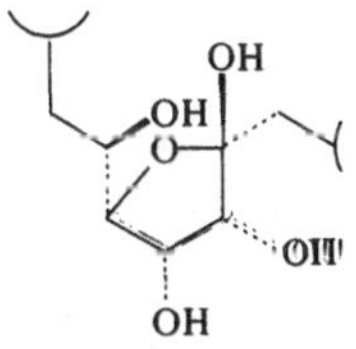

(1 D)

Acknowledgment

Research in the author's laboratory on the structure of palytoxin was supported by grants from the National Science Foundation (CHE 79-25416) and the National Institutes of Health (CA-12623).

References

1. MOORE, R. E., and P. J. SCHEUER: Palytoxin, a New Marine Toxin from a Coelenterate. Science 172, 495 (1971).
2. WILES, J. S., J. A. VICK, and M. K. CHRISTENSEN: Toxicological Evaluation of Palytoxin in Several Animal Species. Toxicon 12, 427 (1974).

3. MOSHER, H. S., F. A. FUHRMAN, H. D. BUCHWALD, and H. G. FISCHER: Tarichatoxin-Tetrodotoxin; A Potent Neurotoxin. Science **144**, 1100 (1964).
4. WALSH, G. E., and R. L. BOWERS: A Review of Hawaiian Zoanthids with Descriptions of Three New Species. Zool. J. Linnean Soc. **50**, 161 (1971).
5. MOORE, R. E., P. HELFRICH, and G. M. L. PATTERSON: The Deadly Seaweed of Hana. Oceanus **25** (2), 54 (1982).
6. HASHIMOTO, Y., N. FUSETANI, and S. KIMURA: Aluterin; A Toxin of Filefish, *Aluteria scripta*, Probably Originating from a Zoantharian, *Palythoa tuberculosa*. Bull. Japan. Soc. Sci. Fish. **35**, 1086 (1969).
7. KIMURA, S., Y. HASHIMOTO, and K. YAMAZATO: Toxicity of the Zoanthid *Palythoa tuberculosa*. Toxicon **10**, 611 (1972).
8. SCHEUER, P. J.: The Chemistry of Toxins Isolated from Some Marine Organisms. Fortschr. Chem. organ. Naturstoffe **22**, 265 (1964).
9. HIRATA, Y., D. UEMURA, K. UEDA, and S. TAKANO: Several Compounds from *Palythoa tuberculosa* (Coelenterata). Pure Appl. Chem. **51**, 1875 (1979).
10. QUINN, R. J., M. KASHIWAGI, R. E. MOORE, and T. R. NORTON: Anticancer Activity of Zoanthids and the Associated Toxin, Palytoxin, against Ehrlich Ascites Tumor and P-388 Lymphocytic Leukemia in Mice. J. Pharm. Sci. **63**, 257 (1974).
11. ATTAWAY, D. H., and L. S. CIERESZKO: Isolation and Partial Characterization of Caribbean Palytoxin. Proceedings of the Second International Coral Reef Symposium, 497 (1974).
12. BERESS, L., J. ZWICK, H. J. KOLKENBROCK, P. N. KAUL, and O. WASSERMANN: A Method for the Isolation of the Caribbean Palytoxin (C-PTX) from the Coelenterate (Zoanthid) *Palythoa caribaeorum*. Toxicon **21**, 285 (1983).
13. MOORE, R. E., and P. J. SCHEUER: Unpublished work.
14. MACFARLANE, R. D., D. UEMURA, K. UEDA, and Y. HIRATA: [252]Cf-Plasma Desorption Mass Spectrometry of Palytoxin. J. Amer. Chem. Soc. **102**, 875 (1980).
15. CHA, J. K., W. J. CHRIST, J. M. FINAN, H. FUJIOKA, Y. KISHI, L. L. KLEIN, S. S. KO, J. LEDER, W. W. MCWHORTER, JR., K.-P. PFAFF, M. YONAGA, D. UEMURA, and Y. HIRATA: Stereochemistry of Palytoxin. 4. Complete Structure. J. Amer. Chem. Soc. **104**, 7369 (1982).
16. MOORE, R. E., R. F. DIETRICH, B. HATTON, T. HIGA, and P. J. SCHEUER: The Nature of the λ263 Chromophore in the Palytoxins. J. Organ. Chem. (USA) **40**, 540 (1975).
17. MOORE, R. E., F. X. WOOLARD, M. Y. SHEIKH, and P. J. SCHEUER: Ultraviolet Chromophores of Palytoxins. J. Amer. Chem. Soc. **100**, 7758 (1978).
18. WALTERS, L. L., and R. E. MOORE: Unpublished work.
19. MOORE, R. E., F. X. WOOLARD, and G. BARTOLINI: Periodate Oxidation of N-(p-Bromobenzoyl)palytoxin. J. Amer. Chem. Soc. **102**, 7370 (1980).
20. MOORE, R. E., and G. BARTOLINI: Structure of Palytoxin. J. Amer. Chem. Soc. **103**, 2491 (1981).
21. UEMURA, D., K. UEDA, Y. HIRATA, C. KATAYAMA, and J. TANAKA: Structural Studies on Palytoxin, a Potent Coelenterate Toxin. Tetrahedron Letters **21**, 4857 (1980).
22. — — — — — Structures of Two Oxidation Products Obtained from Palytoxin. Tetrahedron Letters **21**, 4861 (1980).
23. UEMURA, D., K. UEDA, Y. HIRATA, H. NAOKI, and T. IWASHITA: Further Studies on Palytoxin. I. Tetrahedron Letters **22**, 1909 (1981).
24. — — — — — Further Studies on Palytoxin. II. Structure of Palytoxin. Tetrahedron Letters **22**, 2781 (1981).
25. FLEMING, I., and J. B. MASON: A Product from the Reaction of Pyridine with Acetic Anhydride. J. Chem. Soc. (C) London, 2509 (1969).
26. WOOLARD, F. X., and R. E. MOORE: Unpublished work.
27. SILVERSTEIN, R. M.: Spectrometric Identification of Insect Sex Attractants. J. Chem. Ed. **45**, 794 (1968).

28. MOORE, R. E., G. BARTOLINI, J. BARCHI, A. A. BOTHNER-BY, J. DADOK, and J. FORD: Absolute Stereochemistry of Palytoxin. J. Amer. Chem. Soc. **104**, 3776, 5572 (1982).
29. KLEIN, L. L., W. W. MCWHORTER, JR., S. S. KO, K.-P. PFAFF, Y. KISHI, D. UEMURA, and Y. HIRATA: Stereochemistry of Palytoxin. 1. C85 – C115 Segment. J. Amer. Chem. Soc. **104**, 7362 (1982).
30. KO, S. S., J. M. FINAN, M. YONAGA, Y. KISHI, D. UEMURA, and Y. HIRATA: Stereochemistry of Palytoxin. 2. C1 – C6, C47 – C74, and C77 – C83 Segments. J. Amer. Chem. Soc. **104**, 7364 (1982).
31. FUJIOKA, H., W. J. CHRIST, J. K. CHA, J. LEDER, Y. KISHI, D. UEMURA, and Y. HIRATA: Stereochemistry of Palytoxin. 3. C7 – C51 Segment. J. Amer. Chem. Soc. **104**, 7367 (1982).
32. MOORE, R. E.: Manuscript in preparation.
33. WARREN, R. G., R. J. WELLS, and J. F. BLOUNT: A Novel Lipid from the Brown Alga *Notheia anomala*. Aust. J. Chem. **33**, 891 (1980).
34. MOORE, R. E., and G. BARTOLINI: Unpublished work.
35. DOSKOCILOVA, D., and B. SCHNEIDER: On the Structure and Properties of Vinyl Polymers and Their Models. I. NMR Spectra of *d,l*- and *meso*-Forms of 2,4-Dichloropentane and 2,4-Pentanedioldiacetate: AA^1XX^1 and ABX_2 Systems with Weak Coupling. Collect. Czech. Chem. Commun. **29**, 2290 (1964).
36. MOORE, R. E., J. J. BARCHI, JR., and G. BARTOLINI: Use of Borate Complexation in Assigning Relative Stereochemistry of Acyclic Polyhydroxylated Compounds. J. Org. Chem. **50**, 374 (1985).
37. HUTSON, D. H., and H. WEIGEL: Partial Periodate Oxidation of D-Glucitol and its Borate Complex. J. Chem. Soc. 1546 (1961).
38. MASAMUNE, S., and R. E. MOORE: Unpublished work.
39. CHA, J. K., W. J. CHRIST, and Y. KISHI: On Stereochemistry of Osmium Tetraoxide Oxidation of Allylic Alcohol Systems. Empirical Rule. Tetrahedron **40**, 2247 (1984).
40. MCCONNELL, J. F., A. MCL. MATHIESON, and B. P. SCHOENBORN: Conformation of Iridomyrmecin and Isoiridomyrmecin. Tetrahedron Letters 445 (1962).
41. CHEUNG, K. K., K. H. OVERTON, and G. A. SIM: On the Conformations of δ-Lactones. Chem. Commun. 634 (1965).
42. CARDELLINA II, J. H., R. E. MOORE, E. V. ARNOLD, and J. CLARDY: Structure and Absolute Configuration of Malyngolide, an Antibiotic from the Marine Blue-Green Alga *Lyngbya majuscula* Gomont. J. Org. Chem. **44**, 4039 (1979).
43. KORVER, O.: Optical Rotatory Dispersion and Circular Dichroism of δ-Lactones. Determination of the Absolute Configuration of (+)-5-Decanolide and (+)-5-Dodecanolide. Tetrahedron **26**, 2391 (1970).
44. TIUS, M. A., and A. THURKAUF: Unpublished work.
45. BARCHI, JR., J. J., and R. E. MOORE: Unpublished work.
46. KO, S. S., L. L. KLEIN, K.-P. PFAFF, and Y. KISHI: Synthetic Studies on Palytoxin. Stereocontrolled, Practical Synthesis of the C.101 – C.115 Segment. Tetrahedron Letters **42**, 4415 (1982).
47. WONG, M. Y. H., and G. R. GRAY: 2-Deoxypentoses. Stereoselective Reduction of Ketene Dithioacetals. J. Amer. Chem. Soc. **100**, 3548 (1978).
48. MANCUSO, A. J., S.-L. HUANG, and D. SWERN: Oxidation of Long-Chain and Related Alcohols to Carbonyls by Dimethyl Sulfoxide "Activated" by Oxalyl Chloride. J. Org. Chem. **43**, 2480 (1978).
49. MINAMI, N., S. S. KO, and Y. KISHI: Stereocontrolled Synthesis of D-Pentitols, 2-Amino-2-deoxy-D-pentitols, and 2-Deoxy-D-pentitols from D-Glyceraldehyde Acetonide. J. Amer. Chem. Soc. **104**, 1109 (1982).
50. KATSUKI, T., and K. B. SHARPLESS: The First Practical Method for Asymmetric Epoxidation. J. Amer. Chem. Soc. **102**, 5974 (1980).

51. Nagaoka, H., and Y. Kishi: Further Synthetic Studies on Rifamycin S. Tetrahedron **37**, 3873 (1981).

52. Sondheimer, S. J., H. Yamaguchi, and C. Schuerch: Synthesis of 1,2-Anhydro-3,4,6-tri-O-benzyl-β-D-mannopyranose. Carbohydr. Res. **74**, 327 (1979).

53. Cohen, N., W. F. Eichel, R. J. Lopresti, C. Neukom, and G. Saucy: Synthetic Studies on (2R,4′R,8′R)-α-Tocopherol. An Approach Utilizing Side Chain Synthons of Microbiological Origin. J. Org. Chem. **41**, 3505 (1976).

54. Tamura, M., and J. Kochi: Coupling of Grignard Reagents with Organic Halides. Synthesis 303 (1971).

55. Garegg, P. J., and L. Maron: Inversion at C-2 of 3,4,6-Tri-O-benzyl-α-D-mannopyranosides by Oxidation and Reduction. Acta Chem. Scand., Ser. B **33 B**, 453 (1979).

56. Fraser-Reid, B., R. D. Dawe, and D. B. Tulshian: Stereoselective Routes to Some Unsaturated α- and β-C-Glycopyranosides. Can. J. Chem. **57**, 1746 (1979).

57. Lewis, M. D., J. K. Cha, and Y. Kishi: Highly Stereoselective Approaches to α- and β-C-Glycopyranosides. J. Amer. Chem. Soc. **104**, 4976 (1982).

58. Leder, J., H. Fujioka, and Y. Kishi: Synthetic Studies on Palytoxin. Stereocontrolled Practical Synthesis of the C.23–C.37 Segment. Tetrahedron Letters **24**, 1463 (1983).

59. Weygand, F., and H. Wolz: 2-Desoxy-d-xylose aus d-Glucose über 3-Desoxy-d-galaktose. Chem. Ber. **85**, 256 (1952).

60. Christ, W. J., J. K. Cha, and Y. Kishi: Manuscript in preparation.

61. Oppolzer, W., and M. Petrzilka: An Enantioselective Total Synthesis of Natural (+)-Luciduline. Helv. Chim. Acta **61**, 2755 (1978).

62. Still, W. C., and I. Galynker: Stereospecific Synthesis of the C30–C43 Segment of Palytoxin by Macrocyclically Controlled Remote Asymmetric Induction. J. Amer. Chem. Soc. **104**, 1774 (1982).

63. Vedejs, E., D. A. Engler, and J. E. Telschow: Transition-Metal Peroxide Reactions. Synthesis of α-Hydroxycarbonyl Compounds from Enolates. J. Organ. Chem. (USA) **43**, 188 (1978).

64. Que, L., Jr., and G. R. Gray: [13]C Nuclear Magnetic Resonance Spectra and the Tautomeric Equilibria of Ketohexoses in Solution. Biochemistry **13**, 146 (1974).

65. Koll, P., S. Deyhim, and K. Heyns: Anhydrozucker mit 2,7-Dioxabicyclo[2.2.1]heptan-System. 2,6-Anhydrohexulosen. Chem. Ber. **111**, 2909 (1978).

66. Moore, R. E.: Unpublished work.

67. Doddrell, D., and A. Allerhand: Study of Anomeric Equilibria of Ketoses in Water by Natural-Abundance Carbon-13 Fourier Transform Nuclear Magnetic Resonance. D-Fructose and D-Turanose. J. Amer. Chem. Soc. **93**, 2779 (1971).

(Received September 17, 1984)

Sesterterpenes: An Emerging Group of Metabolites from Marine and Terrestrial Organisms

By P. Crews and S. Naylor
Thimann Laboratories and Center for Coastal Marine Studies, University of California, Santa Cruz, California, U.S.A.

Contents

I. Introduction

Naturally occurring sesterterpenes were first encountered less than thirty years ago, and as a consequence they are often viewed as a rare category of natural products. The first sesterterpenes, isolated during the late 1950's, included ophiobolin A which was present in crude extracts studied by Orsenigo (1) and gascardic acid reported by Brochere-Ferreol and Polonsky (2). No structures were determined initially for either of these. However, the complete tricarbocyclic skeleton for ophiobolin A along with the absolute stereochemistry at eight chiral sites shown in formula (75) (see Table 6) was elucidated in 1965 after an X-ray study by Nozoe and coworkers (3). In the same year structure (i), featuring a polycarbocyclic frame, was announced for gascardic acid by Arigoni and Scartazzini (4, 5). Two out of the five elements of its relative stereochemistry were defined, whereas the remaining centers were pinned down in representation (85) by Clardy and Boeckman only after an X-ray crystallographic study of gascardic acid dicyclohexylammonium salt (6)*. Lagging slightly behind the above was the discovery and characterization of the first acyclic sesterterpene geranylnerolidol (2) in 1968 by Nozoe and coworkers (7).

Additional new sesterterpenes were discovered slowly; by 1972 only 13 compounds were known which represented the aforementioned one acyclic and two carbocyclic frameworks (8). Within three years this number had grown to 31 representatives distributed among just nine different structural families (9)**. Since then, there have been reports of numerous other new

---

* A total synthesis of (85) also established the relative stereochemistry (47).

** We identify nine rather than six structure types among the compounds discussed by Cordell (9).

(i)

sesterterpenes with the total count now standing at 158 compounds comprising 21 major carbon frameworks and 11 modified ones. In view of this diversity, the sestertepene class can no longer be considered as rare. Moreover, the many recent developments in sesterterpene chemistry have prompted the present review. Our discussions jointly considers important chemical as well as biological aspects and also covers some aspects of structure elucidation, especially insights which can be obtained from ^{13}C-NMR data. The literature is covered up to September 1984.

The only previous major review of sesterterpene natural products by CORDELL (9) appeared in 1977; both marine and terrestrial derived metabolites were covered. Although many sesterterpenes are described in an extensive encyclopedia of terpenes (10) and in a handbook listing compounds from marine organisms (11), they can only be accessed by searching through a formula index. Numerous other scant summaries of sesterterpene chemistry can be found in the literature, but they are all part of broader treatments. These include special texts on natural products (12 – 14), a catalog of absolute stereochemistry (15), discussions of the use of ^{13}C-NMR spectroscopy in natural products (16), phytosesterterpenes (17), general reviews of marine natural products (18, 19), antibiotics from marine organisms (20), Australian marine natural products (21, 22), Red Sea natural products (23), marine sponge constituents (24, 26) and a review of chemical approaches to sponge taxonomy (27). Consequently, we have found it useful to cover previously reviewed material along with more recent developments.

Strictly speaking, a sesterterpene ought to have twenty-five contiguous carbons divisible into five isoprene units. Summarized in Section II are, to the best of our knowledge, all of the structural variations observed for sesterterpene natural products. Also included in Section II are natural products which do not contain 25 carbon atoms but are obviously sesterterpene derivatives. The latter fall into several categories which include degraded sesterterpenes with 22 – 24 contiguous carbon atoms, alkylated sesterterpenes with 26 or 27 contiguous carbon atoms, and sesterterpenes containing an additional non-terpenoid unit. We have not considered a large array of noncarbocyclic compounds with 21 contiguous carbon atoms as their relationship to sesterterpenes is still uncertain. Such terpenoids seem to be exclusively of marine origin and have been the subject of prior reviews (19, 25, 28). Finally, other C_{25} compounds which are not of obvious sesterterpene origin have been excluded. Examples are the meroterpenoids such as andibenin-B $C_{25}H_{30}O_6$ (ii) from *Aspergillus* for which a sesterterpenoid origin was proposed originally (29), but which were subsequently shown to arise by a mixed polyketide-terpenoid route (30, 31), quassinoids such as soulameolide $C_{25}H_{32}O_6$ (iii) which is a triterpene degradation product (32), and diumycinol $C_{25}H_{42}O$ (iv), a nonisoprenoid obtained by acid hydrolysis of the antibiotic diumycin (33).

Sesterterpenes occur in widely differing sources and have been obtained from terrestrial fungi, plants and insects as well as from marine sponges and nudibranchs. Though discovery of the first marine sesterterpene followed

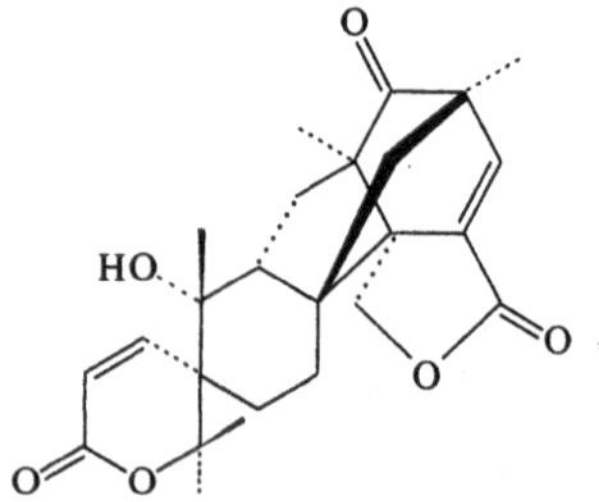

andibenin-B (ii)

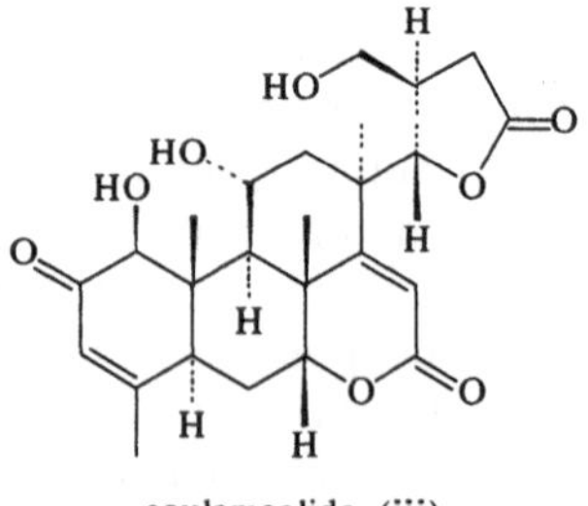

soulameolide (iii)

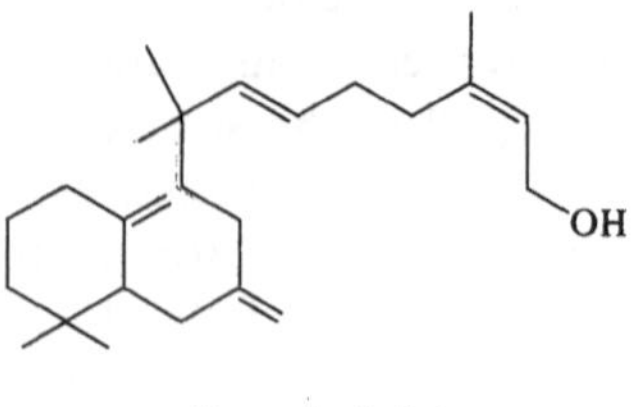

diumycinol (iv)

that of the first terrestrial one by more than a decade, sesterterpenes of marine origin now represent a majority. In fact 66% of the compounds listed in this review are of marine origin. Various possible relationships between the distribution of the various structural types and the phyletics of their organism source are discussed in Section III and summarized in Table 3.

The biogenesis of terrestrial sesterterpenes seems reasonably well understood and is also discussed in Section III. The most definitive information is based upon ^{14}C-mevalonic acid incorporation studies (9). The results imply that a straightforward carbocation-like cyclization of geranylfarnesol phyrophosphate occurs to generate such polycarbocyclics as albolineol (60) and ophiobolin A (75). If the possibility of Wagner-Meerwein alkyl shifts (34) is included, then formation of all of the other 25-carbon carbocyclic skeletons, except those illustrated by tricyclic sesterterpenes (91–94) can be rationalized. By contrast, very little discussion or experimental evidence has appeared to explain the genesis of the variety of non-25-carbon atom sesterterpene derivatives reviewed here.

Many sesterterpenes possess interesting biological activities. Such properties which are summarized in Section IV, have frequently stimulated the interest of synthetic organic chemists. However, the only total syntheses of sesterterpenes reported to date are those of gascardic acid (85) by BOECKMAN and coworkers (47), and ceriferol-I (46) by KATO et al. (35). Other synthetic studies on sesterterpenes remain in the embryonic stage and generally represent approaches to a particular structural class.

Complete structure elucidation of a new isolated sesterterpene frequently presents a challenge. Many sesterterpenes possess relatively complex carbon skeletons with multiple chiral centers as well as polysubstituted double bonds. Only 16 sesterterpenes have been studied by X-ray crystallography, and they represent 11 of the structural sub-categories. This is indicated by the symbol [ ] in Tables 6 and 7. Consequently, most structure elucidations have relied on NMR data and/or chemical degradation or interconversion experiments. Some of the more important spectroscopic techniques used in sesterterpenes analysis are discussed in Section V.

II. Summary of Carbon Skeletons and Structural Variations

1. Conventions

Following the format of DEVON and SCOTT (8) we have organized the sesterterpenes into separate families based upon their carbon skeletons. These are summarized in Tables 1 and 2; each different group is designated by the combination of a boldface letter, a subscript and, in some cases, a superscript. Closely related compounds which have the same number of rings but differs in carbon skeleton are assigned the same boldface letter and a different subscript. For example, the four bicyclic sesterterpenes in Table 1 are designated as (C_1), (C_2), (C_3), (C_4). A superscript in combination with this subscript indicates a relationship between the regular sesterterpenes listed in Table 1 and the modified sesterterpenes which are grouped in Table 2. The following superscript pattern is used: $-$, a degraded skeleton; $+$, an alkylated skeleton, and $\#$, a sesterterpene skeleton with an additional nonisoprenoid substituent. Thus the biogenetic relationship between, for example, the bicyclic compounds such as (52) and (53) is immediately indicated by the symbols C_2 and C_2-. Finally, each structural family in Table 1 and 2 is coded with a T or M to indicate their terrestrial and/or marine origin.

Table 1. *Sesterterpene Carbon Skeleton Classes*

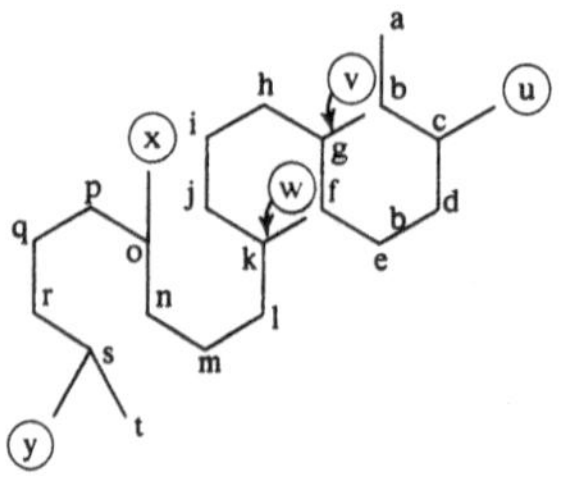

(A) (T, M)

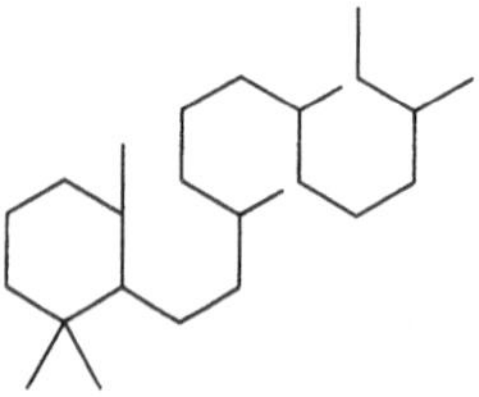

(B₁) (M)

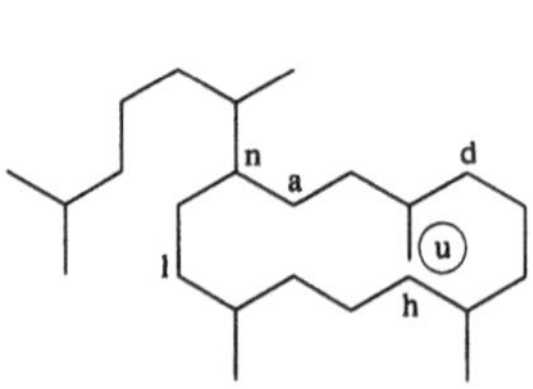

(B₂) (T)

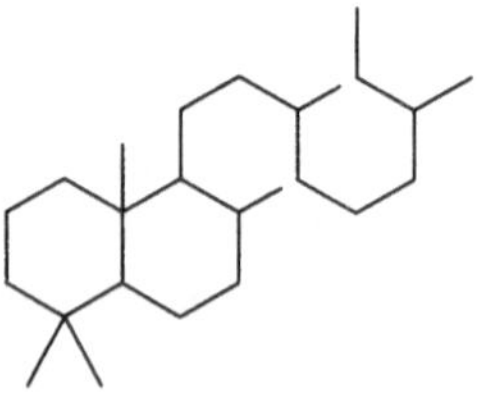

(C₁) (T)

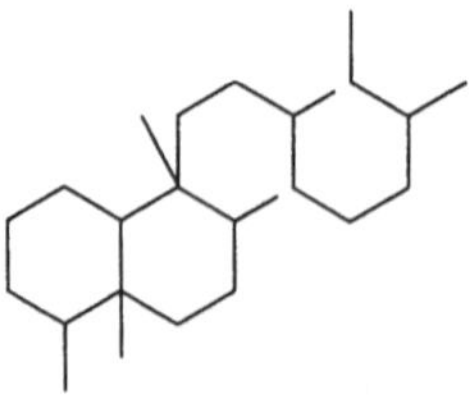

(C₂) (M)

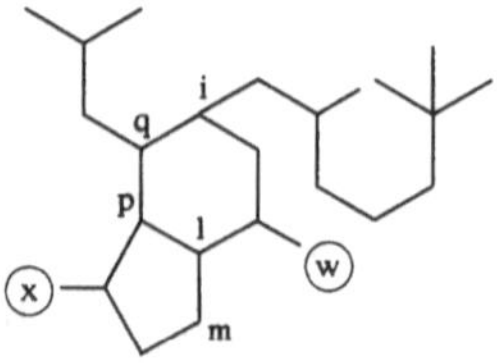

(C₃) (M)

Table 1 *(continued)*

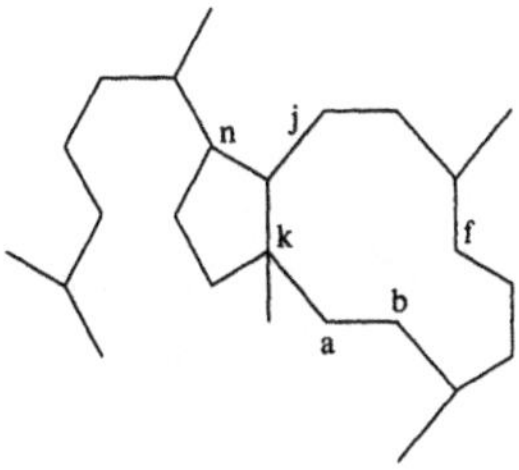

(**C₄**) (T) pre-ophiobolane

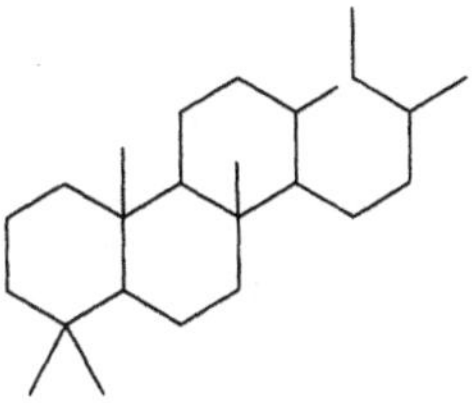

(**D₁**) (T, M) cheilanthane

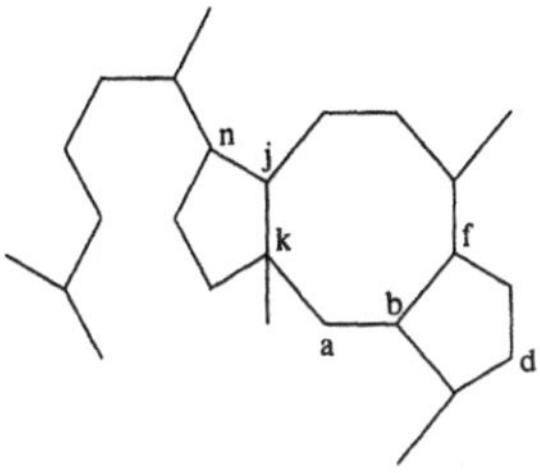

(**D₂**) (T) ophiobolane

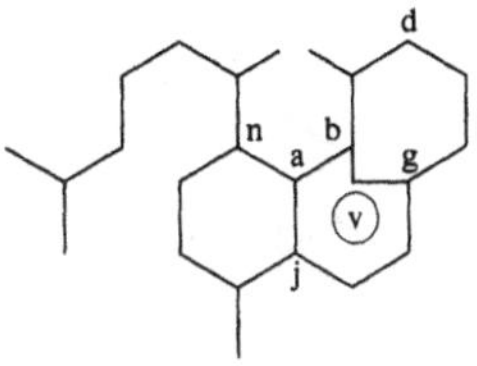

(**D₃**) (T)

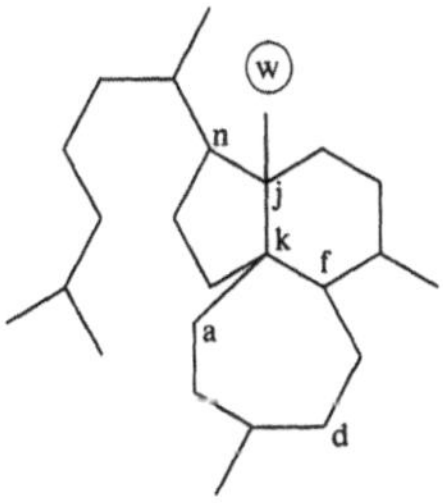

(**D₄**) (T)

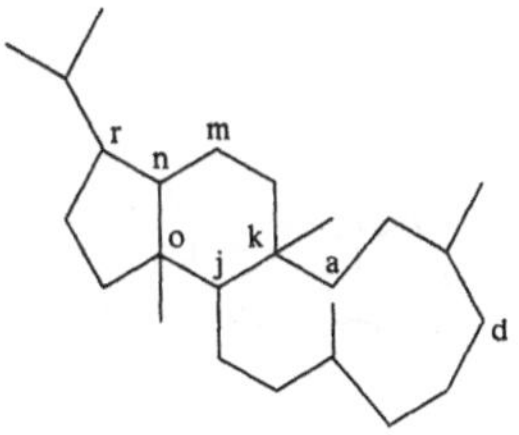

(**D₅**) (T)

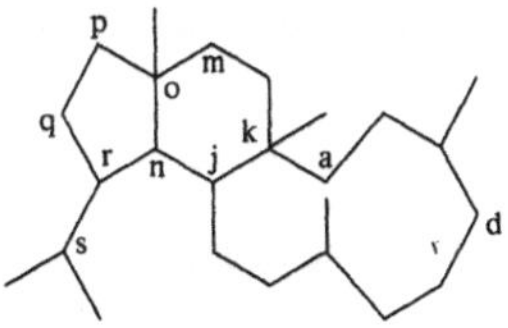

(**D₆**) (T)

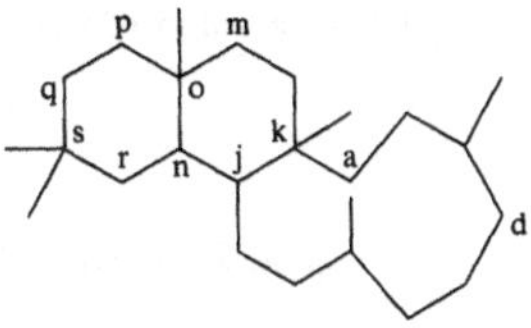

(**D₇**) (T)

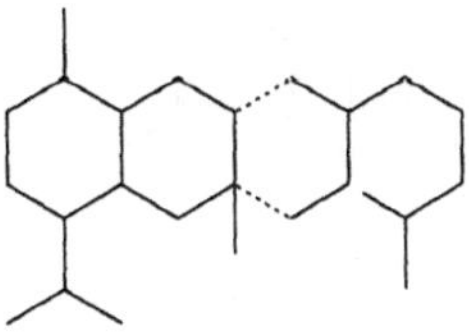

(**D₈**)* (T)

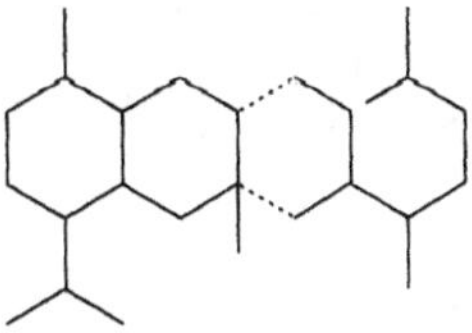

(**D₉**)* (T)

Table 1 *(continued)*

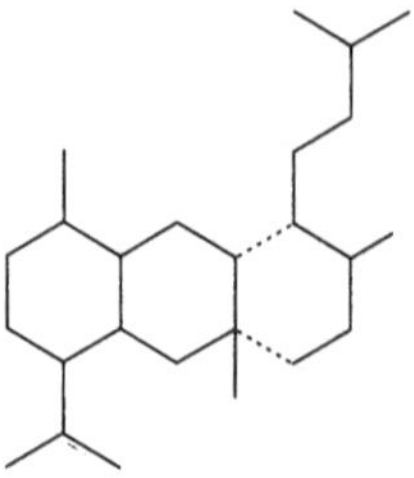

(**D₁₀**)* (T)

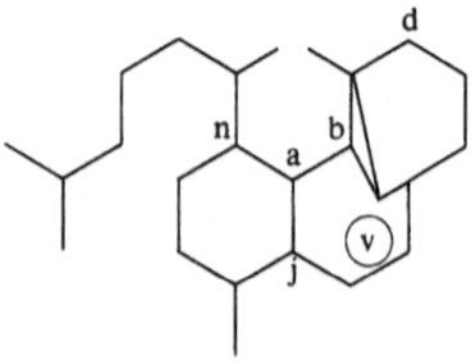

(**E₁**) (T)

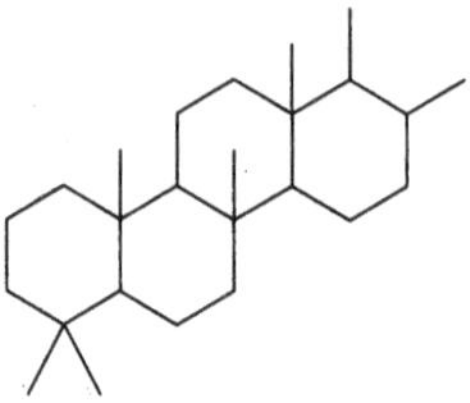

(**E₂**) (M) scalarane

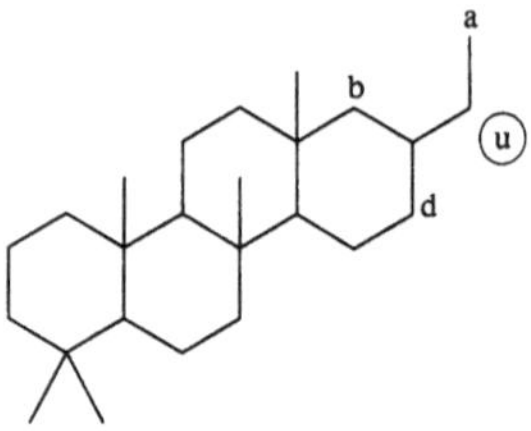

(**E₃**) (M)

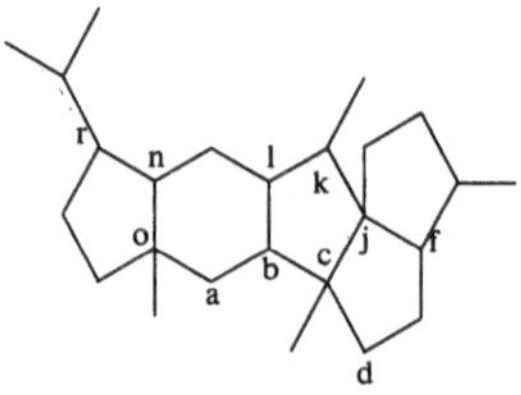

retigerane (**F₁**) (T)

---

* Biosynthesis via union of C_{15} and C_{10} isoprenoids is indicated by dotted lines.

Table 2. *Modified Sesterterpene Carbon Skeleton Classes*

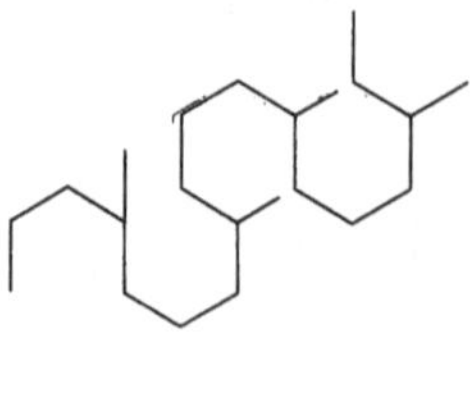

(**A⁻**) (M)

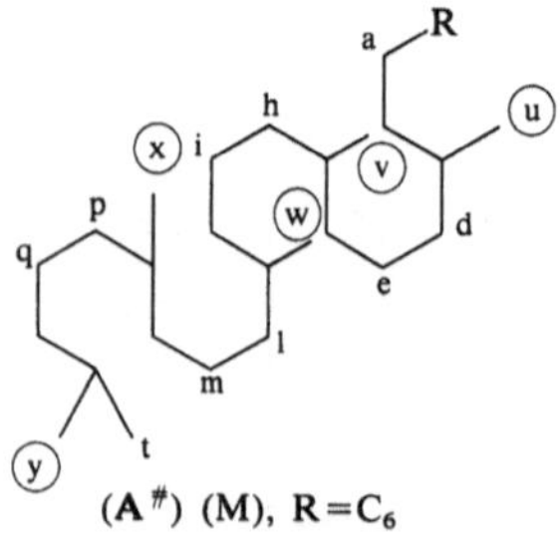

(**A#**) (M), R=C_6

Table 2 *(continued)*

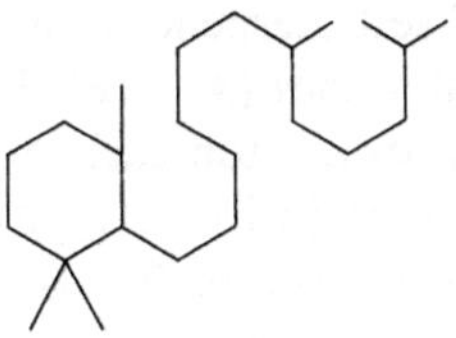

(**B₁⁻**) (M)

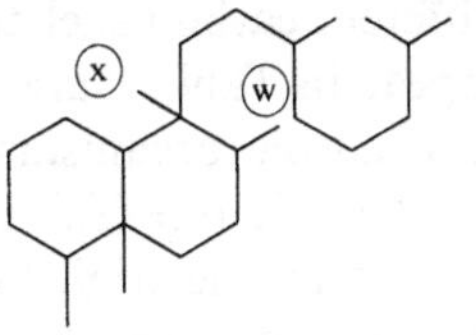

(**C₂⁻**) (M)

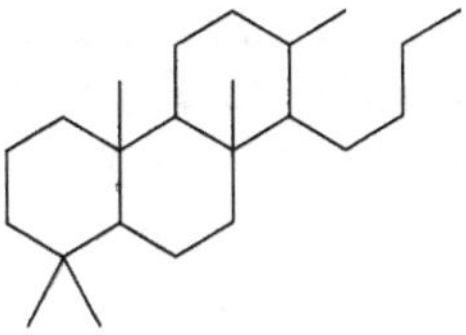

(**D₁⁻**) (M)

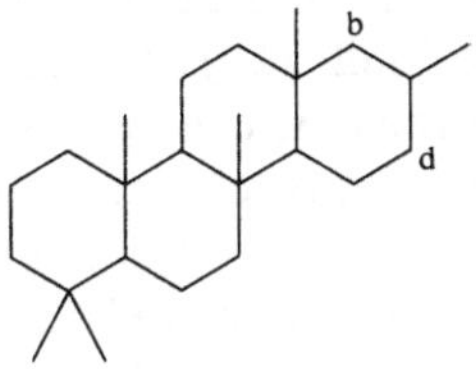

(**E₂⁻**) (M)

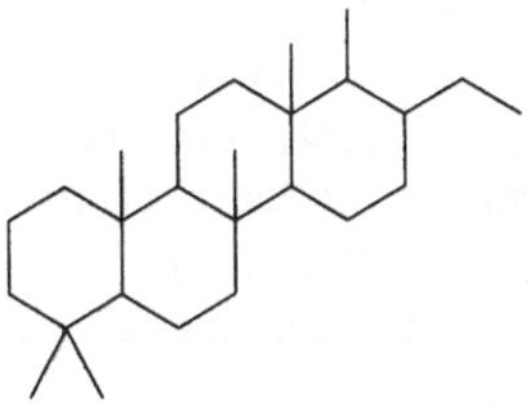

(**E₂⁺**) (M)

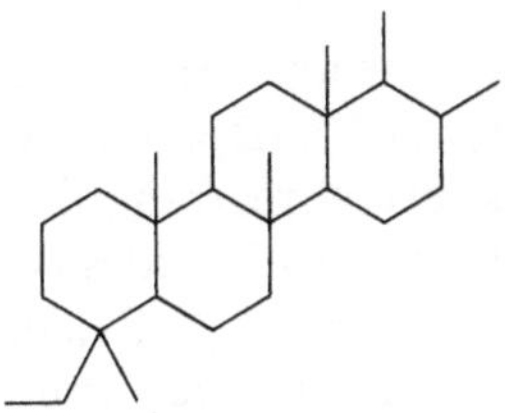

(**E₂⁺′**) (M)

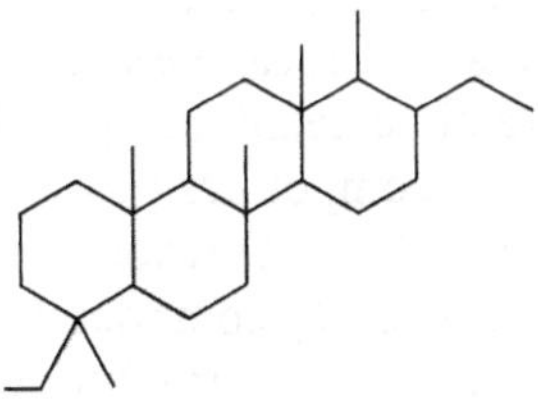

(**E₂⁺⁺**) (M)

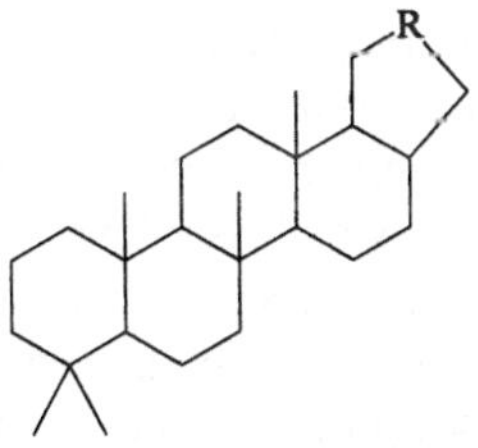

(**E₂ #**) (M)

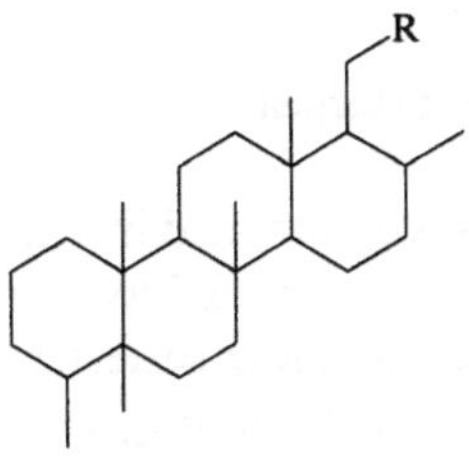

(**E₄ #**) R=C₆ (M)

2. Carbon Skeletons of "Regular" Sesterterpenes

The 21 different carbon skeletons of "regular" sesterterpenes found so far which appear in Table 1 are designated as categories $(A) - (F)$. Both terrestrial and marine organisms elaborate head-to-tail non-carbocyclic [type (A)] compounds (see Table 7). However, the majority of type (A) compounds are of marine origin and only one, geranylfarnesol (3) has been isolated from both a marine (sponge, *Fasciospongia*) and a terrestrial (insect, *Ceroplastes*) source. By contrast, not a single member of a carbocyclic family has been found in both marine and terrestrial source although one family (D_1) contains metabolites isolated from both land and marine organisms. As can be seen in Table 6, structural class (D_1) contains just two compounds, cheilanthatriol (61) from a fern and suvanine (62) from a marine sponge. It is interesting that (61) and (62) have different relative stereochemistry at C-5, C-8, and C-9.

Biogenetic relationship between the families of "regular" cyclic sesterterpenes and the head-to-tail arrangement of geranylfarnesane (A) are outlined by Scheme I. Inspection of the Scheme reveals the following patterns. The terpenoid head-to-tail arrangement is preserved in sesterterpenoid skeletons (B_1), (B_2), (C_1), (C_4), (D_1), (D_2), (D_3), (D_4), (D_5), (E_1), and (E_2), whereas several rearranged isoprenoids can be related back to non-rearranged types. This includes (C_2) from (C_1), (E_3) from (E_2), (D_7) from (D_6), and (D_6) from (C_4). Evidence in support of such biogenetic relationships will be explored further in Section III.

Our symbolism offers some advantages over a somewhat confusing convention used previously for designating different sesterpene categories (9). For example, (37), a type (B_2) compound, and (60), a type (C_4) compound, were both listed under "pre-ophiobolanes". It can be seen from Scheme I that only the skeleton of (60) possesses a direct relationship to that of ophiobolane which is of type (D_2). In addition, compounds possessing several different carbon skeletons were listed as "cheilanthatriol" types. This included not only cheilanthatriol (61) itself, a type (D_1) compound, but also (99). (104), (112), of type (E_2) and (152) of type $(E_2{}^{\#})$. Scheme I conveys that these are all closely related yet biogenetically distinct relatives.

3. Carbon Skeletons of Modified Sesterterpenes

A group of 11 modified sesterterpene skeletons is catalogued in Table 2. Metabolites representing each of these sesterterpene groups have so far been found exclusively in marine organisms. Each of these sesterterpene skeletons, except skeleton (E_4), has obvious precursors among the "regular" families of Table 1, so many biosynthetic relationships suggested by

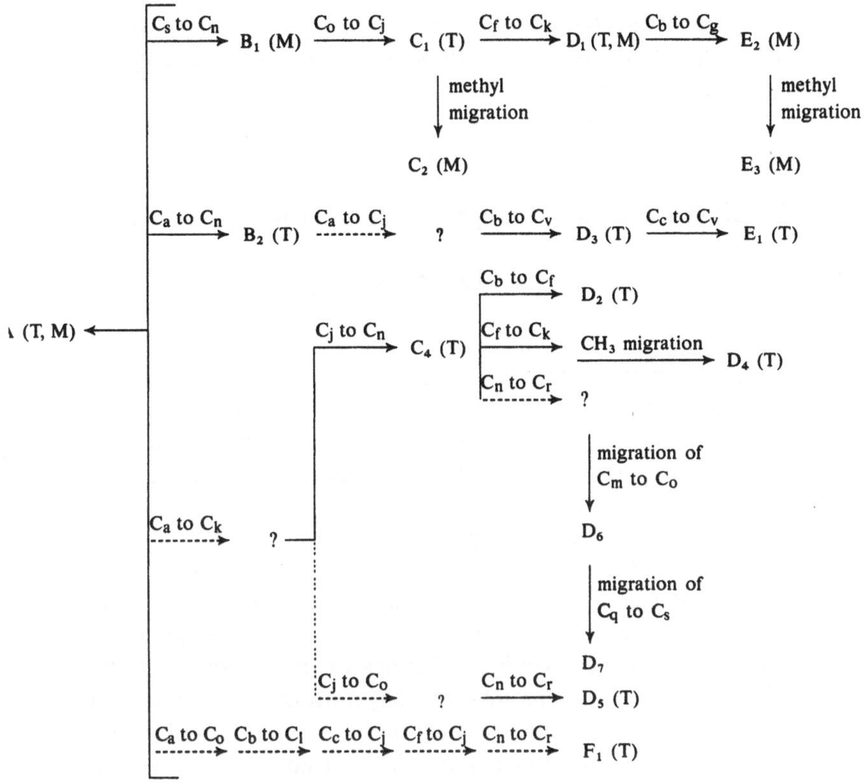

$$C_s \text{ to } C_n \longrightarrow B_1 \text{ (M)} \xrightarrow{C_o \text{ to } C_j} C_1 \text{ (T)} \xrightarrow{C_f \text{ to } C_k} D_1 \text{ (T, M)} \xrightarrow{C_b \text{ to } C_g} E_2 \text{ (M)}$$

methyl migration $\downarrow$ methyl migration $\downarrow$

$$C_2 \text{ (M)} \qquad\qquad E_3 \text{ (M)}$$

$$C_a \text{ to } C_n \longrightarrow B_2 \text{ (T)} \xdashrightarrow{C_a \text{ to } C_j} ? \xrightarrow{C_b \text{ to } C_v} D_3 \text{ (T)} \xrightarrow{C_c \text{ to } C_v} E_1 \text{ (T)}$$

$$C_b \text{ to } C_f \longrightarrow D_2 \text{ (T)}$$

A (T, M) ⟵

$$C_j \text{ to } C_n \longrightarrow C_4 \text{ (T)} \xrightarrow{C_f \text{ to } C_k} \text{CH}_3 \text{ migration} \longrightarrow D_4 \text{ (T)}$$

$$C_n \text{ to } C_r \dashrightarrow ?$$

migration of C_m to C_o $\downarrow$

$$C_a \text{ to } C_k \dashrightarrow ? \qquad D_6$$

migration of C_q to C_s $\downarrow$

$$D_7$$

$$C_j \text{ to } C_o \dashrightarrow ? \xrightarrow{C_n \text{ to } C_r} D_5 \text{ (T)}$$

$$C_a \text{ to } C_o \dashrightarrow C_b \text{ to } C_l \dashrightarrow C_c \text{ to } C_j \dashrightarrow C_f \text{ to } C_j \dashrightarrow C_n \text{ to } C_r \dashrightarrow F_1 \text{ (T)}$$

Scheme 1. Summary of biogenetic relationships between acyclic and regular cyclic sesterterpenes*

---

* For lettering convention see Table 1.

Scheme 1 should also apply to this group. The alkylated scalaranes including types (E_2^+), $(E_2^{+\prime})$, and (E_2^{++}) probably arise by transfer of a methyl group of methionine (*164*) to a carbon in a type (E_2) compound.

4. Heteroatom Substituent Patterns

Oxygen is a prominent sesterterpene heteroatom. With only one exception [see (**64**)] each metabolite is oxygenated at one or more locations. Nitrogen is the other heteroatom found in sesterterpenes but it is present only sparingly and is restricted to the marine metabolites, as can be seen by scanning Table 6 which reveals only five compounds with nitrogen from structure categories (D_1) and (E_2). While halogen substituents often abound in marine terpene metabolites (*11, 19, 36*) none have been discovered to date among the sesterterpenes.

III. Biosynthetic and Comparative Biochemical Observations

1. Introduction

Organisms from rather diverse phyletic groups are sources of sesterpenoids. A summary of the distribution of structure classes vs. species is presented in Table 3. The table indicates that the range of terrestrial organisms which elaborate sesterterpenes is considerably greater than the range of marine organisms. Marine sesterterpenes have been isolated mainly from sponges, the exception being seven metabolites restricted to families (A), (D_1^-) and (E_2^+) which are found in nudibranchs and they may actually be derived from the sponge diets of the nudibranchs (*37, 67*).

Acyclic sesterterpenes abound in marine sponges (Table 7), but are poorly represented in each major group of terrestrial organisms. Circumstantial evidence (see below) indicates that acyclic sesterterpenes found in terrestrial organisms are formed by normal head-to-tail isoprenoid biosynthesis (*9*). Carbocyclic sesterterpenes are found in both marine and terrestrial organisms as summarized in Table 1 and 2, but strikingly little overlap exists among the sesterterpene skeletons from marine and terrestrial organisms. To be more more specific, *all carbocyclic* compounds from marine organisms, except those based on skeleton (C_3) (in III.2) possess a common biosynthetic feature, which is the formation of one six-membered ring by connection of C_s to C_n. In other words, skeleton (B), appears to be a common biosynthetic starting point for these metabolites. By contrast, a variety of ring sizes are found in metabolites from terrestrial organisms and no common cyclization path appears to be evident for them.

As noted previously (*9*) experimental evidence shows that certain terrestrial organisms elaborate cyclic sesterterpenes by cyclization of normal head-to-tail isoprenoid precursors. By contrast, no information is available on the biosynthesis of either cyclic or acyclic sesterterpenes in marine sponges or nudibranchs. In fact it has been suggested, based upon unpublished results, that sponges are probably not capable of *de novo* terpene synthesis (*25, 28*).

References, pp. 262—269

Table 3. *Sesterterpene Carbon Skeletons and Their Sources*

Genera	Structure types elaborated
A. Marine Sponges	
Family Spongiidae (Order, Dictyoceratida)	
Spongia	$(A), (E_2), (E_3)$
Carteriospongia (= Phyllospongia)	$(A), (A-), (E_2+), (E_2++)$
Lendenfeldia	(E_2+)
Family Thorectidae (Order, Dictyoceratida)	
Fasciospongia	(A)
Ircinia	$(A), (A\#), (C_3), (D_1)$
Thorecta	(A)
Luffarariella	(B_1)
Cacospongia	$(E_2), (E_2\#), (E_3)$
Hyrtios	$(E_2), (E_2-)$
Family Dysideidae (Order, Dictyoceratida)	
Dysidea	$(A\#), (E_2), (E_2+), (E_2+'),$ $(E_2++), (E_2\#)$
Family Hymeniacidonidae (Order, Halichondrida)	
Prianos	$(B_1-), (C_2-)$
Family Clathriidae (Order, Poecilosclerida)	
Microciona	$(E_4\#)$
Family Latrunculiidae (Order, Hadromerida)	
Sigmosceptrella	(C_2-)
B. Marine Nudibranchs	
Family Chromodorididae (Suborder, Doridacea)	
Chromodoris	(E_2+)
Family Cadlinidae (Suborder, Doridacea)	
Cadlina	$(A), (D_1-)$
C. Terrestrial Fungi	
Family Moniliaceae (Order, Moniliales)	
Cochliobolus*	$(A), (D_2)$
Helminthosporium*	(D_2)
Aspergillus	(D_5)
Family Hypocreaceae (Order, Hypocreales)	
Cephalosporium	(D_2)
D. Terrestrial Lichens	
Family Stictaceae (Order, Lecanorales)	
Lobaria	(F_1)
E. Terrestrial Plants	
Family Polypodiaceae (Order, Polypodiales)	
Cheilanthes (fern)	$(D_1), (D_2)$
Family Malvaceae (Order, Malvales)	
Gossypium (cotton flower buds)	$(D_8), (D_9), (D_{10})$
Family Solanaceae (Order, Tubiflorae)	
Solanum (potato leaves)	(A)

Table 3 *(continued)*

Genera	Structure types elaborated
Family Stachyoideae (Order, Tubiflorae)	
Salvia (sage)	(C_1)
F. Terrestrial Insects	
Family Lacciferidae (Order, Hamiptera)	
Gascardia	(D_4)
Family Phylloxeridae (Order, Homoptera)	
Ceroplastes (scale insects)	(A), (B_2), (C_4), (D_2), (D_3), (D_6), (D_7), (E_1)

* *Cochliobolus* is the perfect stage of *Helminthosporium* (syn = *Ophiobolus*)

2. Biogenesis

Speculations about sesterterpene biogenesis commonly begin with the acyclic head-to-tail sesterterpene geranylfarnesol (**3**), its pyrophosphate, or some other acyclic biological equivalent. It is reassuring to find that (**3**) has been observed in both terrestrial and marine sources. Cordell (*9*) reviewed evidence which shows that *Cochliobolus* cultures can convert ^{14}C-labeled mevalonate to ophiobolane type (D_2) sesterterpenes such as ophiobolins A (**75**), B (**74**), C (**73**), and F (**72**). The same genus also effects tricyclization of ^{3}H-labeled all-*trans*-geranylfarnesol and all-*trans*-geranylfarnesyl pyrophosphate to ophiobolin F (**72**). While no experimental information is available on the biogenetic pathways to marine carbocyclic terpenes, it can be imagined that the routes employed by terrestrial and marine organisms to the related D_1 tricyclic compounds chelanthatriol (**61**) from a fern and suvanine (**62**) from a sponge are parallel.

Many hypotheses about sesterterpene biogenesis can be found in the literature. Inspection of Tables 1 and 2 and Scheme 1 provides a summary of what bonds must interconnect if a geranylfarnesane were a precursor for the various carbocyclic frameworks. Past mechanistic proposals have often been accompanied by drawings showing stereochemical details and these are best appreciated by consulting the original literature.

In analogy to biosynthetic cyclizations in the diterpenoid series (*39*), we speculate that protonation at C_b of a type (**A**) pyrophosphate leads initially to monocyclic skeleton (**B$_1$**), which may be transformed successively to bicyclic (**C$_1$**), then to (**D$_1$**) and finally to (**E$_1$**). Such a hypothesis has been advanced to explain the biogenesis of cheilanthatriol (**61**) and scalarin (**112**) (*9*). Straightforward cyclization of a geranylfarnesyl pyrophosphate connecting C_a to C_n has been diagrammed as leading to type (**B$_2$**) compounds such as albocerol (**37**) (*9*). By contrast, the biosynthesis of the bicyclic (**C$_3$**) ircinianin (**58**) has been proposed to occur by an internal Diels-Alder cyclization of a type (**A**) $\Delta^{13,15}$ dehydro derivative of fasciculatin (**10**) (*40*). Bicyclization of a geranylfarnesyl pyrophosphate initiated by loss of OPP at C_a due to attack from C_k followed by bond formation between C_j and C_n would lead to the type (**C$_4$**) skeleton of albolineal (**60**) (*41*). Further cyclization of the latter by connection of C_b with C_f has been envisioned as leading to the ophiobolane (**D$_2$**) skeleton (*41*). Beginning with a 2-(*Z*)-geranylfarnesyl pyrophosphate the sequence of events involves connection of C_k with C_a, C_n with C_j, migration of Me_w to C_j and cyclization by connection of C_f with C_k which explains (*9*) the biogenesis of gascardic acid (**85**) a type (**D$_4$**) compound. Tricyclization of all-*trans*-geranylfarnesyl pyrophosphate *via* bond formation of C_k with C_a, C_o with C_j and C_r with C_n presumably occurs in the generation of the (**D$_5$**) skeleton of stellatic acid (**86**) (*42*), whereas cyclization of 2-(*Z*)-geranylfarnesyl pyrophosphate *via* connection of C_k with C_a, C_n with C_j, migration of C_m to C_o and cyclization by connection of C_r with C_n provides an explanation for the unusual skeletons of type (**D$_6$**). Rearrangement of type (**D$_6$**) by migration of C_q to C_s rationalizes formation of (**D$_7$**) (*43*). In yet another variation of this theme, cyclization of a geranylfarnesyl pyrophosphate precursor *via* connection of C_o with C_a, C_r with C_n, C_b with C_l, C_j with C_o, and C_f with C_i represents a route to the type (**F$_1$**) skeleton of retigeranic acid (**158**) (*44*).

3. Biomimetic Synthesis

Occasionally insights can be derived from biomimetic syntheses. However, very little work has been done along these lines in the sesterterpene area (*89*). In one noteworthy exception, HERZ showed that synthetic (**C$_1$**) type bicyclic derivatives (**v**) and (**vi**) underwent stereospecific cationic cyclization to give tetracyclic type (**E$_2$**) products, (**vii**) and (**viii**) (*45*). In the only other publication of this type, a Diels-Alder reaction between hemigossypolone (**ix**) and myrcene (**x**) yielded heliocide H$_2$ (**91**). This was presumed to mimic the proposed biosynthetic route for categories (**D$_8$**–**D$_{10}$**) (*46*).

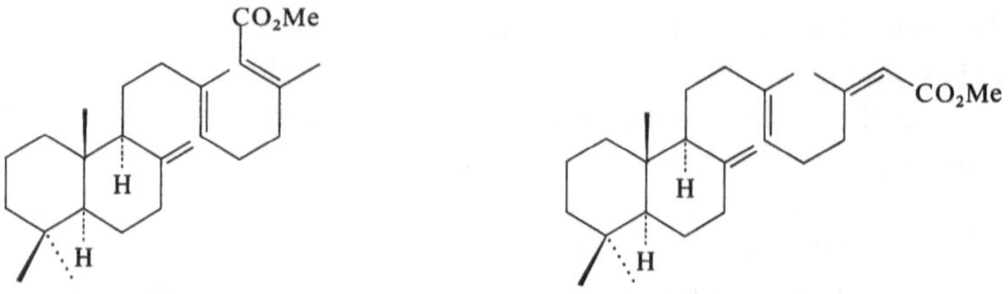

(v)

(vi)

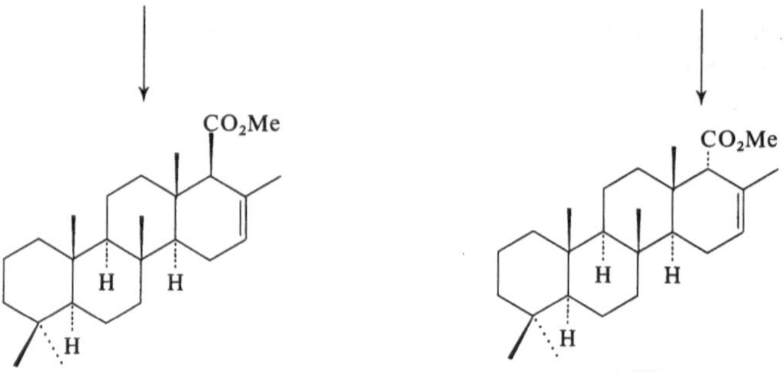

(vii)

(viii)

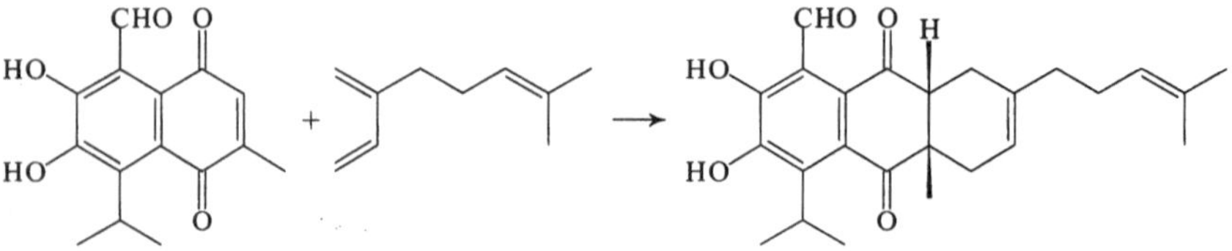

(ix)   +   (x)   →   Heliocide H$_2$ (**91**)

4. Comparative Biochemistry

The relevance of sesterterpenes to taxonomy was recently considered by BERGQUIST and WELLS in their extensive discussion of sponge chemotaxonomy (27). Instances were identified in which distribution of terpene metabolites correlated with taxonomic conclusions based upon more classical criteria. Terpene types such as (**E$_2$+**) with limited species distribution seem to provide the most useful chemical taxonomic markers. Conversely sesterterpenes with wide distribution such as those of type (**A**) are presumably of little or no taxonomic value. While their analysis may have been somewhat premature owing to the relatively small number of sesterterpenes known at the time, it is interesting to reexamine their conclusions in light of the data of Table 3. For example, it was argued that the most distinctive genera based upon unique sesterterpene patterns are *Lendenfeldia*, *Ircinia*, *Dysidea*, and *Carteriospongia*. Interestingly, Table 3 shows that among the Dictyoceratida, *Lendenfeldia*, *Ircinia*, and *Luffariella* are indeed distinctive. Moreover large differences exist in the classes of carbocyclic sesterterpenes found in the sponge orders Dictyoceratida, Halichondrida, Poecilosclerida and Hadromerida.

Upon extending the survey to terrestrial organisms, it is seen that class (**D$_2$**) sesterterpenes which are not found in marine organisms are broadly distributed in terrestrial organisms. While such wide distribution renders them useless as chemotaxonomic markers, the correlation between phyletics and relative stereochemistry of (**D$_2$**) metabolites is of interest. Taking into account only those results which are based on X-ray analyses, insect-derived (**D$_2$**) compounds such as (**70**) (from *Ceroplastes*) which have H-2 (β), H-6 (α), H-10 (β), C-14 (α), C-22 (α) differ in relative stereochemistry from fungi-derived metabolites such as (**75**) (from *Cochliobolus*) for which H-2 (β), H-6 (β), H-10 (α), C-14 (β) and C-22 (β) and (**77**) (from *Cephalosporium*) for which H-2 (β), H-10 (α), C-14 (β), C-22 (β).

IV. Synthesis and Biological Activity

1. Synthesis

Sesterterpenes have been targets for synthesis in order to elucidate unknown elements of stereochemistry or in response to the challenge of devising new routes to an intricate polycycle. The first successful total synthesis of a carbocyclic sesterterpene was that of gascardic acid (**85**) which was undertaken in response to a desire to establish its complete stereochemistry (*47*). The synthesis began with bicyclic intermediate (**xi**), prepared by using known methodology, Claisen rearrangement of which followed by oxidation led stereospecifically to (**xiia**). A subsequent three-step sequence yielded (**xiii**); it began with compound (**xiib**) which was converted to the C$_6$-C$_7$-β-epoxide by iodolactonization followed by base treatment and subsequently rearranged to (**xiii**). Wittig olefination of (**xiii**) followed by Dieckmann cyclization yielded (**xiv**). NaBH$_4$ reduction of this followed by base treatment of the corresponding mesylate yielded a mixture of epimeric methyl gascardates [(**85**) and 18-*epi*-(**85**)] which were separated by HPLC.

(**xi**)   (**xii**)

a) R=CHO
b) R=CO$_2$H

(xii)

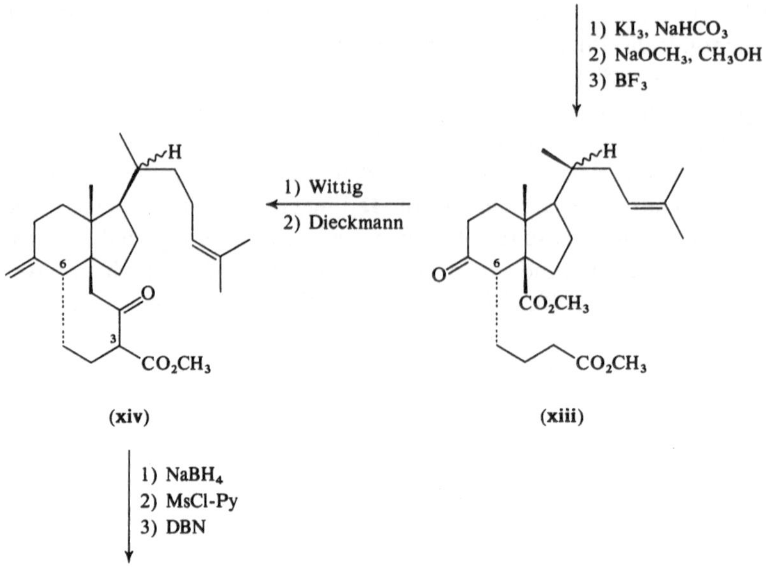
1) KI_3, $NaHCO_3$
2) $NaOCH_3$, CH_3OH
3) BF_3
1) Wittig
2) Dieckmann
H
H
6
6
3
O
O
CO_2CH_3
CO_2CH_3
CO_2CH_3
1) $NaBH_4$
2) MsCl-Py
3) DBN

(xiv) (xiii)

(**85**) + 18-epi – (**85**)

A total synthesis of ceriferol-I (**46**), another sesterterpene having questionable stereochemical assignments, was carried out by KATO and coworkers (*35*). Their work was completed about the same time as a spectroscopic study by NAYA (*49*) which addressed the same uncertainty (see Section V p. 230). The key intermediate was the macrocyclic (**xv a**)

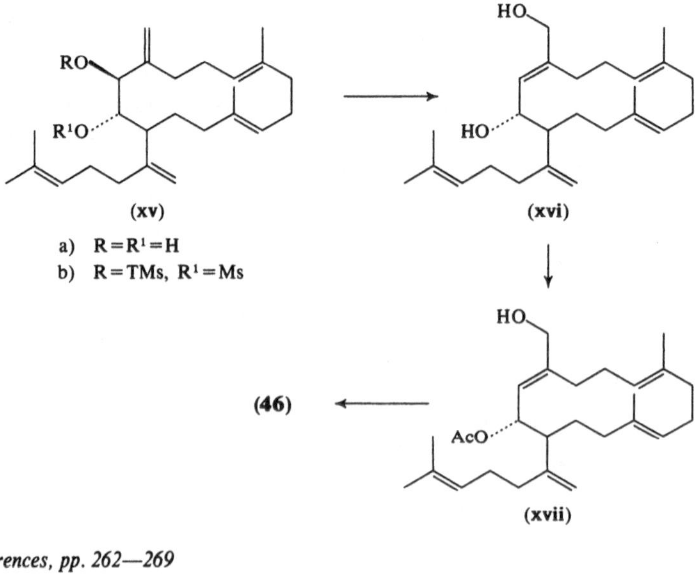
RO
R^1O
HO
HO
HO
AcO
(46)

(xv) (xvi)

a) $R=R^1=H$
b) $R=TMs$, $R^1=Ms$

(xvii)

prepared by cationic cyclization of an acyclic precursor. Base induced rearrangement of the trimethylsilyloxy mesylate (xvb) gave, after hydrolysis, the diol (xvi). This was selectively esterified to yield (xvii); deacetylation of the latter gave ceriferol-I (46). A parallel reaction sequence produced another metabolite, ceriferol (42).

Synthesis was also used by KATO and coworkers (50) in an attempt to resolve a structural ambiguity for cericerol-I (43). That the three ring double bonds of cericerol-I had $Z/E/E$ stereochemistry was based upon the conversion of cericerol-I to a hydrocarbon cericerene for which formula (xviiia) had been proposed earlier (51). However, the arguments used to correlate Z or E stereochemistry with specific double bonds of cericerene were not unequivocal. KATO et al. converted (xixa) to (xviiia) by way of (xixb) and found that the properties of the synthetic material did not match those of cericerene obtained from natural product cericerol-I. As a result, the Z double bond of cericerene must be located at C-2 as in (xviiib), and the previous structure (xxa) (Scheme 2) proposed for cericerol-I (51) is incorrect. The two step synthesis consisted of a cationic cyclization initiated by $SnCl_4$, dehydrochlorination of the resultant reaction mixture by $LiBr/LiCO_3$ to afford (xixb) which was subsequently reduced. The results of KATO and coworkers are equally consistent with any one of three formulas (xxb), (xxc) or (xxd) for cericerol-1 which are summarized in Scheme II, not just (xxc $\equiv$ (43). However (xxb) can be ruled out on other grounds (see below). NAYA and coworkers (49) have also obtained spectroscopic data which are relevant but which do not resolve this ambiguity which is further discussed in Section V. Finally, for reasons similar to those stated above, the same ambiguity exits for structures (36), (41), (44), and (45).

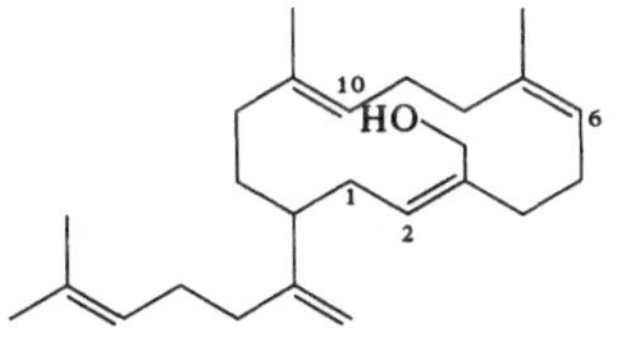

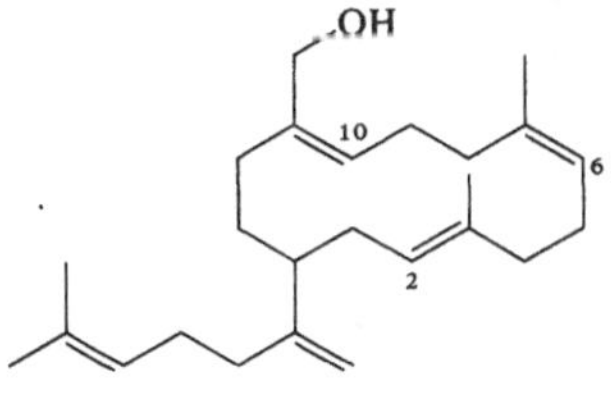

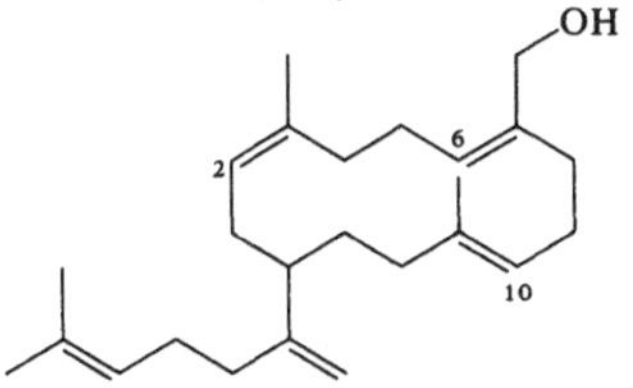

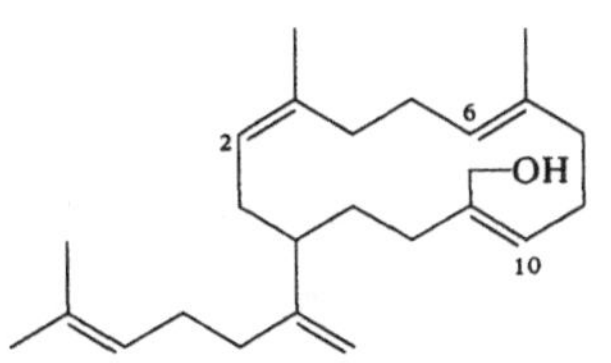

Scheme 2. Possible structures for cricerol-I (43) based upon cericerene candidates (xviiia) or (xviiib)

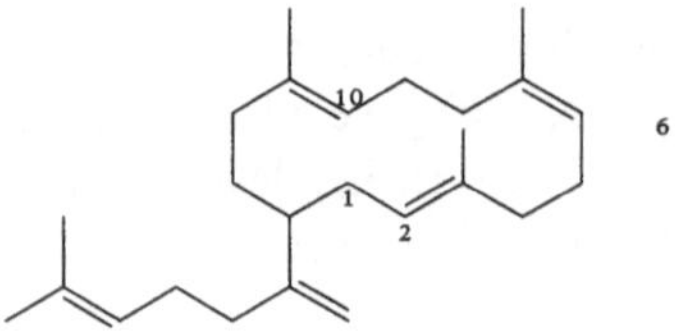

(xviii a) Cericerene

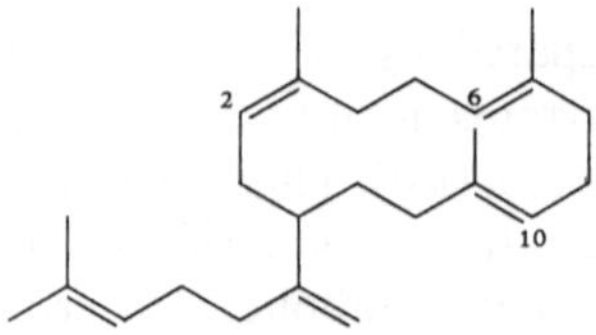

(xviii b)

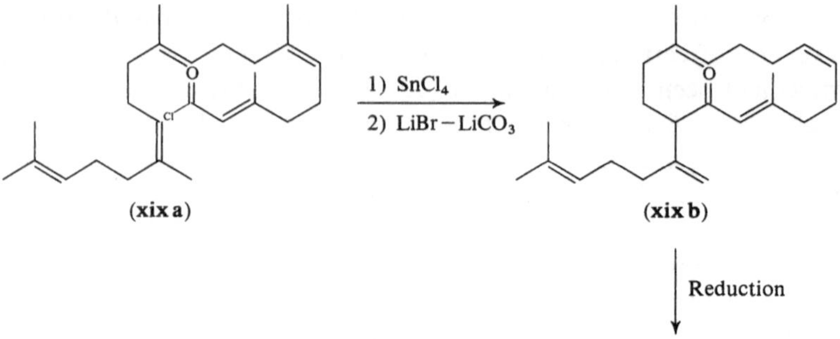

(xix a)

1) SnCl$_4$

2) LiBr−LiCO$_3$

(xix b)

Reduction

(xviii a)

Syntheses of several acyclic sesterterpenes so far not found in nature have been reported (48, 52, 53). The 2Z,6E,10E,14E isomer of geranylfarnesol [(xxi)] has been prepared by a two step five carbon homologation sequence (52). The ^{1}H and ^{13}C chemical shifts are diagnostic of the double bond stereochemistry. This can be seen by comparing the chemical shifts in (xxi) shown below with those of the farnesol isomers shown in Table 5. A set of geranylfarnesols (xxii b) of unspecified configuration was prepared by regiospecific microbial oxidation of (xxii a) (53) for examination as potential antiulcer agents.

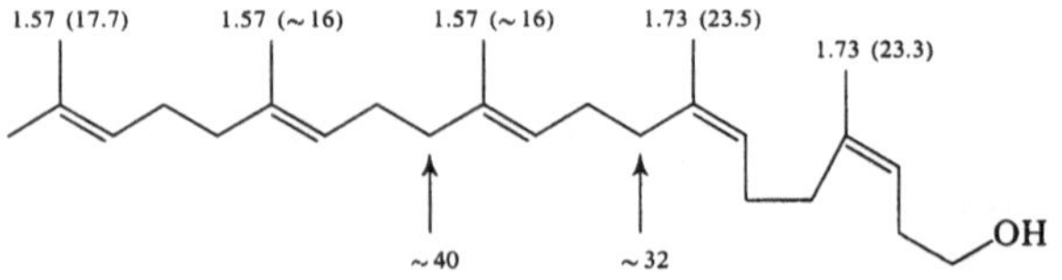

(xxi)

(xxii)

a) R=CH$_3$
b) R=CH$_2$OH, CHO or CO$_2$H

The ring systems of type **(D₁)** have also been a synthetic target. In two separate recent total syntheses of **(64)** the proposed structure and stereochemistry were confirmed. One of these utilized methyl isocopalate **(xxiiia)** as a convenient starting material. Wittig reaction upon the aldehyde derived from **(xxiiia)** yielded **(xxiv)** which was eventually reduced to provide **(64)** and its 13-methyl epimer (*55*). The other employed a similar set of reactions with methyl copolate **(xxiiib)** serving as starting material (*162*). Another example dealing with system D₁ was an approach to the total synthesis of cheilanthatriol **(61)** (*56*). It began from tricyclic dienedione **(xxv)** which was converted to the **(xxvi)** and **(xxvii)** which were considered to be key intermediates for future efforts aimed at derivatives of **(61)**.

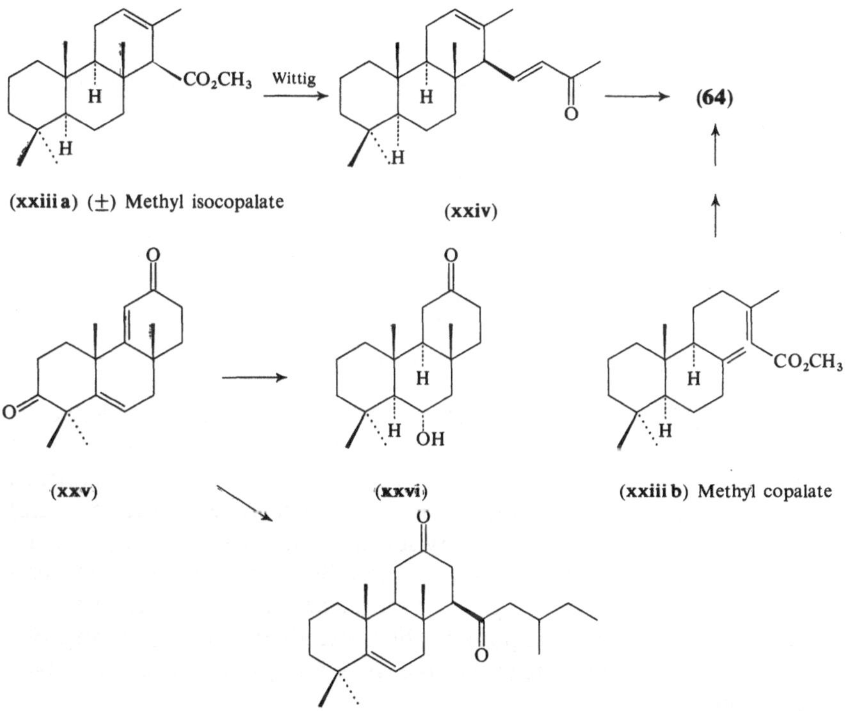

(xxiiia) (±) Methyl isocopalate   **(xxiv)**

(xxv)   **(xxvi)**   **(xxiiib)** Methyl copalate

(xxvii)

Synthesis of the tricyclic ophiobolane skeleton has been a goal of several research groups (*57–62*). As has been mentioned previously, two stereochemical patterns are found in this series including rings AB *cis* and rings BC *trans* in metabolites such as **(72)** from *Cochliobolus* vs. rings AB *trans* and rings BC *trans* in metabolites such as **(70)** from *Ceroplastes*. BOECKMAN and coworkers (*57*) have prepared the ring system present in each of these

(xxviii)　　　　　　　　(xxixa)　　　　　　　(xxxa) Cochliobolis type

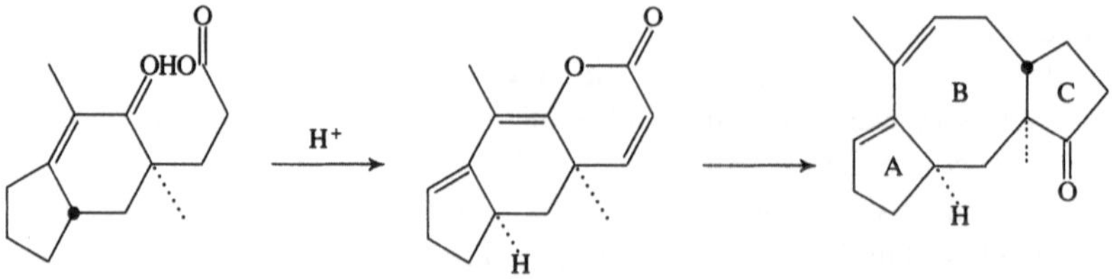

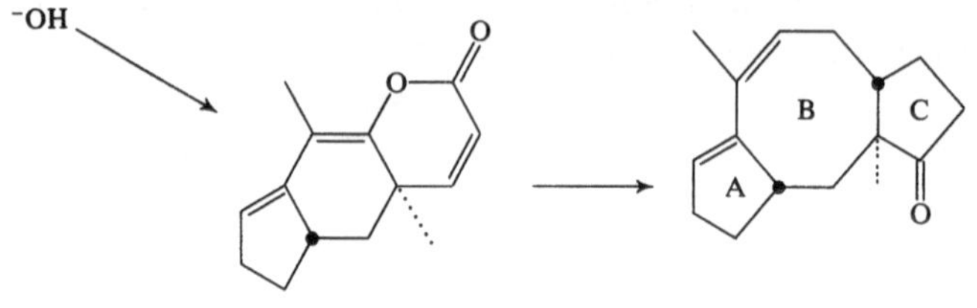

(xxixb)　　　　　　　(xxxb) Ceroplastes type

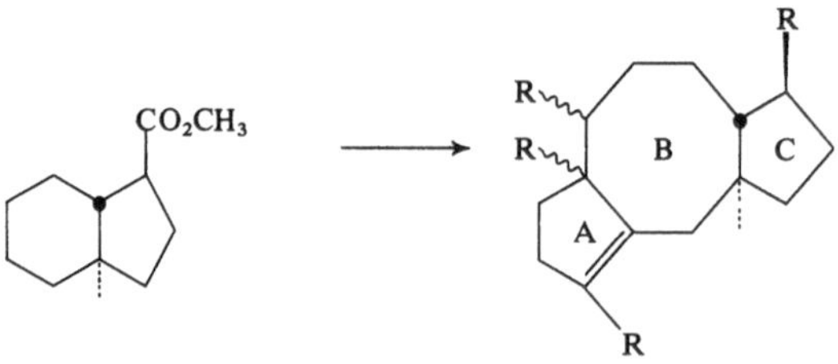

(xxxi)　　　　　　　　　　　(xxxii)

classes from the common intermediate (xxviii). Lactonization under acidic conditions produced (xxixa) as a major product while lactonization of (xxviii) under basic conditions produced (xxixb). These were transformed in several steps to (xxxa) with a *Cochliobolis* type skeleton and to (xxxb) with a *Ceroplastes* type skeleton. By starting from bicyclic (xxxi) DAUBEN and coworkers (58) prepared (xxxii) whose BC ring junction stereochemistry matched that of the *Ceroplastes* type metabolites. Two other syntheses have resulted in bicyclic compounds which resemble the *trans*-A/B rings of a *Ceroplastes* type metabolite (59) or the *cis*-A/B rings of a *Cochliobolis* type metabolite (60).

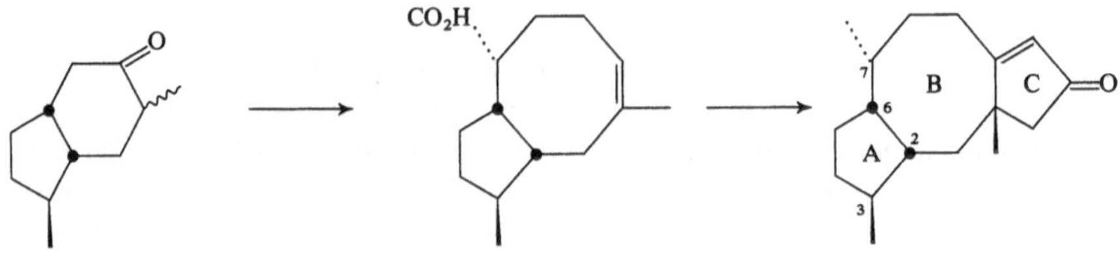

(xxxiii)　　　　　　　　　(xxxiv)　　　　　　　　　　(xxxv)

DUTTA and coworkers (61, 62) have converted the cis-fused bicyclo-nonane (xxxiii) to a key intermediate (xxxiv) whose stereochemistry was confirmed by X-ray analysis. This compound was subsequently elaborated to compound (xxxv) whose stereochemistry at C-2, C-3, C-6 and C-11 matches that of ophiobolin F (72).

Two different research groups have explored approaches to the pentacyclic (F₁) skeleton. In work aimed eventually at the total synthesis of retigeranic acid (158), HUDLICKY and coworkers (63) converted (+)-pulegone to (xxxvii) in which the C/D/E rings of (158) and their substituents are properly assembled. DAUBEN and coworkers (64) have reported an analogous accomplishment by converting the relatively simple hydrindene skeleton (xxxviii) by way of a key intermediate (xxxvix) to the tricyclic derivative (xl).

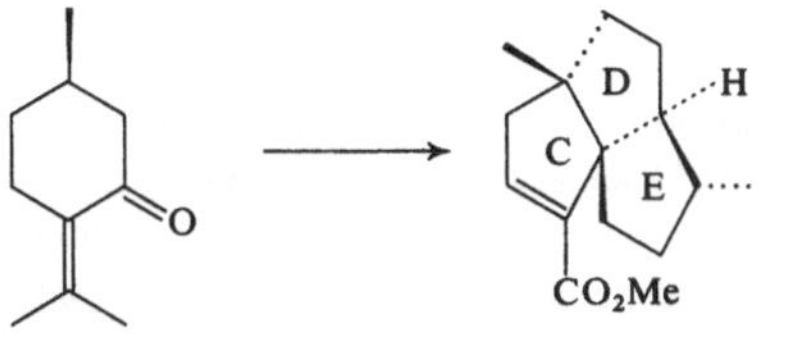

(+) pulegone
(xxxvi)   (xxxvii)

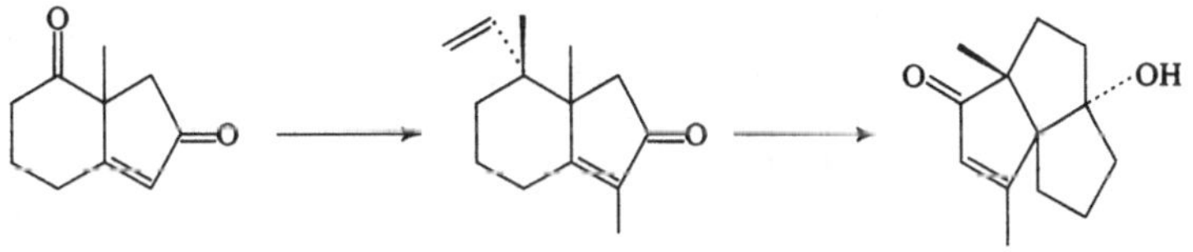

(xxxviii)   (xxxix)   (xl)

2. Biological Activity

Interesting physiological properties have sometimes been described for organisms which yield sesterterpenes. These range from marine sponges which are toxic (65 – 69), nudibranchs with a pleasant fruity odor (70), and fungi which produce extracts having antimicrobial and phytotoxic activity (71 – 74), to cotton varieties which are resistant to worms (75) or insects producing secretions which defend against weather and enemies (76 – 78). Many recent studies, especially those involving marine organisms, have attempted to link a specific biological activity to an individual sesterterpene metabolite.

Table 4 summarizes the biological activities known for sesterterpenes. Antimicrobial activity is the most commonly recorded property but antiinflammatory activity and inhibition of cell division is also of interest. To the best of our knowledge only two sesterterpenes ophiobolin (**75**) (as an antibiotic) and monoalide (**31**) (as an antiinflammatory agent) have been the subjects of patents.

Table 4. *Summary of Sesterterpenes with Biological Activity*

Structure type	Compound	Activity	Ref.
(A)	(21), (20)	antibacterial vs. *Diplococcus* (MIC = 1 mcg/ml) and *S. aureus* (MIC = 1 mcg/ml)	(20)
(A)	(11)	antibacterial vs. *S. aureus*	(86)
(A)	(11), (14)	antibacterial vs. *S. aureus* and *Bacillus subtilis* (MIC = 3 – 6 ppm)	(87)
(A)	(6)	toxic to numerous preditory marine organisms including a sea star (5 mg/l) an abalone larvae (1 mg/l) and a bryozoan, and brine shrimp (10 mcg/l)	(67)
(A)	(1)	toxic to brine shrimp (10 mcg/l)	(58)
(A)	(31)	inhibition of phorbol-induced inflammation	(163)
(B)	(28-31)	antibacterial vs. *Bacillus subtilis* and *S. aureus*	(112)
(B)	(31)	antibacterial vs. *Streptomyces pyrogenes*	(113)
(B)	(32)	inhibition of sea urchin cell division (16 mcg/ml)	(68)
(B)	(32)	antibacterial vs. *Streptococcus* (MIC = 2.5 mcg/l), *S. aureus* (MIC = 12 mcg/ml), *Corynebacterium diphteriae*	(91)
(C)	(52)	growth inhibition of *S. aureus*	(119)
(C)	(55)	antibacterial vs. *Streptococcus* (MIC = 1.0 mcg/ml)	(91)
(C)	(53), (55)	from an extract fraction toxic to the fish *Lebistes reticulatus* (LD_{50} = 5 mg/l)	(80)
(D_2)	(75)	growth inhibition of *Trichophyton interdigitale* (1.5 – 2.5 mcg/ml)	(71, 74)
(D_2)	(77)	weak growth inhibition of *S. aureus*	(72)
(D_3)	(81-84)	kairomones between the parasitic wasp *Anicetus beneficus* and the scale insect *Ceroplastes rubens*	(136)
(D_8)	(91)	growth inhibition of *Heliothis virescens* (50% GI at 3 mmoles/kg)	(46)
(D_{10})	(93)	growth inhibition of *Heliothis virescens* (50% GI at 3 mmoles/kg)	(46)
(E_1)	(95)	kairomones between the parasitic wasp *Anicetus beneficus* and the scale insect *Ceroplastes rubens*	(136)
(E_2)	(98)	*in vitro* citotoxicity vs. L-1210 cells (ED/50 = 0.6 mcg/ml)	(139)

Table 4 *(continued)*

Structure type	Compound	Activity	Ref.
(E₂)	(100)	toxic to numerous preditory marine organisms including a sea star (5 mg/l) an abalone larvae (1 mg/l) a hydroid, and brine shrimp (10 mcg/l)	*(67)*
(E₂)	(103)	toxic to brine shrimp (10 mcg/l)	*(67)*
(E₂)	(106)	toxic to brine shrimp (10 mcg/l), abalone larvae (1 mg/l), and gametes of the kelp *Macrocystis pyrifera*	*(67)*
(E₂)	(113)	anti-inflammatory activity (50 mcg/ml)	*(69)*
(E₂)	(118)	anti-inflammatory and antifungal	*(147)*
(E₂)	(120), (121), (131)	inhibition of platelet aggregation	*(149)*
(E₂)	(126), (132)	growth inhibition of *Vibrio anguillarum* (at 100 mcg/disk)	*(37)*
(E₂)	(136)	anti-inflammatory (18% at 10 mcg/disk)	*(148)*

V. Spectroscopic Analysis

1. Introduction

Establishing the structure of a sesterterpene can be involved because of the relative complexity of the carbon skeleton. Only 16 sesterterpene structures have been established by X-ray analyses even though more than fifty compounds have been reported as crystalline. Most structure elucidations of new sesterterpenes have relied on NMR data and/or results from chemical degradations or interconversions. We will focus on some of the prominent spectroscopic properties used in their structural analysis because of their applicability to future characterizations of sesterterpenes.

No revisions have been needed in the gross structure attributed to a sesterterpene as initially published. However, in the absence of an X-ray analysis difficulties were sometimes encountered in assignment of the correct relative stereochemistry at chiral sites or across double bonds. Examples of such dfficulties will be considered below.

The low field (e.g. < 100 Mz) ^{1}H-NMR spectra of the sesterterpenes listed in Table 6 are generally complex. Not unexpectedly, sesterterpenes give ^{13}C-NMR resonances which can be completely assigned, especially when NMR spin echo procedures are employed $(79, 80)$. Reliable ^{13}C-NMR chemical shift data can provide a powerful and rapid means for characterization of new members of a series and for pinpointing differences in stereochemistry or the existance of isomers. In order to broaden the information base ^{13}C-NMR data have been included with each structure in Table 6, when available. In cases where chemical shifts were unassigned we have attempted to complete this task. Also, while comparing shifts of closely related metabolites it has been possible to eliminate inconsistencies by making reassignments. Both of these situations are noted by superscript a in Table 6. We have also attempted to correct minor errors which have appeared in the actual drawings of sesterterpene structures. Finally, some structures have been labeled as "tentative" when insufficient characterization evidence was presented.

2. Acyclic Sesterterpenes

The R(H)C = C(Me)R array is a ubiquitous structural feature of an acyclic sesterterpene. The assignment of the double bond geometry can be greatly simplified by utilizying NMR data, but availability of data for model compounds is important. The proton NMR chemical shifts (81) of the methyl groups in the four isomeric farnesols as well as the corresponding ^{13}C shifts (82) are especially diagnostic and are summarized in Table 5. The observed ^{1}H-NMR methyl shifts of (3) are (ppm): 1.70, 1.65, 1.58, 1.58, 1.58, 1.58. Comparison of these values with those in Table 5 clearly indicates that the E stereochemistry first assigned to the C-2, C-3 double bond of (3) (9, 83) was inappropriate. Fortunately this was corrected to Z in a subsequent publication (84). Such information has been useful in assigning the double bond stereochemistry in other acyclic sesterterpenes. Thus the ^{1}H-Me shifts of (5) (δ 1.66, 1.58, 1.58, 1.58, 1.58) were consistent with the double bond stereochemistry as drawn (85). Some investigators have been cautions in applying ^{1}H-NMR shifts for assignment of the trisubstituted double bond stereochemistry. For example no C-7 and C-11 stereochemistry was initially assigned to variabilin (11) (86, 87); however chemical shifts of the two vinyl methyls were originally given as δ 1.67 and 1.62 (86). In a very recent study (28) chemical shifts of δ 1.55 and 1.58 were observed for (11) and δ 1.65 and 1.68 for isomer (12). These values, together with ^{13}C-methyl shifts of δ 15.6 and 15.8 for (11) and δ 23.3 and 23.3 for (12), permitted assignment of the C-7 and C-11 double bond stereochemistry as shown in Table 6. In view of the above, it seems reasonable to complete previously unassigned trisubstituted double bond stereochemistry for (14) (Me's = δ 1.55, 1.59). Unfortunately, neither ^{1}H- or ^{13}C-NMR shifts can be utilized in the stereochemical analysis of the trisubstituted double bond conjugated with the tetronic and group which is present in eleven acyclic sesterterpenes, so this feature remains unspecified in Table 6.

Table 5. Methyl ^{1}H-NMR and ^{13}C-NMR Chemical Shifts for Tri-Substituted Double Bond Isomers

(xli a)

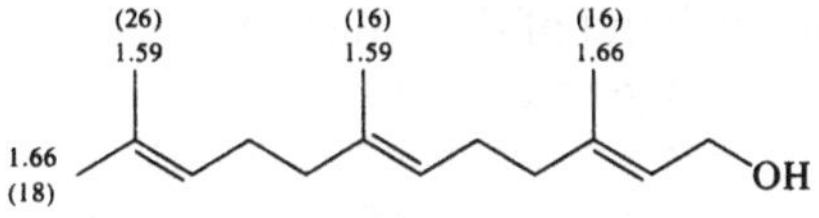

(xli b)

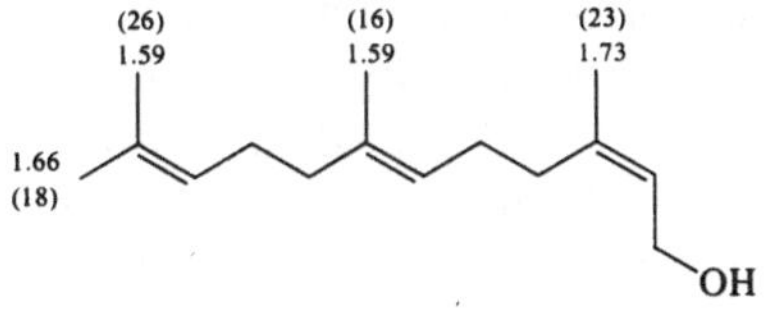

(xli c)

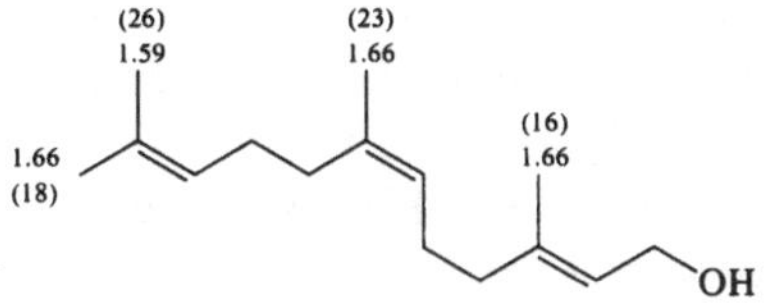

(xli d)

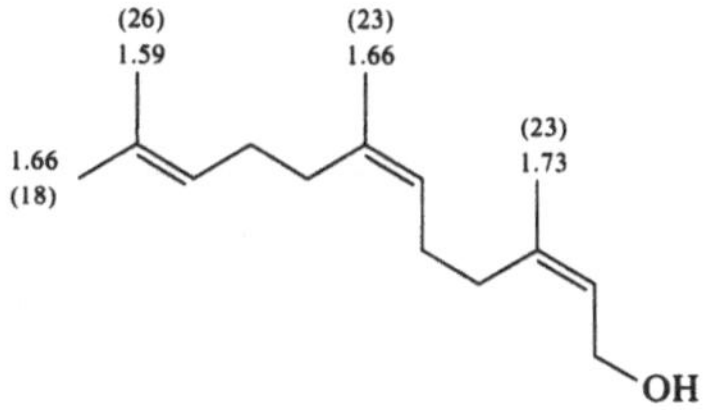

3. Monocarbocyclic Sesterterpenes

The R(H)C=C(Me)R array is also a common structural feature of monocyclic sesterterpenes since at least one such group can be found in every compound known in this class. The ^{13}C-NMR methyl shifts listed in Table 5 are also of value for evaluating the stereochemistry of both the trisubstituted cyclic and acyclic double bonds present in this family. Manoalide (31) is illustrative in that the chemical shift of Me-23 (δ 16.3) is in agreement with E stereochemistry of the C-10,11 double bond. Likewise, the methyl shifts of δ 9.8, 15.4, and 22.5 for the vinyl methyls attached to the rings of 13-methoxycericerene (39) immediately reveal that the three trisubstituted ring double bonds are of the E,E,Z type, but locating each of these bonds within the ring system can be difficult, as illustrated by the case of a ceriferic acid (38) and cericerol-I (43).

The macrocyclic sesterterpene ceriferic acid (38) was first isolated in 1979 (77); analysis of its ^{13}C-NMR methyl shifts indicated that each of the three R(H)C=C(Me)R groups had E geometry. Argument for E geometry at the conjugated double bond was based on reduction of the carboxyl to a methyl group which had a ^{13}C shift of δ 22.4 (77). In spite of this evidence it was subsequently concluded on the basis of the CD spectrum of (38) that the conjugated double bond of (38) must have Z geometry. However, in a later publication this suggestion was withdrawn and the double bond stereochemistry was revised back to E (49). This was based in part on ^{1}H-NMR difference decoupling of ceriferol-I (46) and ceriferic acid-I (47) in the presence of a shift reagent and analysis of their SFORD ^{13}C-NMR spectra. In view of structure (46) for ceriferol-I the possibility (xxb) in Scheme 2 for cericerol-1 can be ruled out. Also consistent with the structure proposed for ceriferol-I was its reduction to cericerene in the ^{13}C-NMR spectrum of which a signal at δ 22.5 could be unambiguously assigned to the methyl attached at C-3. This additional verification of the Z stereochemistry across the C-2, C-3 double bond in cericerene indicated that the E stereochemistry initially assigned to the C-2, C-3 double bond in several compounds, (35) (36), (39), (44), (45), each of which were reduced to cericerene, was incorrect (51, 78, 90, 91).

(46)   (xviii b) (Cericerene)

Another difficulty in structure elucidation of monocarbocyclic sesterterpenes may arise in connection with the stereochemistry of substituents. In illustration, an error was initially made in the stereochemistry of the peroxide rings of muqubilin (32) and sigmosceptrellin-B (55) (91). Use of ^{13}C-NMR shifts in the case of (32) (68) and X-ray crystallography in the case of (55) (92) subsequently resulted in revision of the C-6 stereochemistry.

4. Tri- and Tetracarbocyclic Sesterterpenes

Determination of ring junction stereochemistry is one of the major problems in structure analysis of a polycyclic sesterterpene. Utilization of NMR data can be a powerful short cut, especially when ring methyl substituents are present and their ^{13}C-NMR chemical shifts are available. Crews and Kho-Wiseman mapped out the chemical shifts and substituent increments which allow prediction of methyl ^{13}C-NMR shifts for axial or equatorial methyls flanked by various substituents (93). In excellent agreement with such predictions are the chemical shift trends observed for sesterterpene methyls as a function of changes in ring junction stereochemistry.

For example, it is useful to compare ^{13}C-NMR methyl shifts for bicyclic systems differing in stereochemistry of the ring junction. A series of compounds with similar features whose stereochemistry is known from X-ray analysis includes (54) with a *trans* A/B ring junction, the quinone sesquiterpene ilimaquinone (xlii) with a *trans* A/B ring junction, and the quinone sesquiterpene arenarol (xliii) with *cis* A/B stereochemistry. The ^{13}C-NMR chemical shifts of the various methyl groups are included (*94, 95*), the assignments having been made by reference to clerodanes which were the subject of a previous ^{13}C-NMR study (*96*). Using these shifts as reference, comparison of methyl and ring carbon shifts of (54) and (52), (56), (57) shows that they all possess analogous stereochemistry.

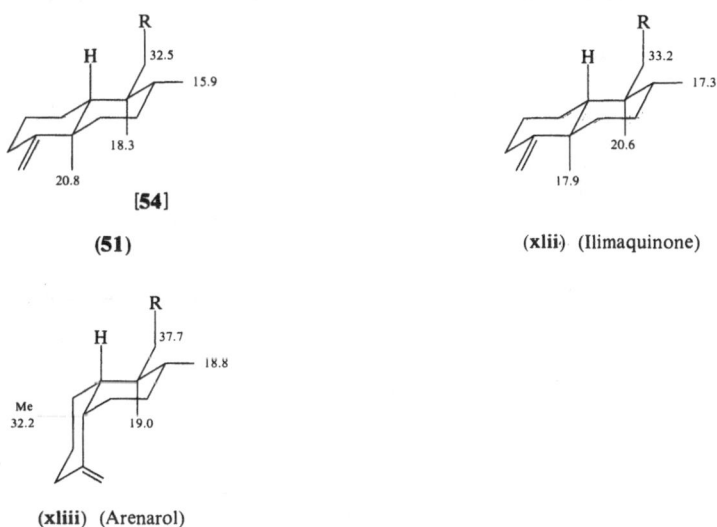

(51) [54]

(xlii) (Ilimaquinone)

(xliii) (Arenarol)

A parallel trend can be observed for (**D₁**) type tricyclic sesterterpenes. The differences in the C-22 and C-23 shifts of (61) (δ 17.7 and 18.1) vs. those of (62) (δ 26.8 and 17.7) reflect the change from *trans* to *cis* A/B stereochemistry (*38*). The same variation of methyl and ring ^{13}C-NMR shifts with stereochemistry exists in type (**E**) tetracyclic sesterterpenes. Tracking the chemical shift variation for Me-23 in scalaranes of type (**E₂**) is particularly revealing as shown by the following compounds in Scheme 3: in (113), C-23 at δ 14.7 is influenced by an equatorial substituent at C-12 and none at C-18; in (101), C-23 at δ 16.9 is influenced by an equatorial substituent at C-12 and an axial substituent at C-18; in (99), C-23 at δ 15.3 and in (106), C-23 at δ 8.8 whose shifts are influenced by an axial substituent at C-12 and an equatorial one at C-18. Other data in Scheme 3 illustrate the sensitivity of the methine shifts at C-9 and C-14 to the stereochemistry of substituents at C-12 and C-18. A change in the stereochemistry of C-12 acetate from axial to equatorial shields C-9 by 6 ppm and C-14 by 3 ppm. A change in the stereochemistry of the substituent at C-18 from equatorial (or none) to axial causes an upfield shift of 4 ppm at C-14.

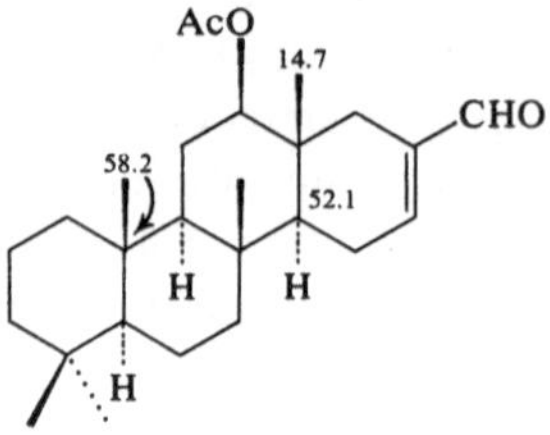

(113) Hyrtial

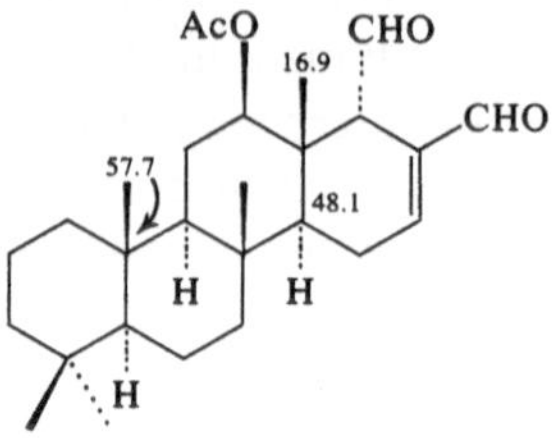

(101) 12,18-Diepiscalaradial

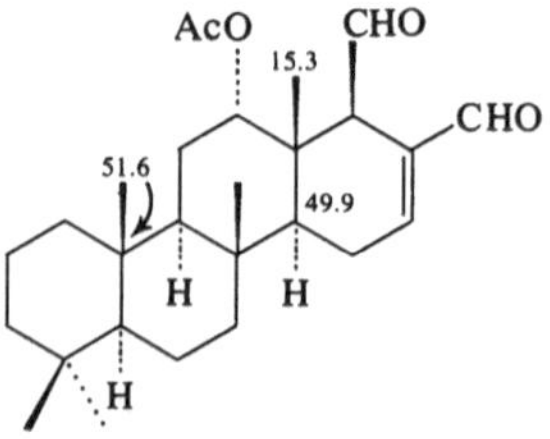

(99) Scalaradial

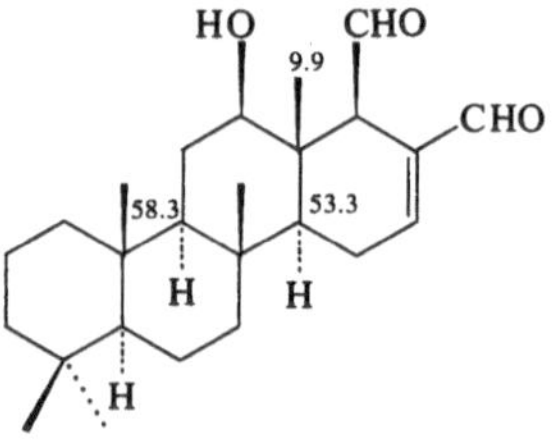

(96) 12-Deacetyl-12-episcalaradial

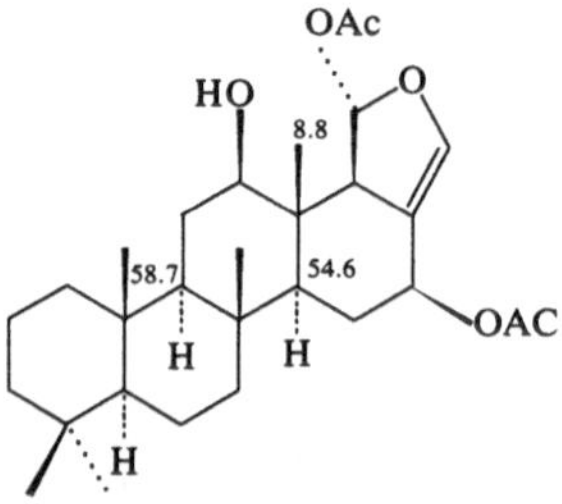

(106) Heteronemin

Scheme 3. ^{13}C-NMR Chemical shift variations vs. stereochemistry in scalaranes

5. Pentacarbocyclic Sesterterpenes

Only one pentacarbocyclic sesterterpene, retigeranic acid [158], appears in this review. Its structure was established by X-ray analysis and little information is available on its spectroscopic properties. For example only two ^{1}H-NMR resonance were listed (44).

6. Absolute Stereochemistry

A structure elucidation cannot be considered complete until all features of relative and absolute stereochemistry have been assigned. Relative stereochemistry has been reported for almost all the sesterterpenes reviewed here; by contrast, absolute stereochemistry has been specified for just a few. Discussions have appeared concerning the relative merits of various methods for determining absolute configurations in organic molecules (15, 97, 98). KLINE and BUCKINGHAM (15) have emphasized that the only completely reliable results are based on use of the BIJVOET method which involves a statistical comparison of diffraction patterns expected for Friedel pairs (97), while chiroptical methods, the Fredga method of quasi racemates, and other empirical methods such as asymmetric synthesis were stated to be less so. We have previously pointed out (36) that occasionally errors have occurred in determining absolute stereochemistry by comparison of Friedel paris even when heavy atoms are present in the molecule under study*.

Absolute configurations for (70), (75), (77), (86), (147), (158) have been deduced by X-ray crystallographic comparison of statistical data for Friedel pairs. Absolute stereochemistry has been established for a few sesterterpenes by a method stated to be "non empirical" involving circular dichroic exciton coupling (101). NAYA and coworkers (43) established the relative stereochemistry of a ketone derived from floridenol (89) by X-ray crystallography and used the exciton chirality CD method to deduce the absolute stereochemistry of the 5-bromobenzoate of 5α-hydroxyfloridenol (90). Similarly, a combination of X-ray crystallography and application of the ORD octant rule provided absolute configurations for (50) and (51) (65, 66).

VI. Physical and Spectroscopic Tables

This section contains physical constants which are divided into two Tables. The carbon skeletons of the different families in Table 6 have been numbered differently by various workers and in some cases a numbering convention has been proposed in the literature. We have used all such existing chemes; for the others we tried to adopt the most often utilized numering scheme. A numbered structure appears at the beginning of each new family in Table 6. The conventions and symbols used in these tables were outlined in Section II.1.

---

* The difficulty of establishing absolute configuration of a complex natural product by X-ray or chiroptical methods is further illustrated by revisions which have been published recently (99, 100).

Table 6. *Summary of Structures and Carbon-13 NMR Chemical Shifts*

A.	Acyclic Sesterterpenes

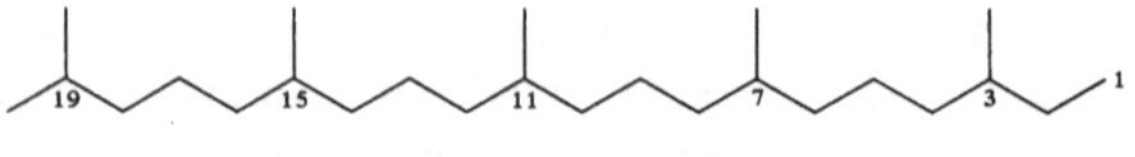

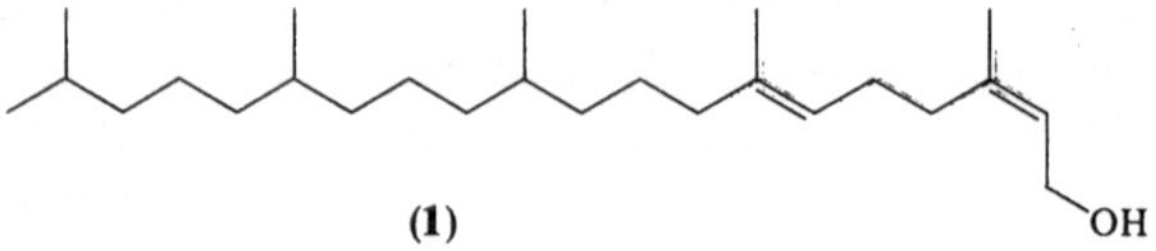

(1)

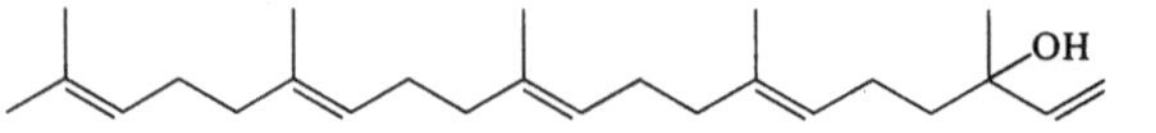

(2) Geranylnerolidol

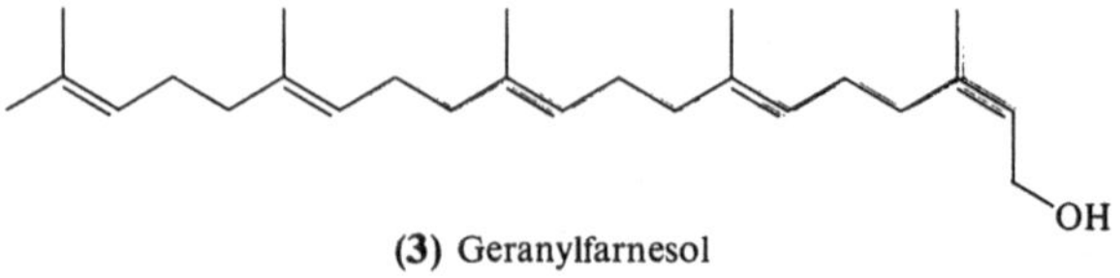

(3) Geranylfarnesol

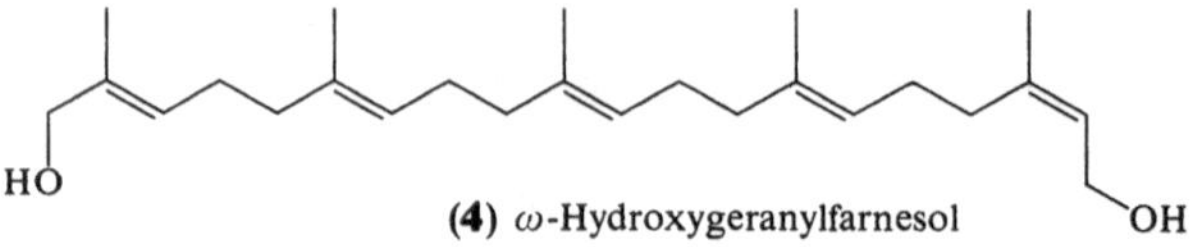

(4) ω-Hydroxygeranylfarnesol

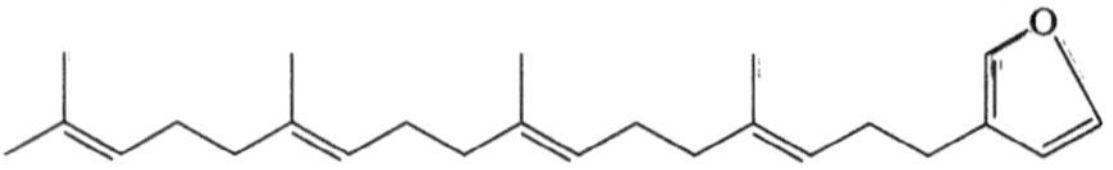

(5) Furospinosulin-1

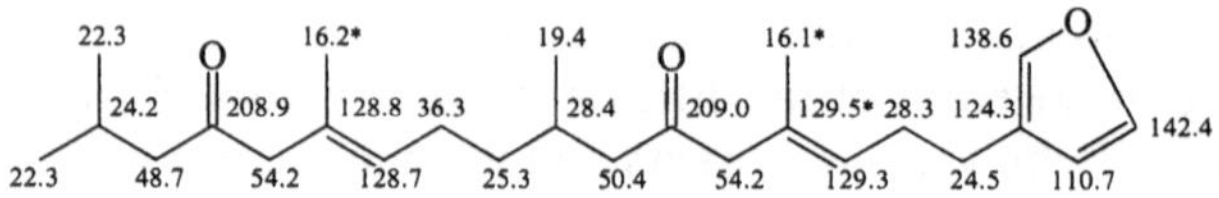

(6)[a] Idiadione

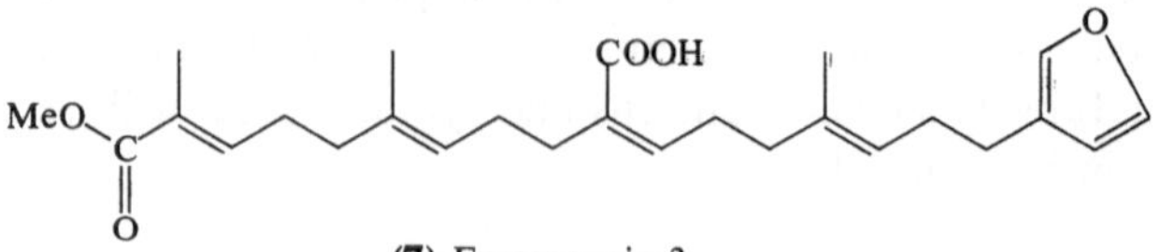

(7) Furospongin-3

[a] Our ^{13}C-NMR assignments

Table 6 *(continued)*

A.   Acyclic Sesterterpenes (continued)

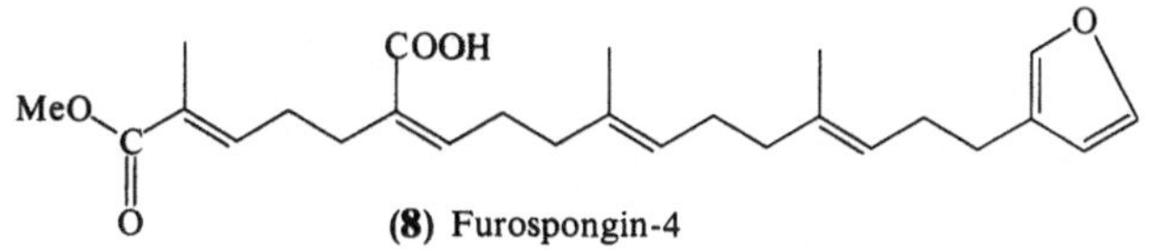

(8) Furospongin-4

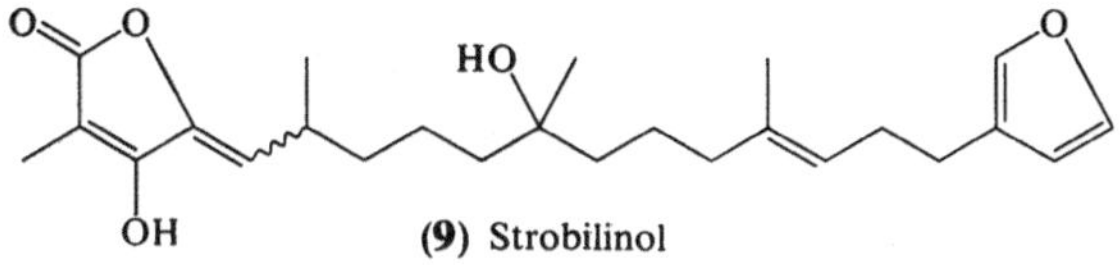

(9) Strobilinol

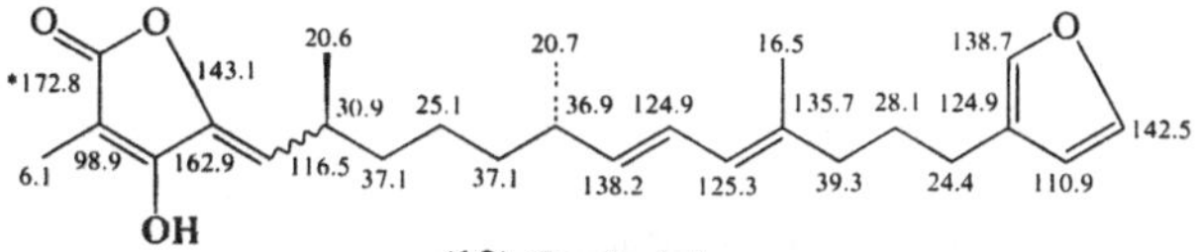

(10) Fasciculatin

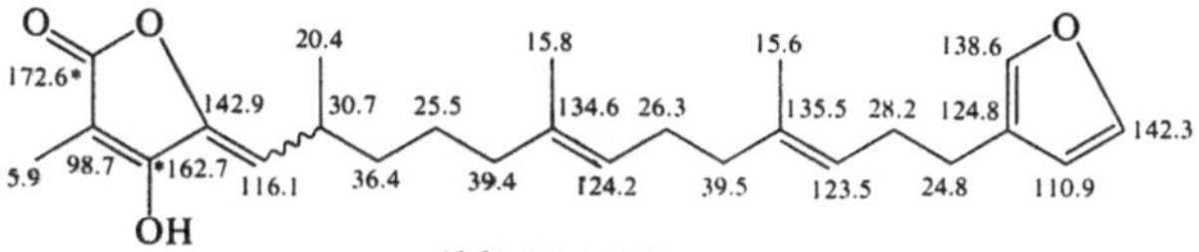

(11) Variabilin

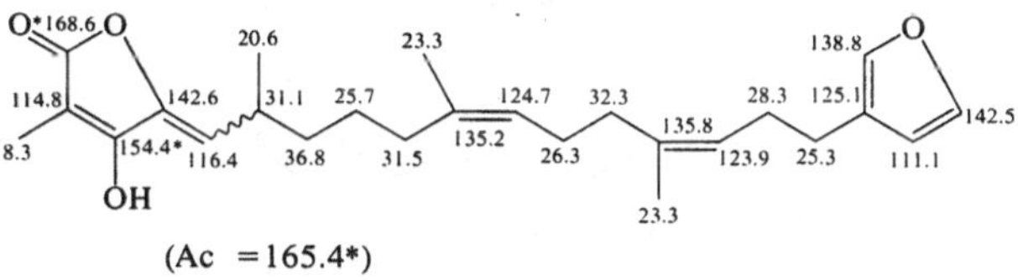

(Ac = 165.4*)
(CH$_3$ = 20.6)

(12) ^{13}C of Acetate

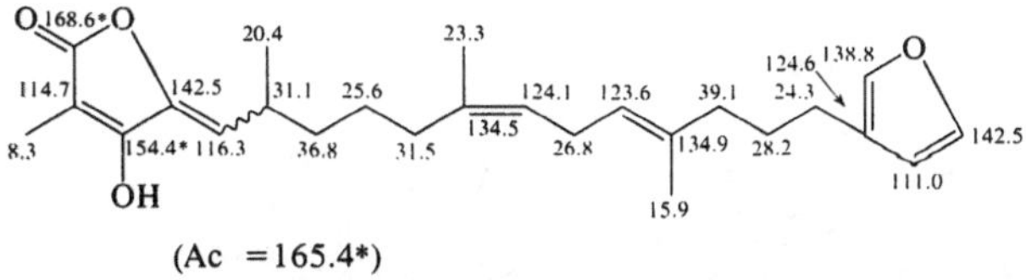

(Ac = 165.4*)
(CH$_3$ = 20.6)

(13) ^{13}C of Acetate

Table 6 *(continued)*

A.   Acyclic Sesterterpenes (continued)

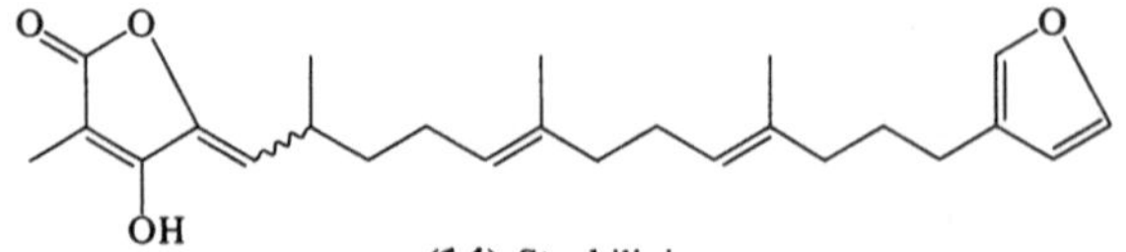

(14) Strobilinin

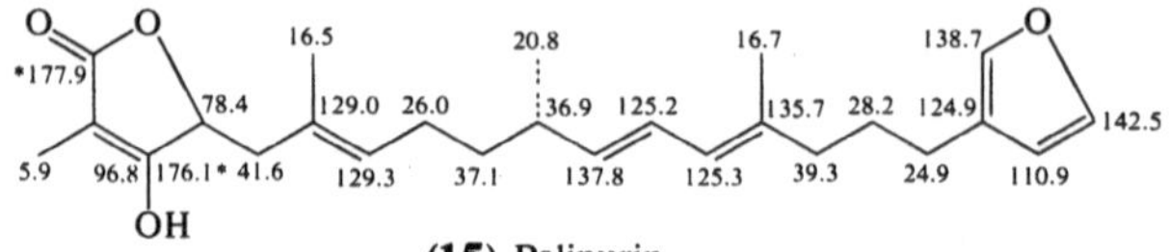

(15) Palinurin

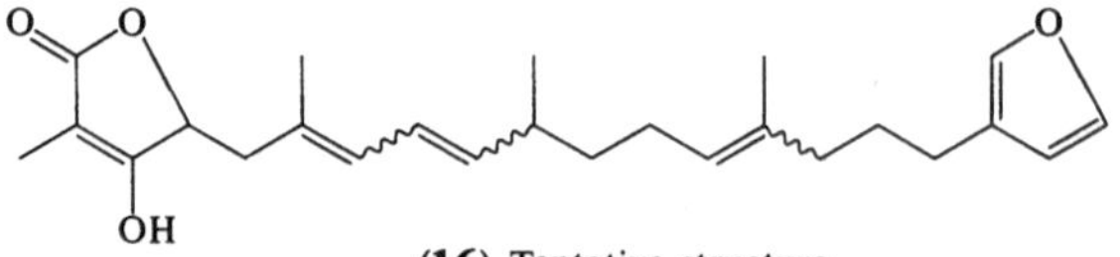

(16) Tentative structure

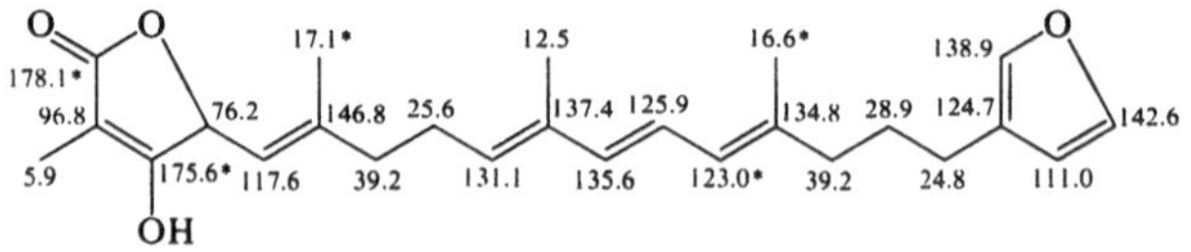

the triene regiochemistry may also be double bonds across
C-6, C-7; C-8, C-9; and C-10, C-11.

(17) Tentative structure

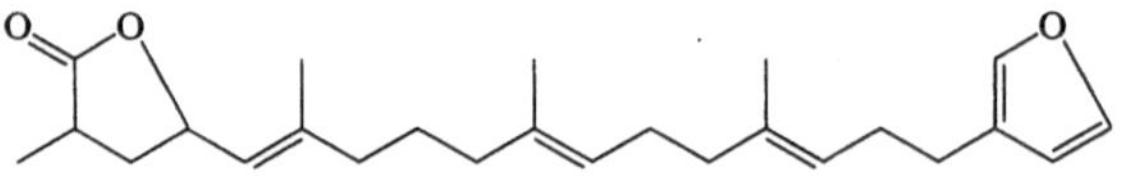

(18) Tentative structure

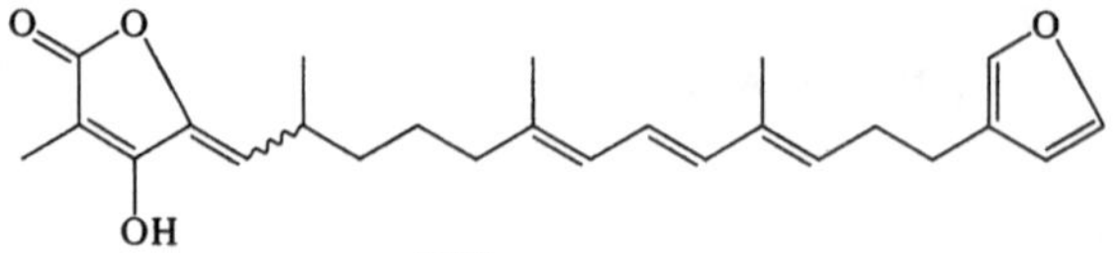

(19)

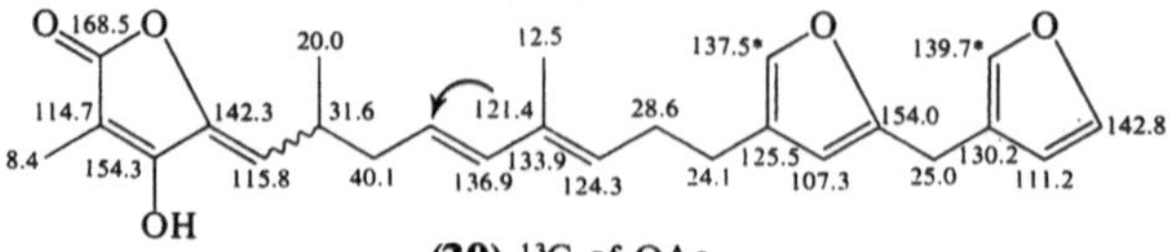

(20) ^{13}C of OAc

Table 6 *(continued)*

A. Acyclic Sesterterpenes (continued)

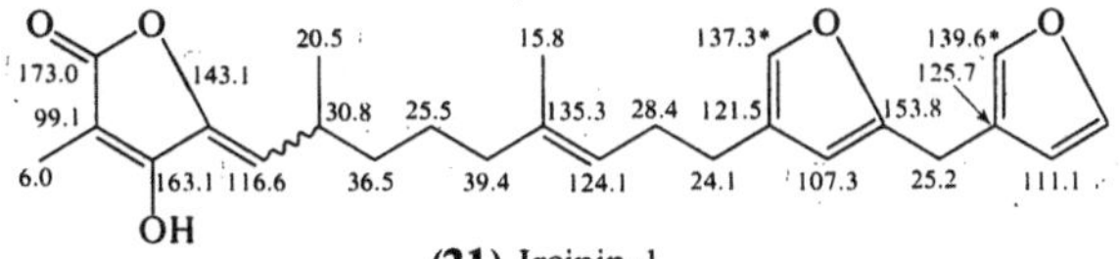

(21) Ircinin-1

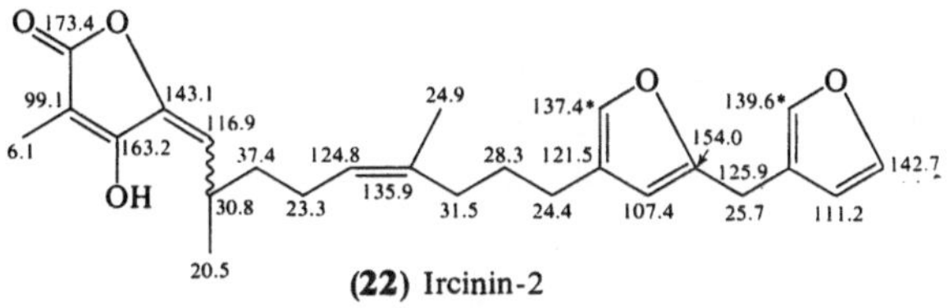

(22) Ircinin-2

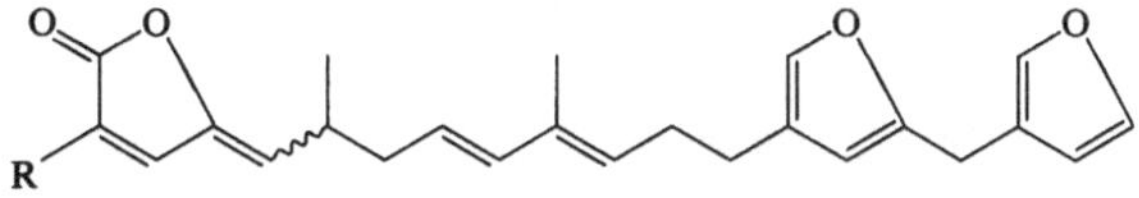

(23) R=CH₃ Ircinolide
(24) R=CH₂OH 24-Hydroxyircinolide

A−. Acyclic Degraded Sesterterpenes

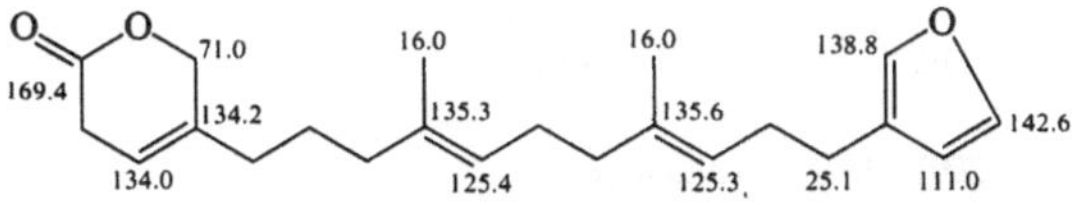

(25)ᵃ Furodendin (Incomplete ¹³C assignment)

A#. Derivatives of Acyclic Sesterterpenes with Mixed Biogenesis

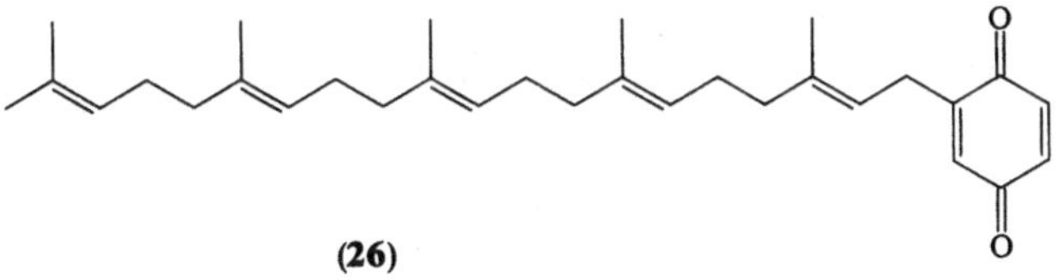

(26)

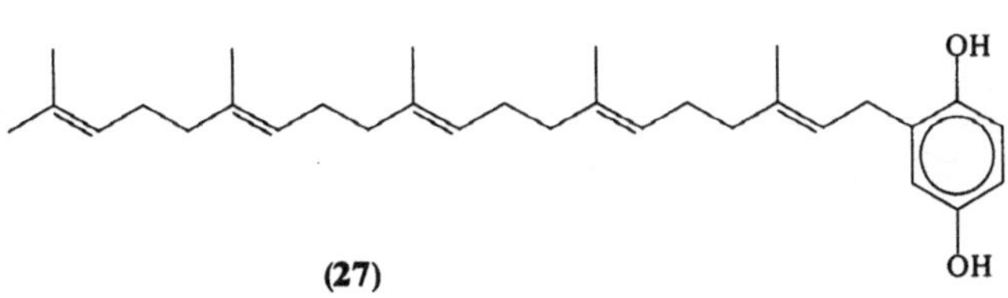

(27)

ᵃ Our ¹³C-NMR assignments

Table 6 *(continued)*

B₁.   Monocyclic Sesterterpenes

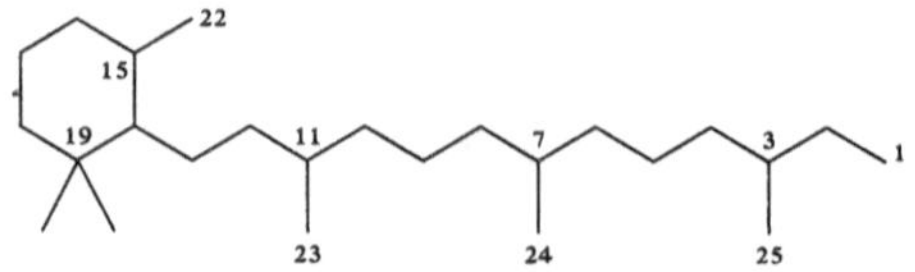

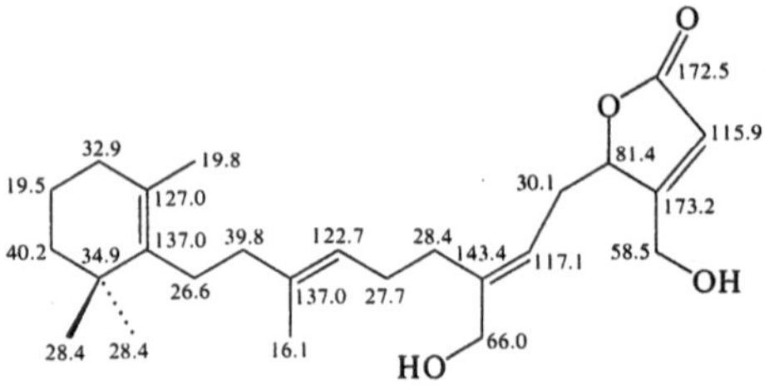

(28) (E)-Neomanoalide

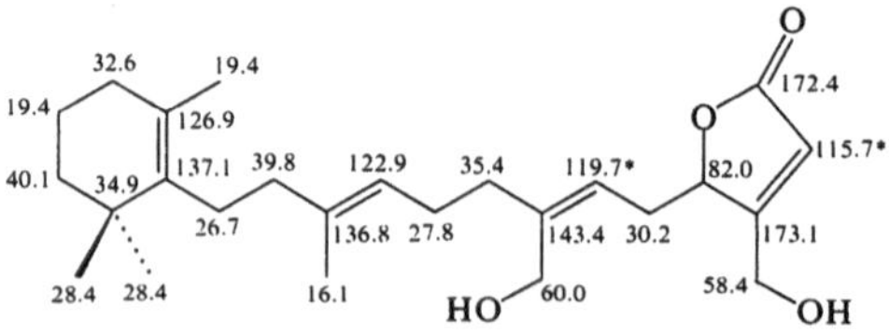

(29) (Z)-Neomanoalide

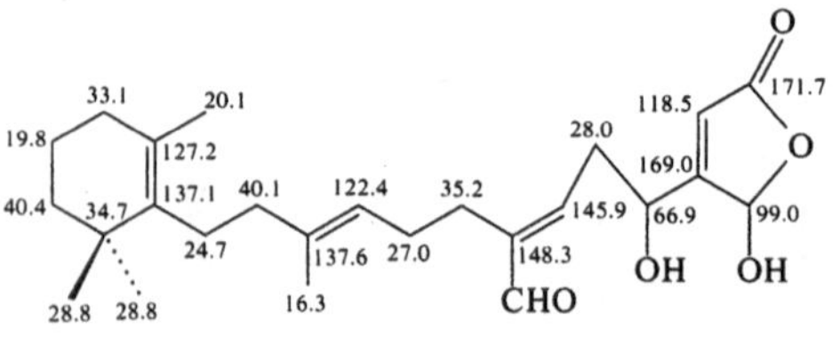

(30) Seco-manoalide

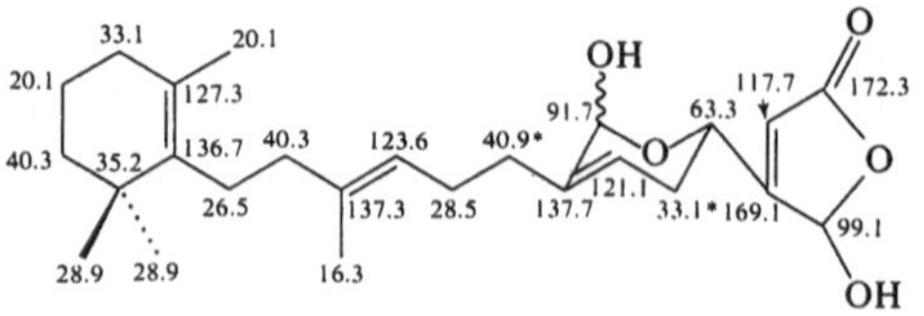

(31) Manoalide

References, pp. 262—269

Table 6 *(continued)*

B₁ —.   Monocyclic Norsesterterpenes

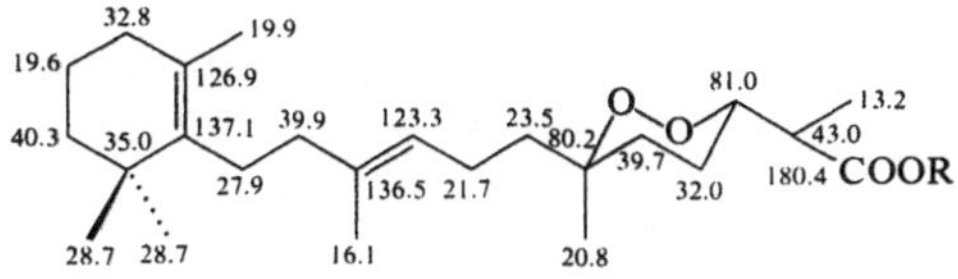

(32) R=H Muqubilin ¹³C data shown
(33) R=Me Muqubilin ester

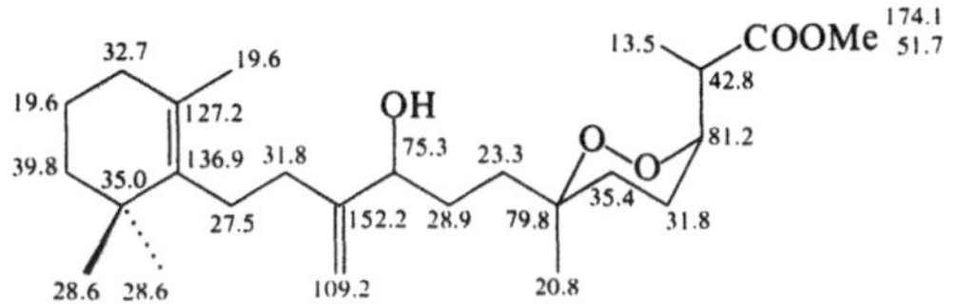

(34) Tentative structure

B₂.   Monocyclic Sesterterpenes

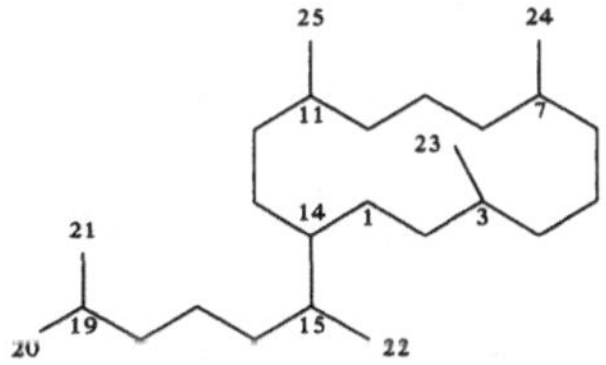

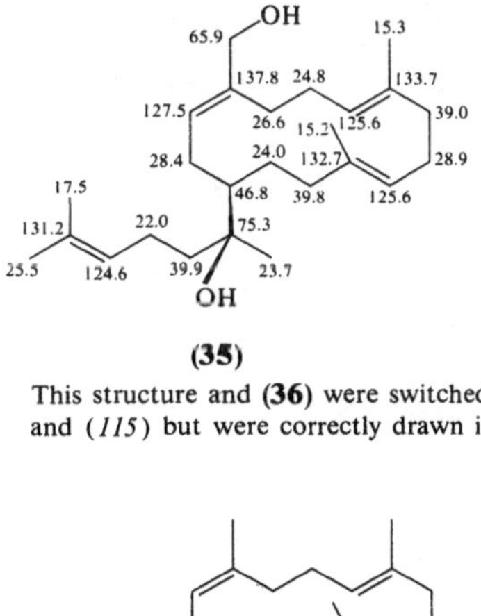

(35)

This structure and **(36)** were switched in (*90*)
and (*115*) but were correctly drawn in (*49*).

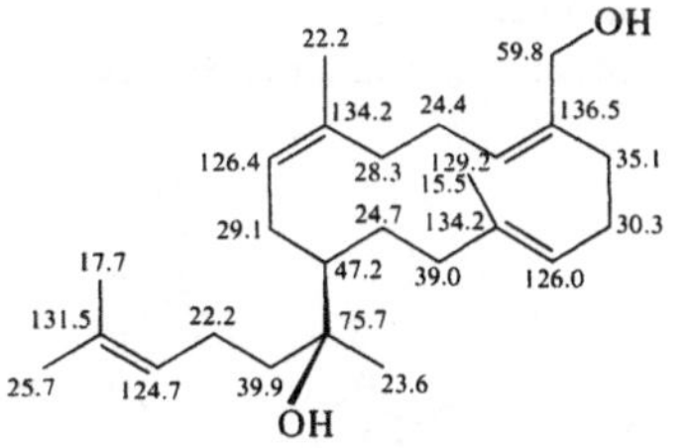

(36) Cericerol-II

This structure and **(35)** were
switched in (*90*) and (*115*) but were
correctly drawn in (*49*)

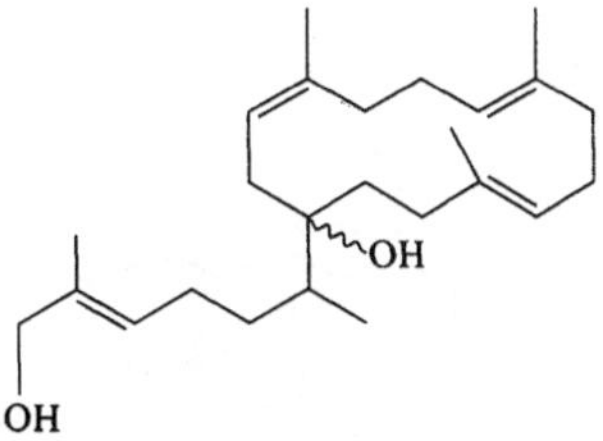

(37) Albocerol

Table 6 *(continued)*

B₂.   Monocyclic Sesterterpenes (continued)

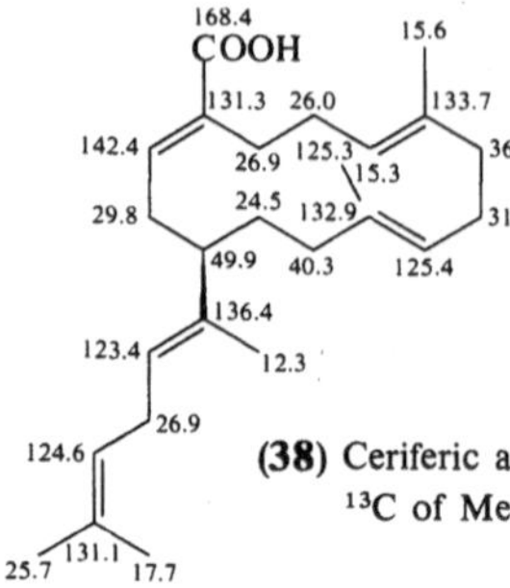

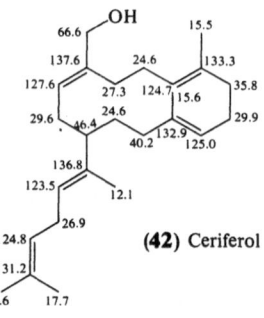

(38) Ceriferic acid
¹³C of Me ester

A typographical error appeared for
this structure in *(51)*, but it is
correctly drawn in *(49)*

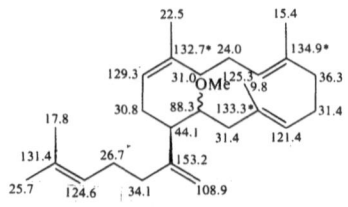

(39) 13-Methoxycericerene

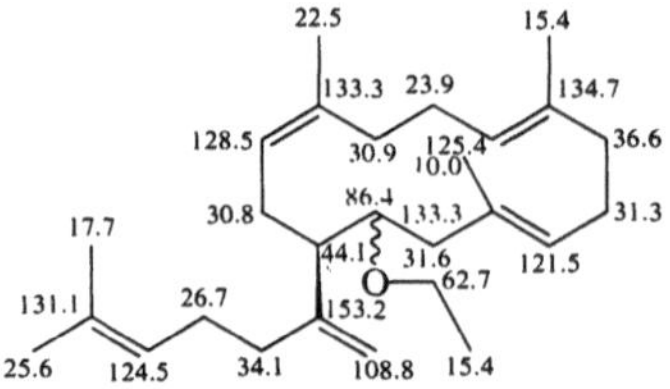

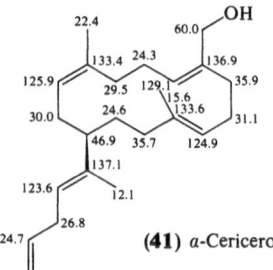

(40) 13-Ethoxycericerene

(41) α-Cericerol-I

(42) Ceriferol

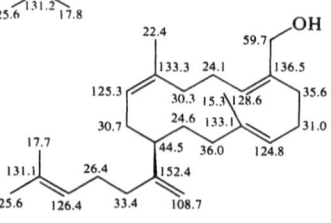

(43) Cericerol-I
Tentative structure (see Scheme 2)

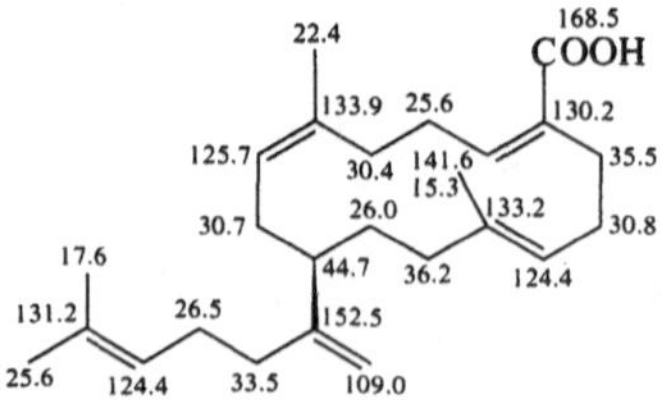

(44) Cerienoic acid
¹³C of Me ester

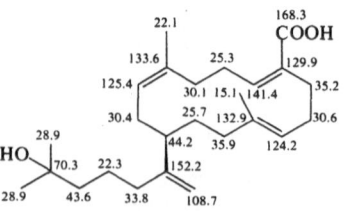

(45) ¹³C of Me ester

Table 6 *(continued)*

B₂. Monocyclic Sesterterpene (continued)

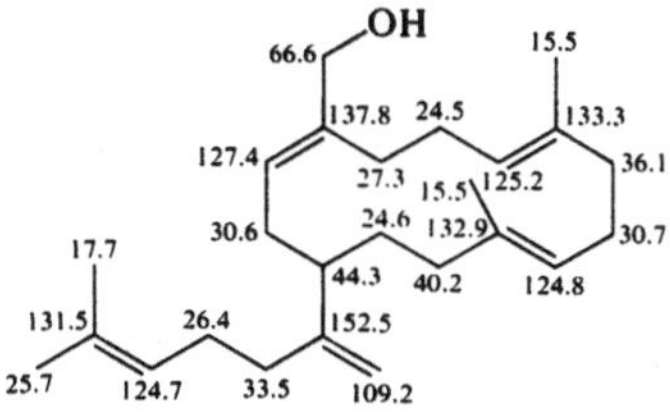

(46) Ceriferol-I

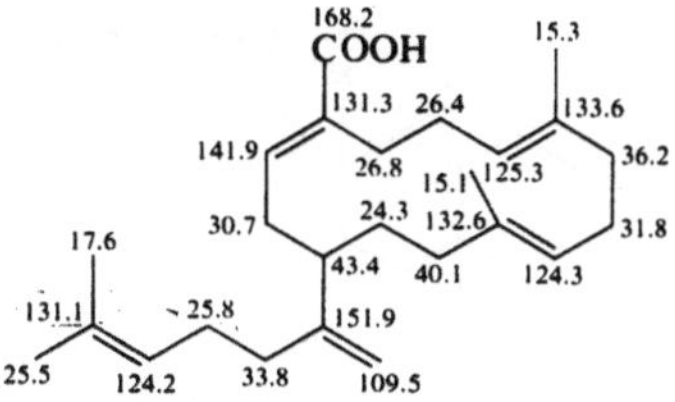

(47) Ceriferic acid-I
¹³C of Me ester

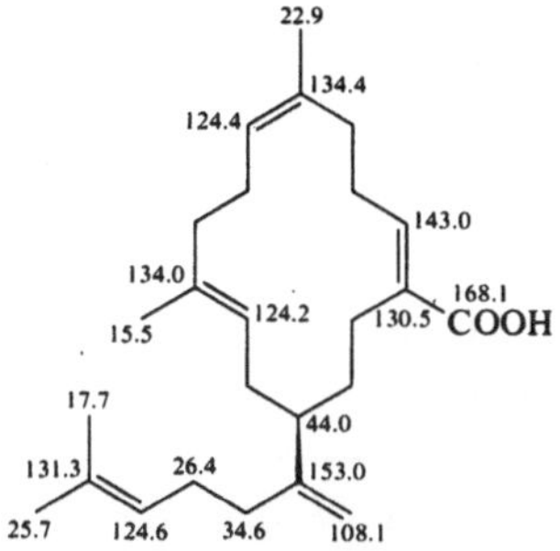

(48)ᵃ ¹³C of Me ester
(incomplete ¹³C assignment)

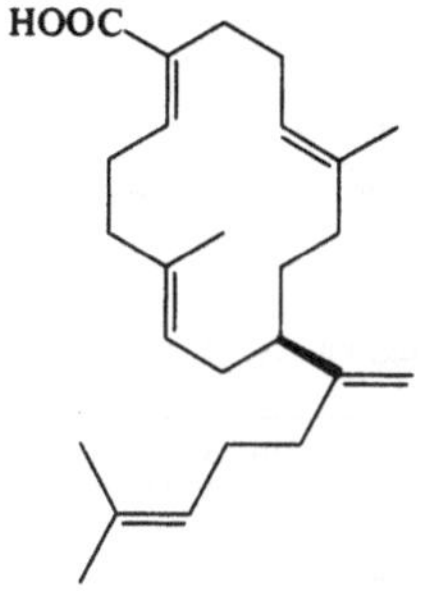

(49)

C₁. Bicyclic Sesterterpenes

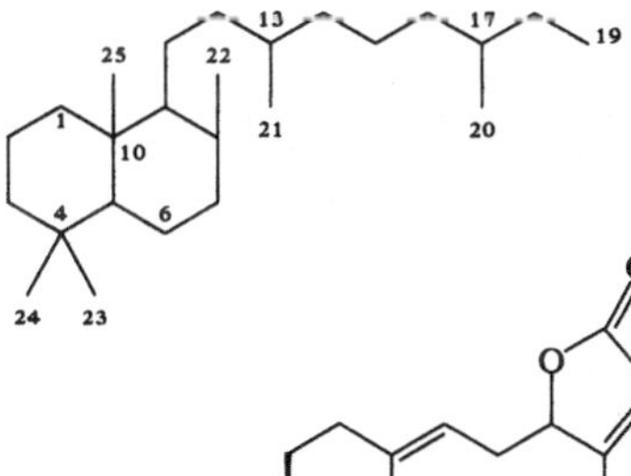

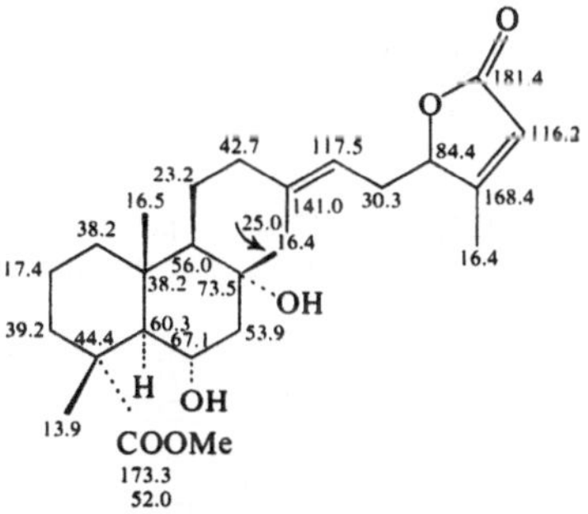

(50) Salvileucolide methyl ester

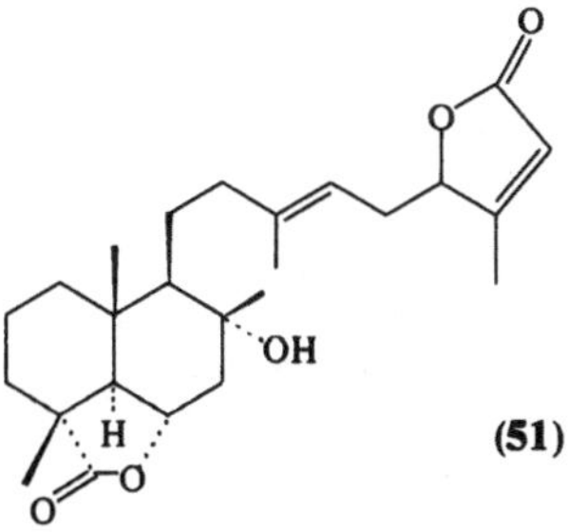

(51) Salvileucolide-6,23-lactone

ᵃ Our ¹³C-NMR assignments

Table 6 *(continued)*

C₂. Bicyclic Sesterterpenes

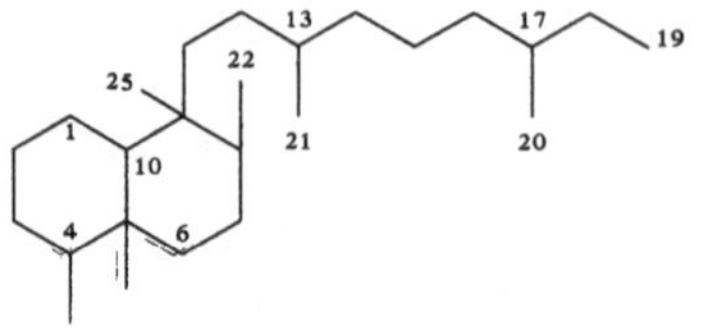

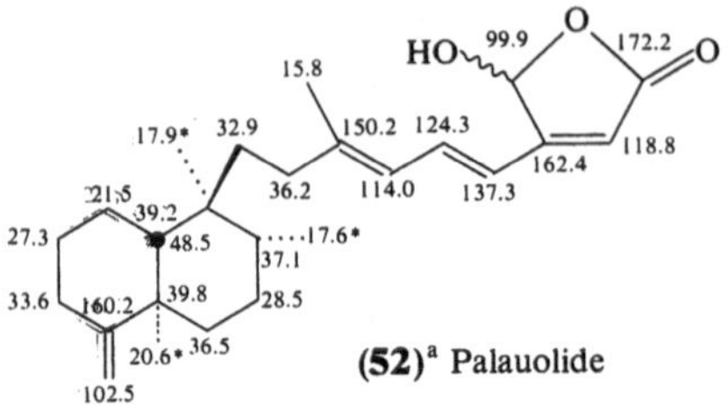

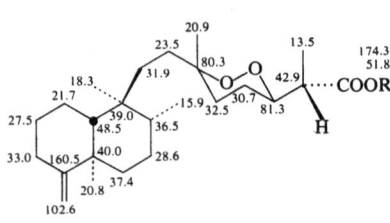

(52)[a] Palauolide

¹³C assignments were made by comparison with (53)–(54)

C₂–. Bicyclic Norsesterterpenes

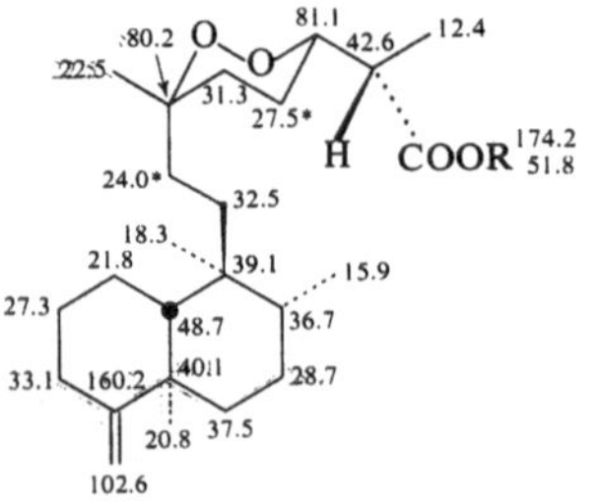

(53) R=H Sigmosceptrellin-A
[54][a] R=Me ¹³C data shown

(55) R=H Sigmosceptrellin-B
(56)[a] R=Me ¹³C data shown

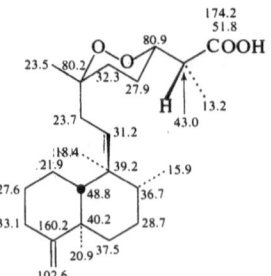

(57)[a] Sigmosceptrellin-C
¹³C data is for the Me ester

[a] Our ¹³C-NMR assignments

Table 6 *(continued)*

C₃.   Bicyclic Sesterterpenes

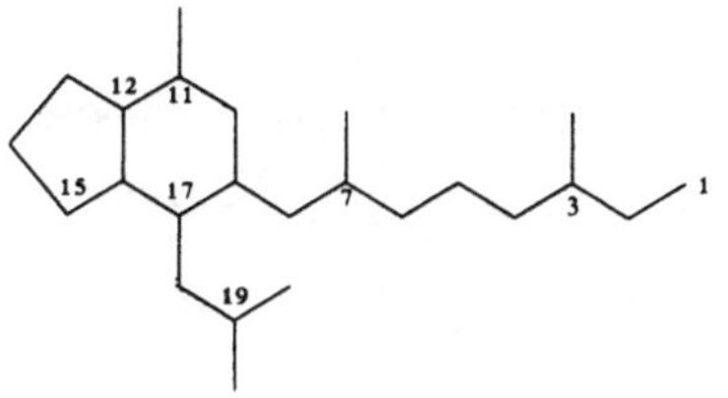

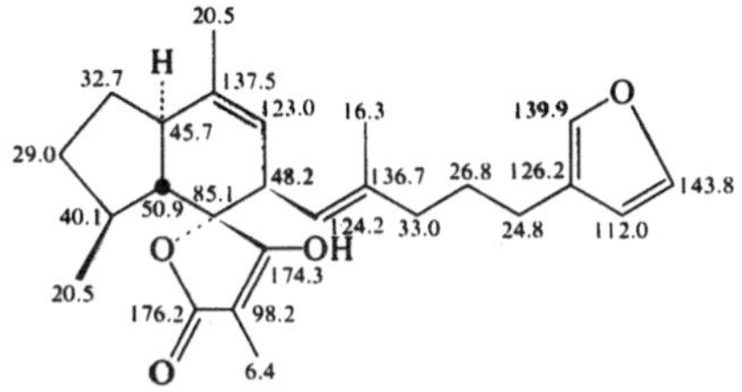

[58] Ircinianin

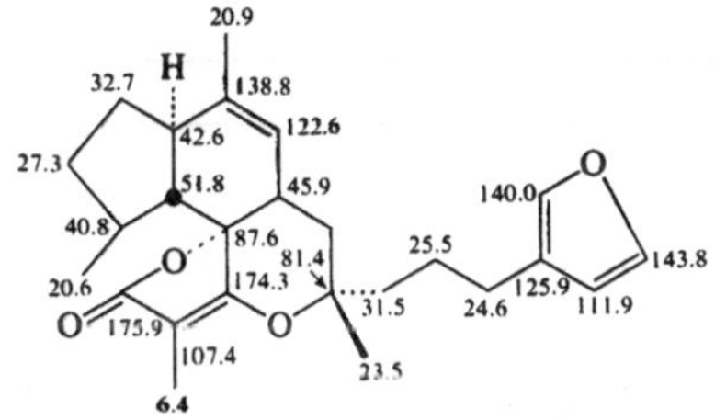

(59) Wistarin

C₄.   Bicyclic Sesterterpenes

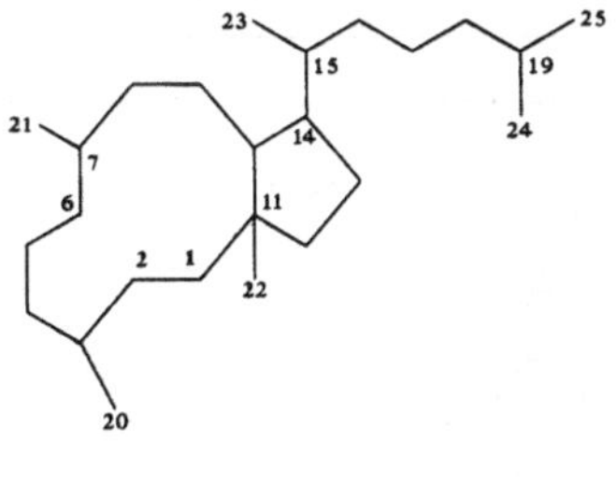

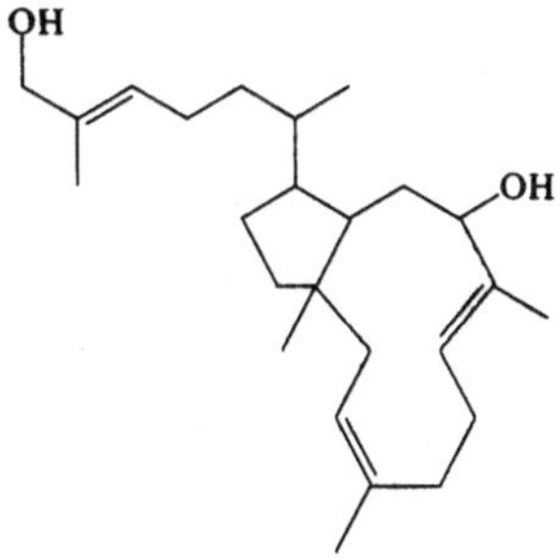

(60) Albolineol

Table 6 *(continued)*

D₁.   Tricyclic Sesterterpenes

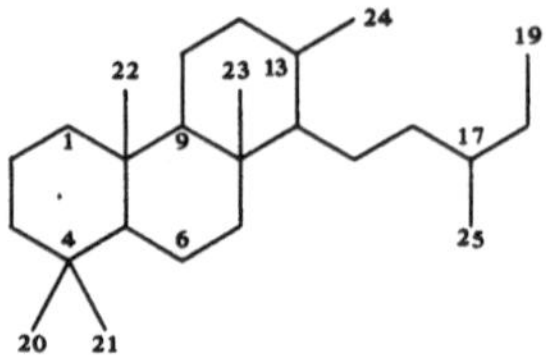

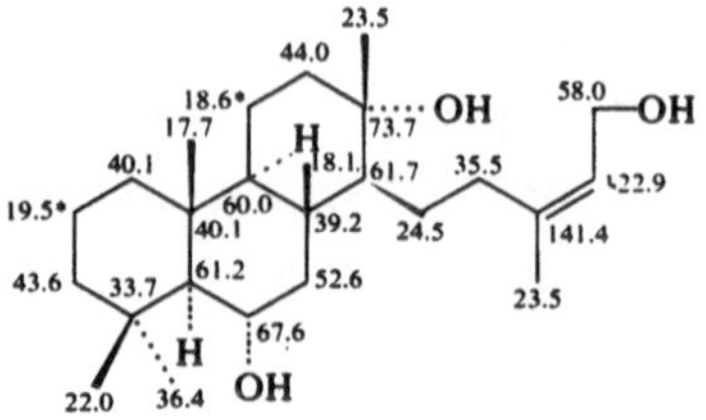

(61) Cheilanthatriol

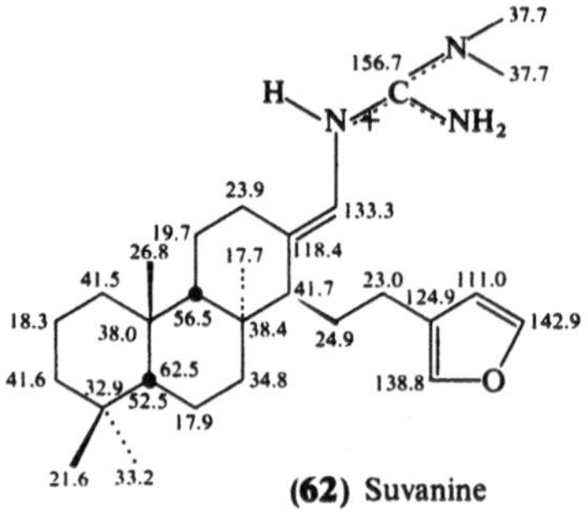

(62) Suvanine

D₁−.   Tricyclic Sesterterpenes

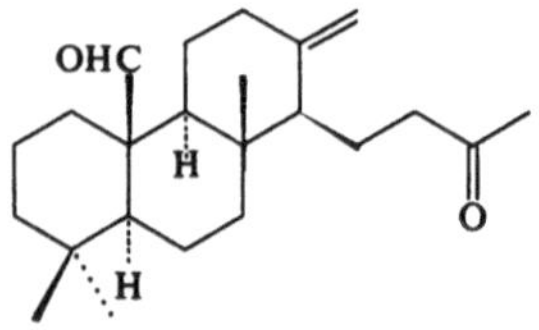

[63] Luteone

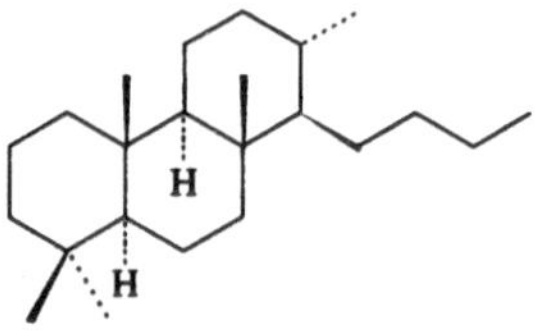

(64)

may not be a sesterterpene as it occurs with a
series of higher homologs [see (*163*)]

D₂.   Tricyclic Sesterterpenes

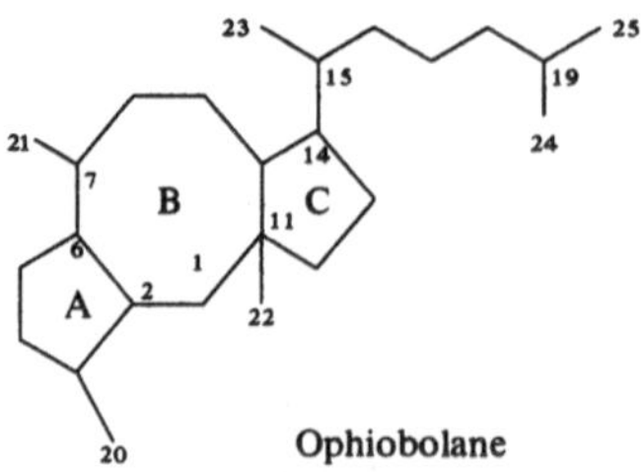

Ophiobolane

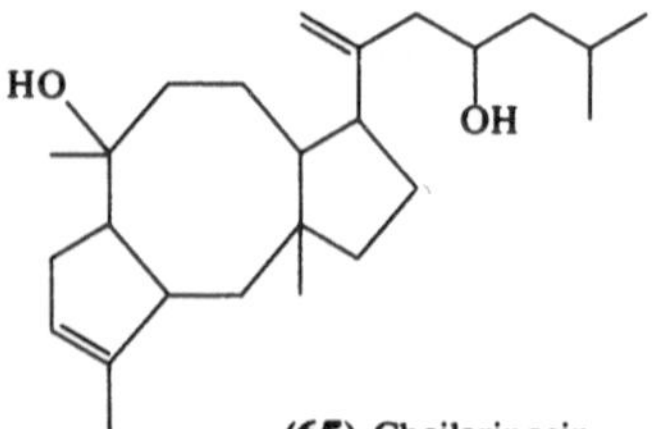

(65) Cheilarinosin

Table 6 *(continued)*

D₂. Tricyclic Sesterterpenes (continued)

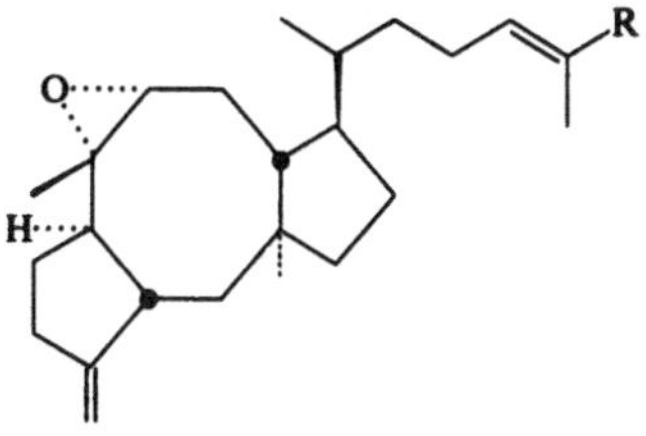

(66) R=COOH Ceralbic acid I
(67) R=CH₂OH Ceralbol

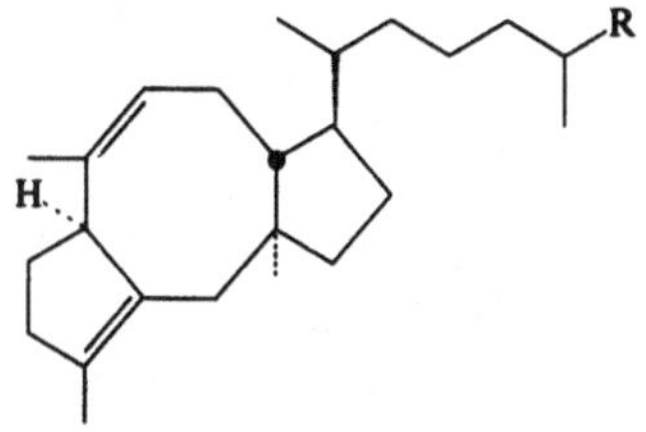

(68) R=CH₂OH Ceroplastol-II
(69) R=COOH Albolic acid

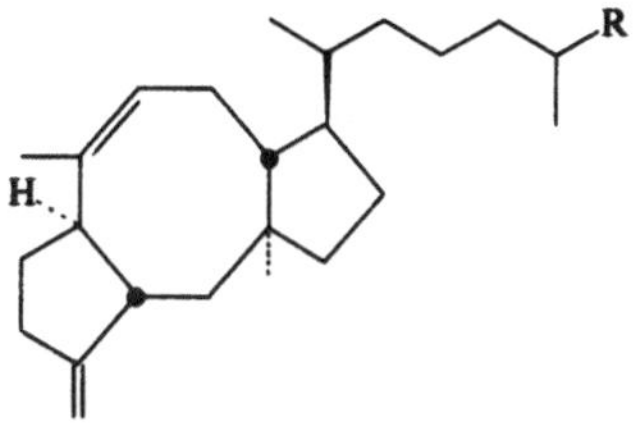

[70] R=CH₂OH Ceroplastol-I
(71) R=COOH Ceroplasteric acid

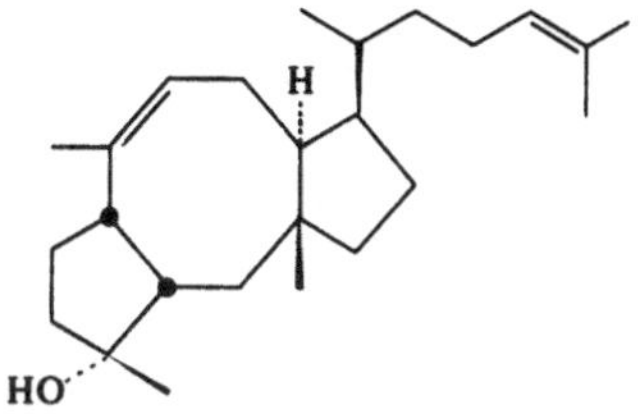

(72) Ophiobolin-F

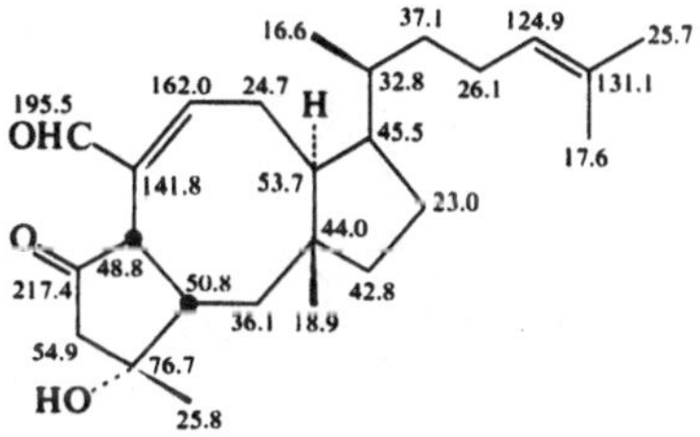

(73) Ophiobolin-C

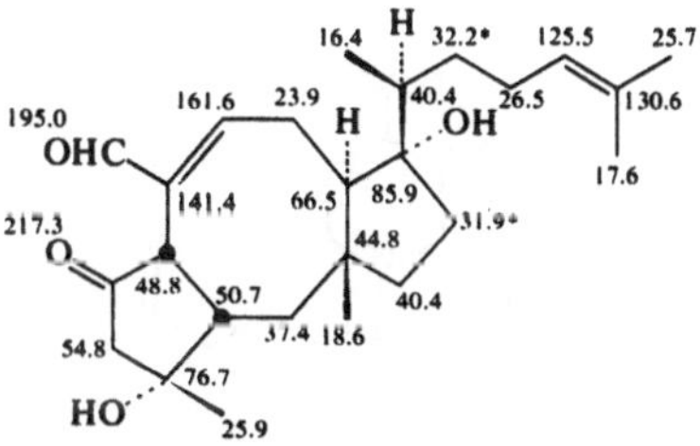

(74) Ophiobolin-B

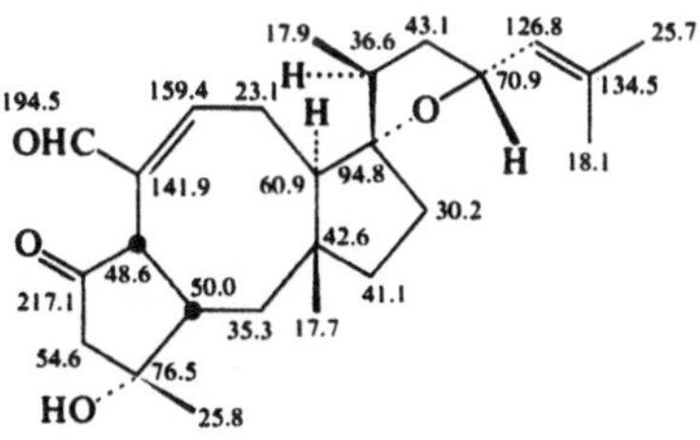

[75] Ophiobolin-A

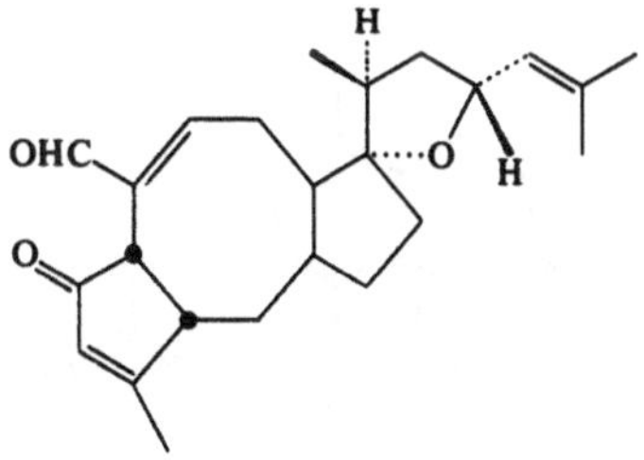

(76) Anhydroophiobolin-A

Table 6 *(continued)*

D₂.   Tricyclic Sesterterpenes (continued)

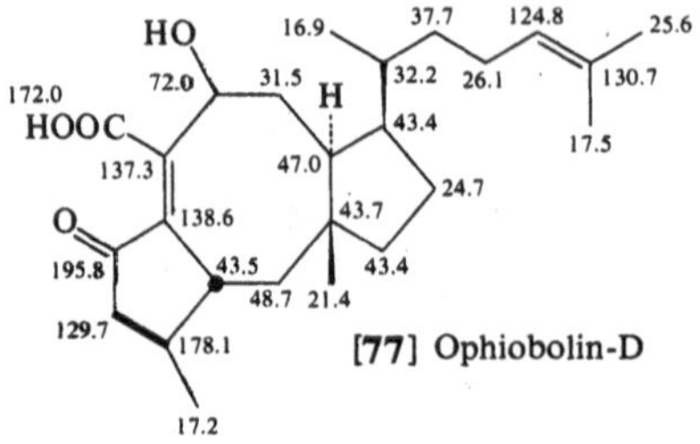

HO
16.9   37.7   124.8   25.6
172.0   72.0   31.5   H   32.2   26.1   130.7
HOOC   43.4   17.5
137.3   47.0
43.7   24.7
O   138.6   43.7
195.8   43.5   43.4
48.7   21.4
129.7   178.1   **[77]** Ophiobolin-D
17.2

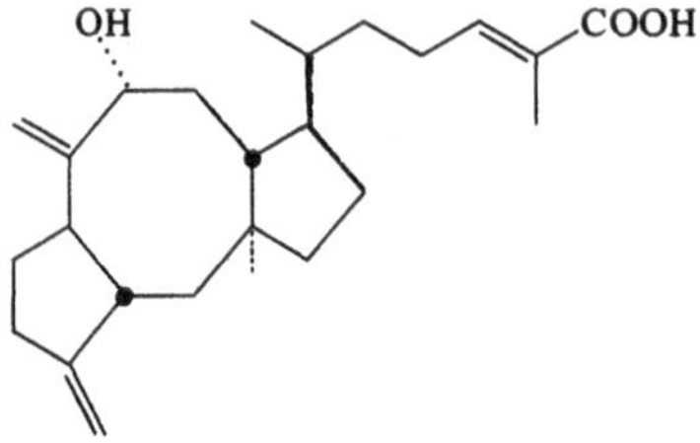

OH   COOH

(78) Ceralbic acid II

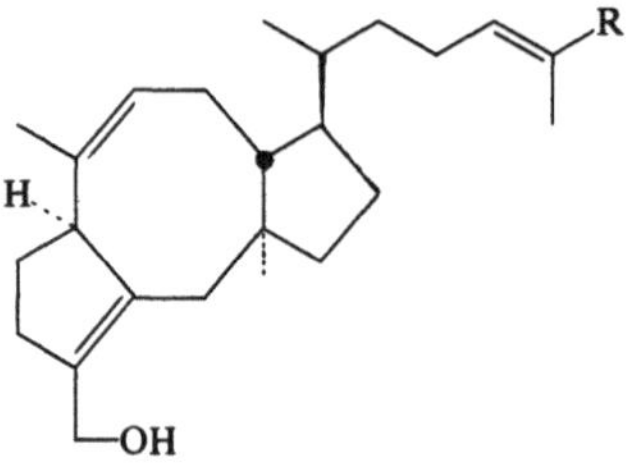

H.   R

OH

(79) R=COOH Ceroplastolic acid
(80) R=CH₂OH Ceroplastodiol

D₃.   Tricyclic Sesterterpenes

25
10
7   24   14   21
2   1   19
3   15   20
23   22

H
H   R

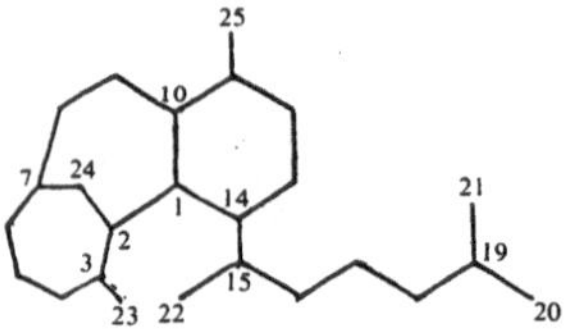

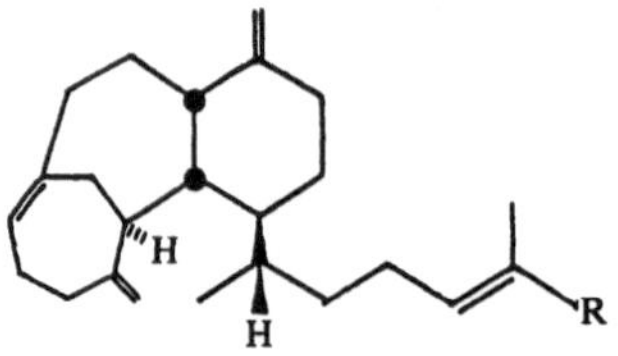

(81) R=COOH Cerorubenic acid-II
(82) R=CH₂OH Cerorubenol-II

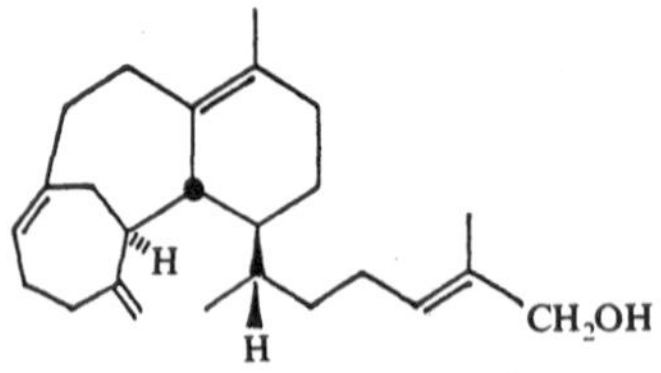

H
H   CH₂OH

(83) Cerorubenol-I

References, pp. 262—269

Table 6 *(continued)*

D₃.   Tricyclic Sesterterpenes (continued)

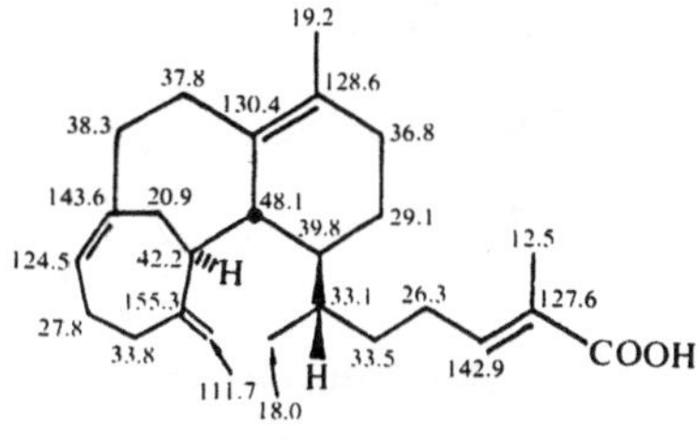

(84) Cerorubenic acid-I
¹³C of Me ester

D₄.   Tricyclic Sesterterpenes

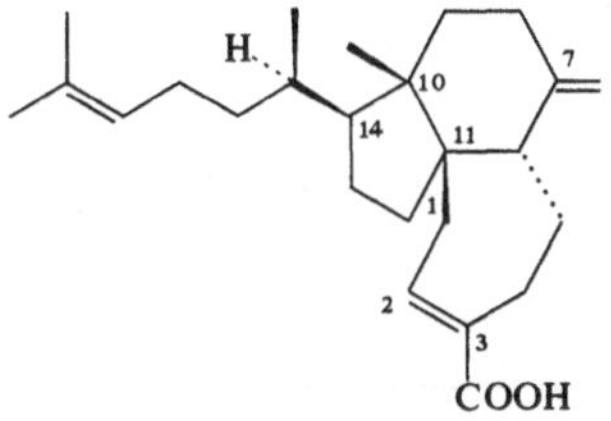

[85] Gascardic acid

D₅.   Tricyclic Sesterterpenes

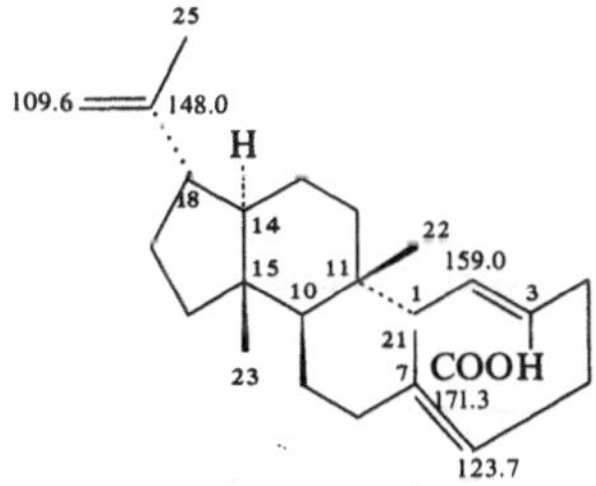

[86] Stellatic acid

D₆.   Tricyclic Sesterterpenes

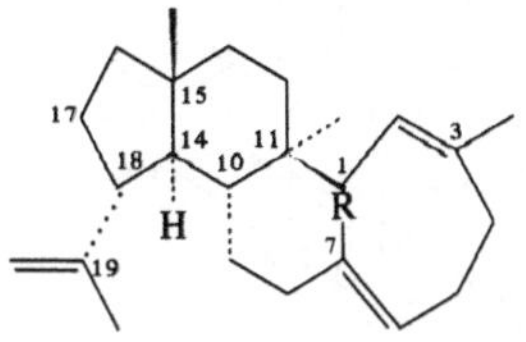

(87) R=CH₂OH Flocerol
(88) R=COOH Floceric acid

Table 6 (continued)

D₇.   Tricyclic Sesterterpenes

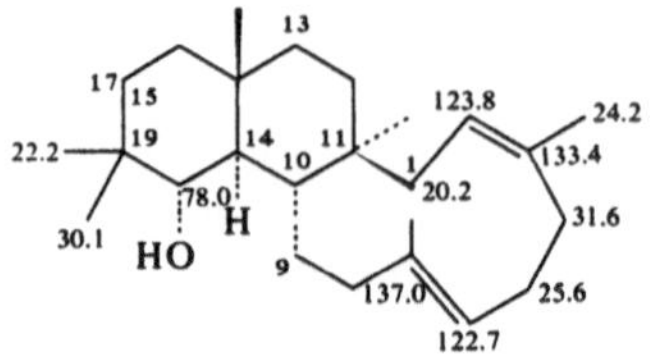

[89] Floridenol

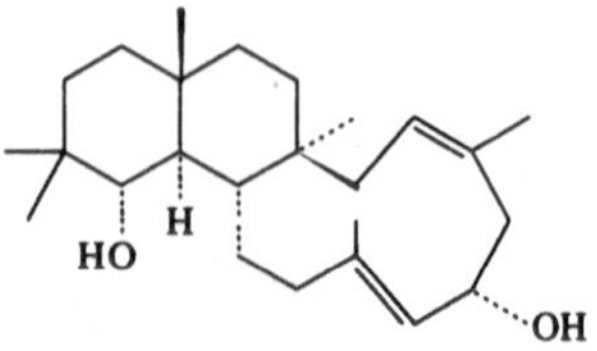

(90) 5a-Hydroxyfloridenol

D₈.   Tricyclic Sesterterpenes

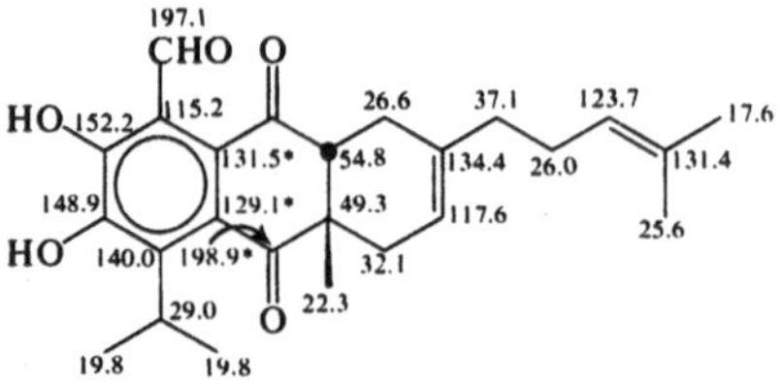

[91] Heliocide H₂

D₉.   Tricyclic Sesterterpenes

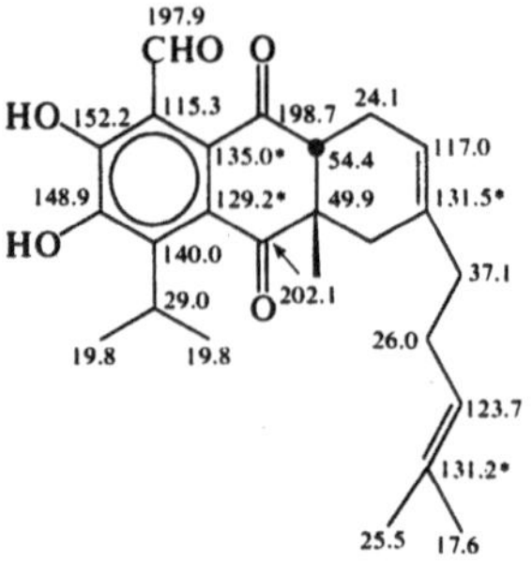

(92) Heliocide H₃

D₁₀.   Tricyclic Sesterterpenes

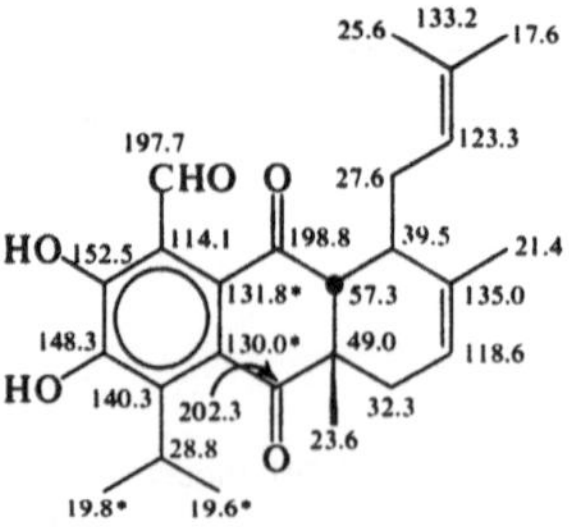

(93) Heliocide H₁

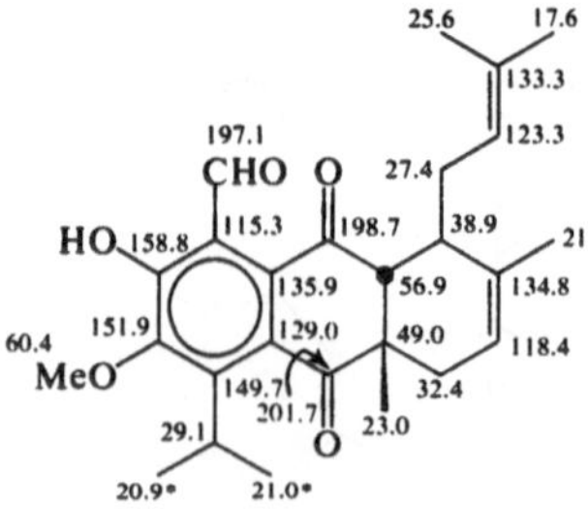

(94) Heliocide B₁

Table 6 *(continued)*

E₁.   Tetracyclic Sesterterpenes

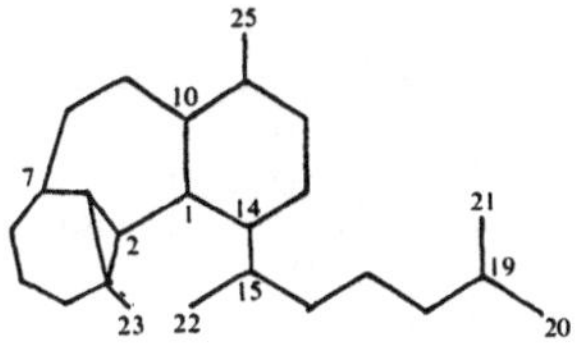

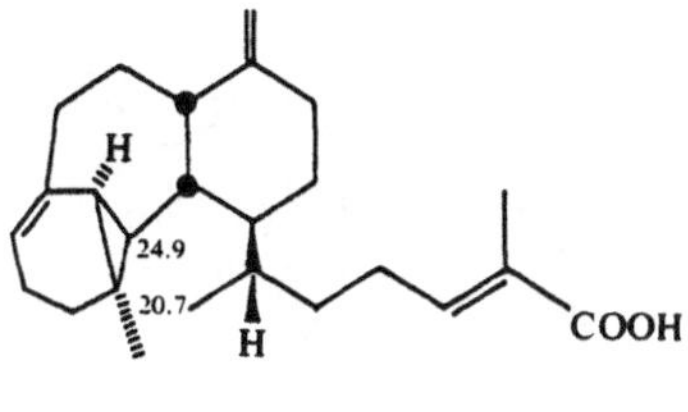

(95) Cerorubenic acid-III

E₂.   Tetracyclic Sesterterpenes

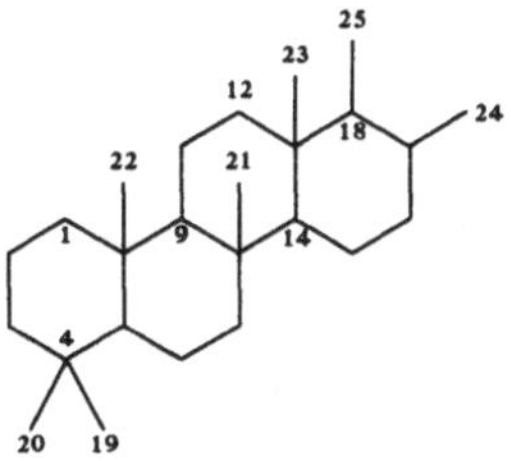

Scalarane

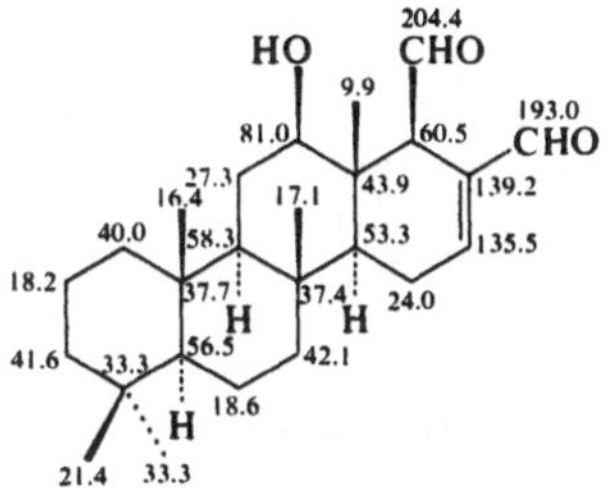

(96) 12-Deacetyl-12-episcalaradial

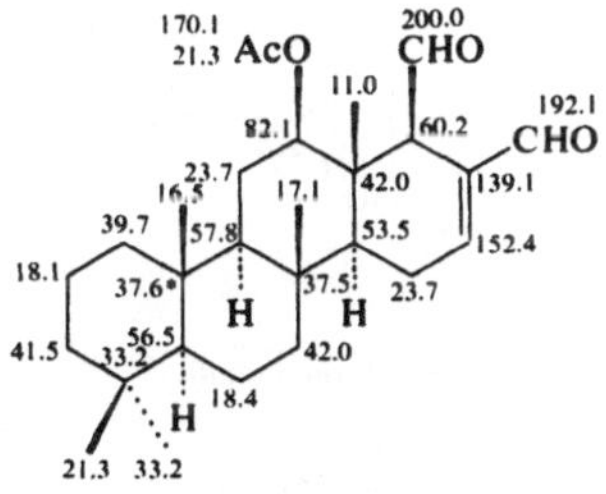

(97) 12-Episcalaradial

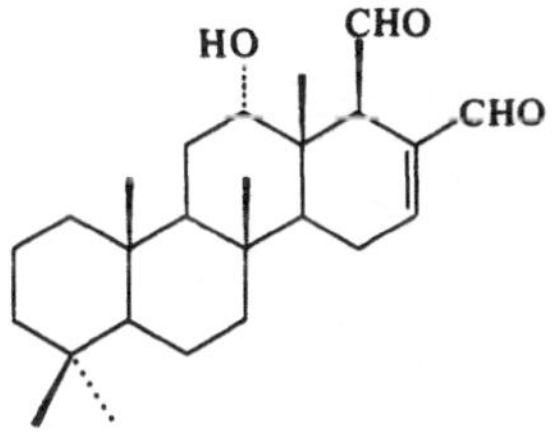

(98) Desacetylscalaradial

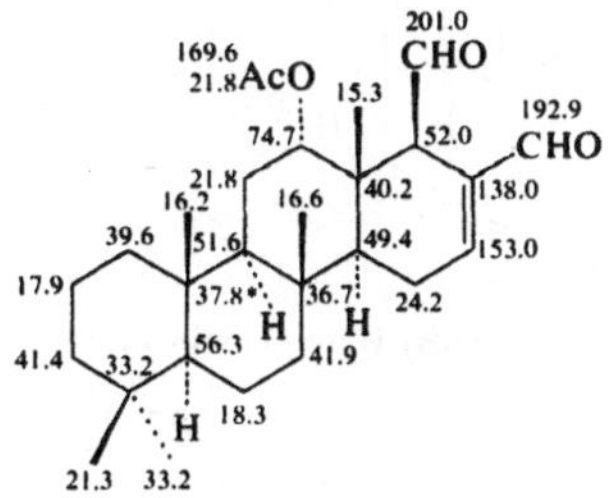

(99) Scalaradial

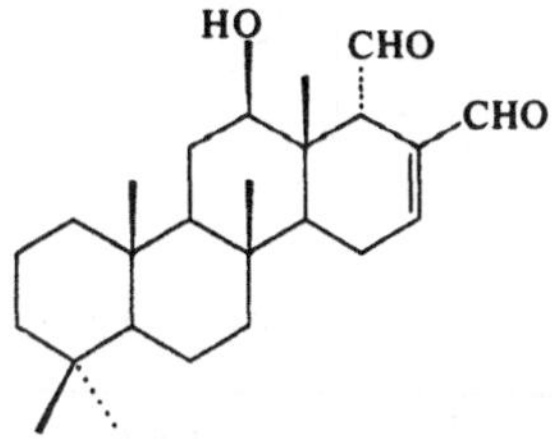

(100)

Table 6 *(continued)*

E₂.   Tetracyclic Sesterterpenes (continued)

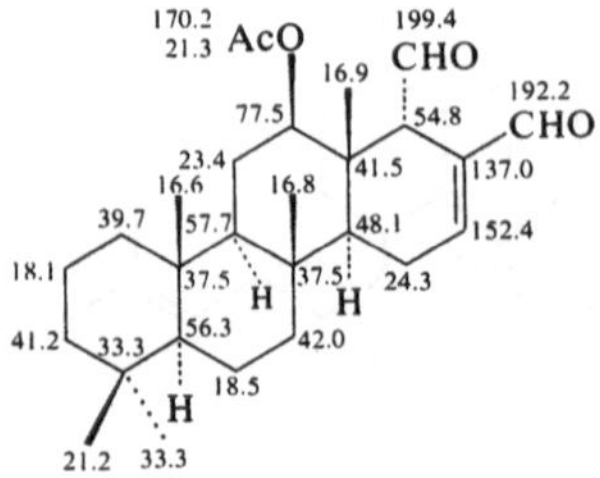

(101) 12,18-Diepiscalaradial

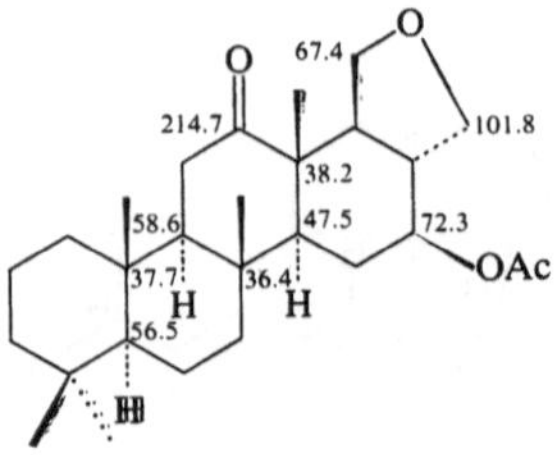

(102) Scalardysin-A

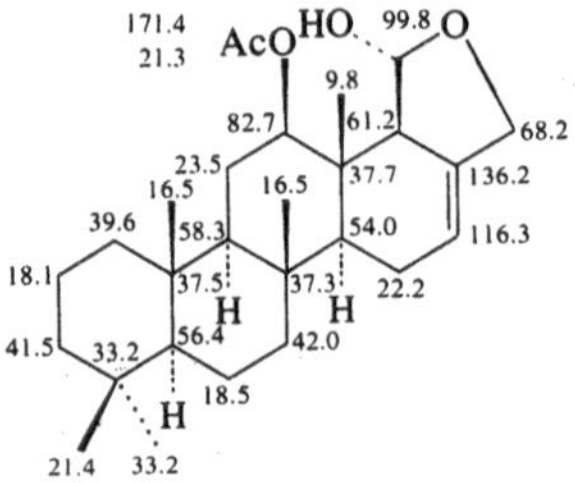

(103) 12-Epi-deoxoscalarin

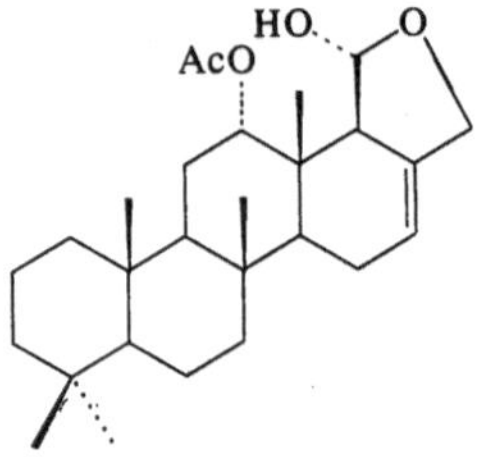

(104) Deoxoscalarin

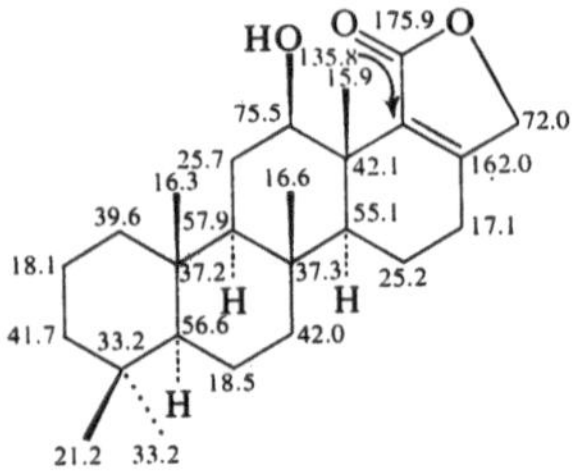

(105)[a] Scalarolide

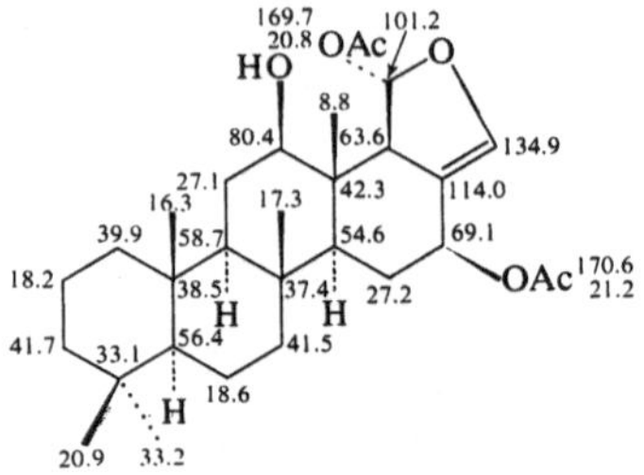

(106) Heteronemin

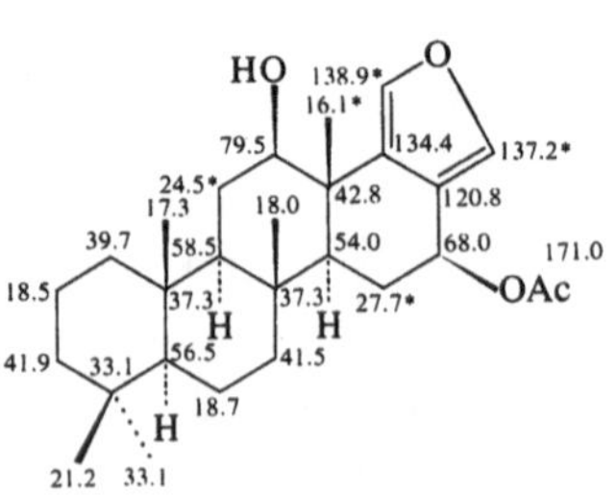

(107)[a] Scalarafuran

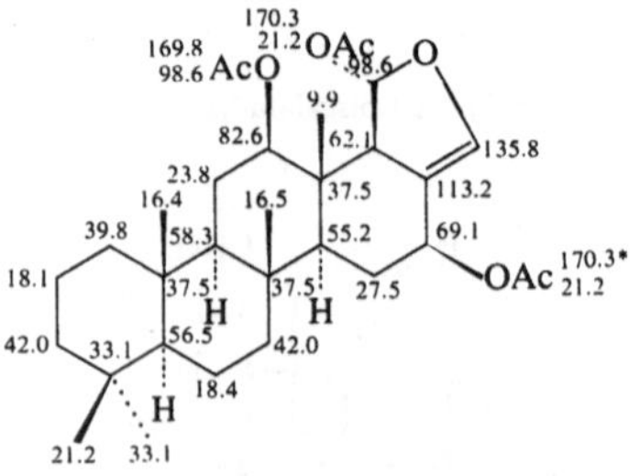

(108) Heteronemin acetate

[a] Our ¹³C-NMR assignments

Table 6 *(continued)*

E₂.   Tetracyclic Sesterterpenes (continued)

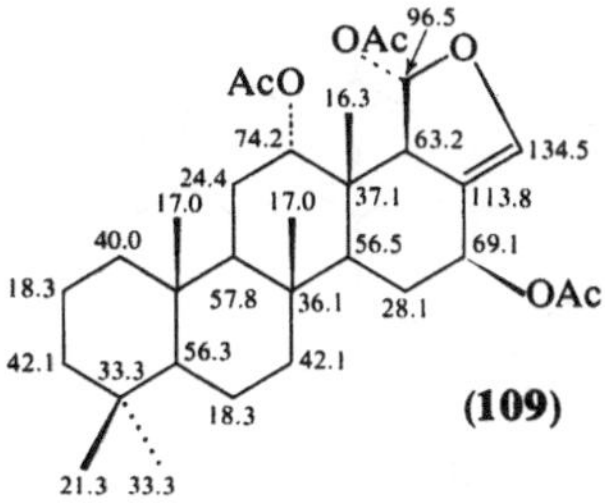

(109)

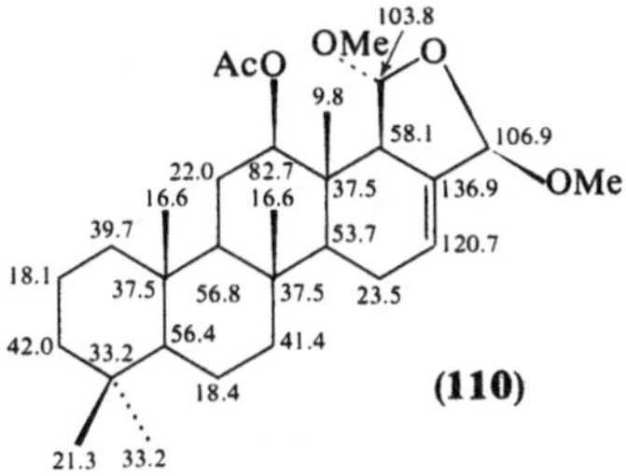

(110)

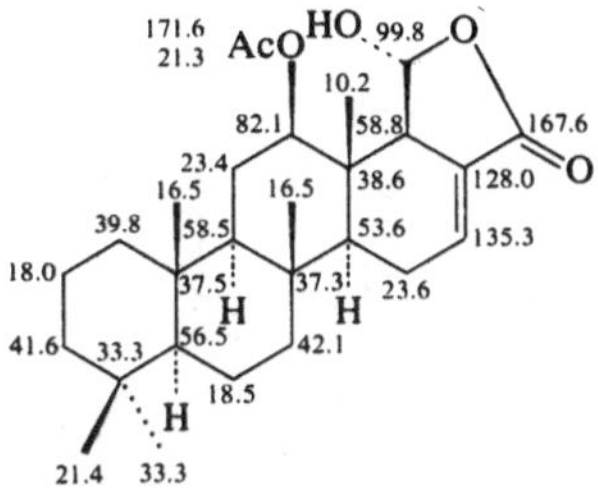

(111) 12-Episcalarin

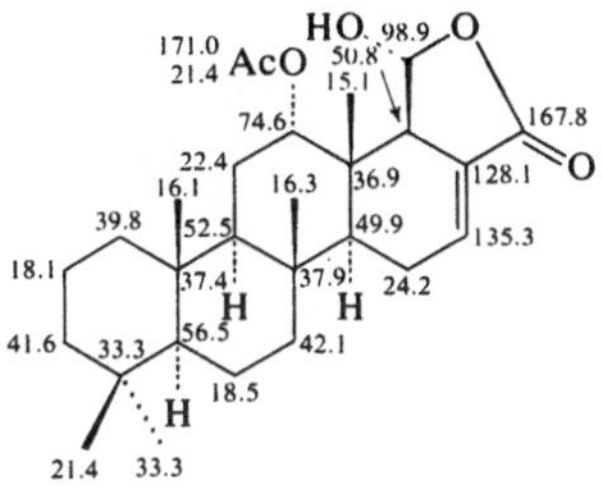

(112) Scalarin

E₂−.   Tetracyclic Norsesterterpenes

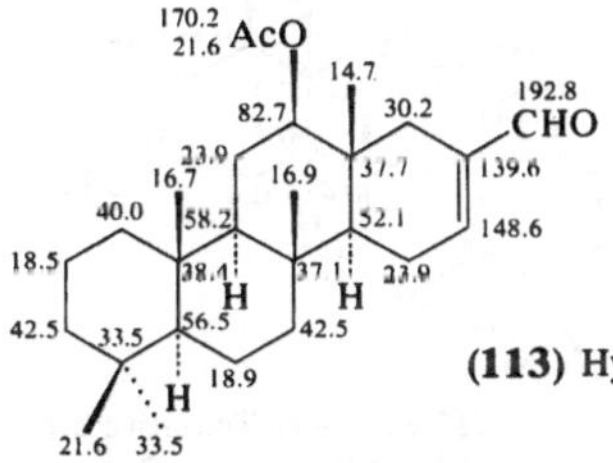

(113) Hyrtial

E₂+.   Tetracyclic Homosesterterpenes

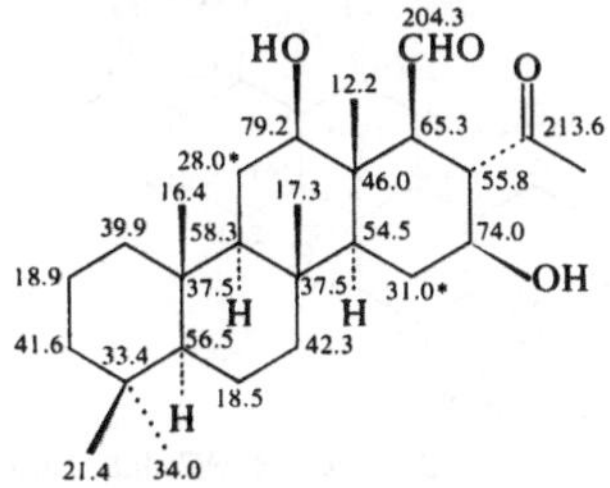

(114)[a]

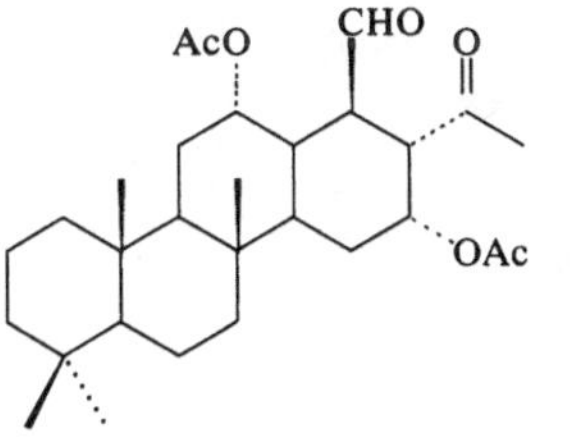

(115) Scalarherbacin-A acetate

[a] Our ¹³C-NMR assignments

Table 6 *(continued)*

E$_2$+. Tetracyclic Homosesterterpenes (continued)

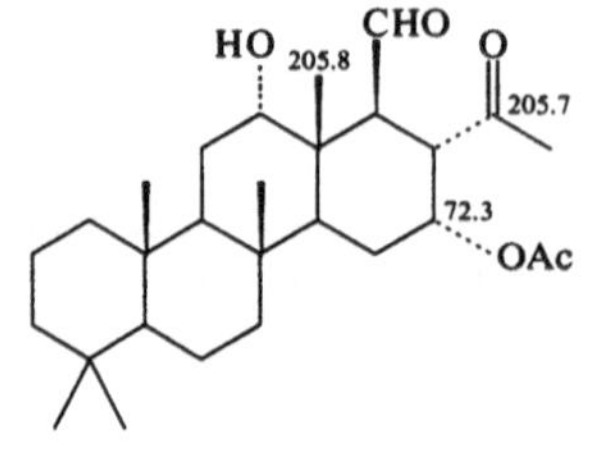

(116) R=H Scalarherbacin-A

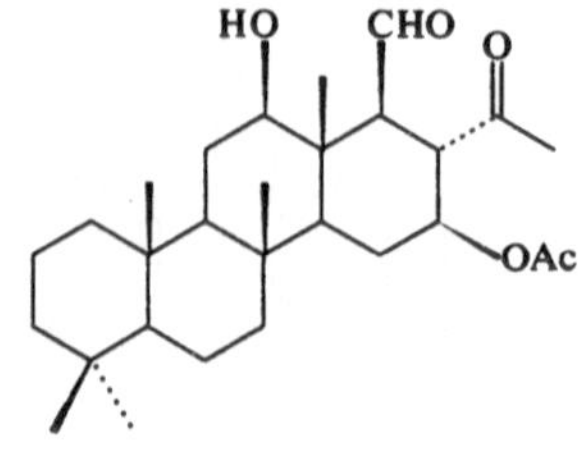

(117)

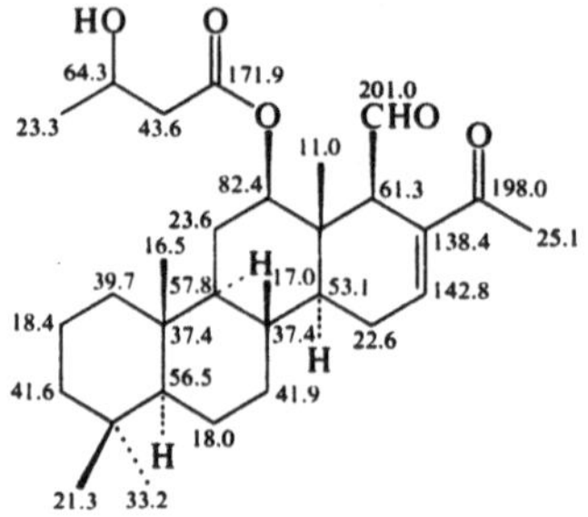

(118)[a] (Incomplete ^{13}C assignment)

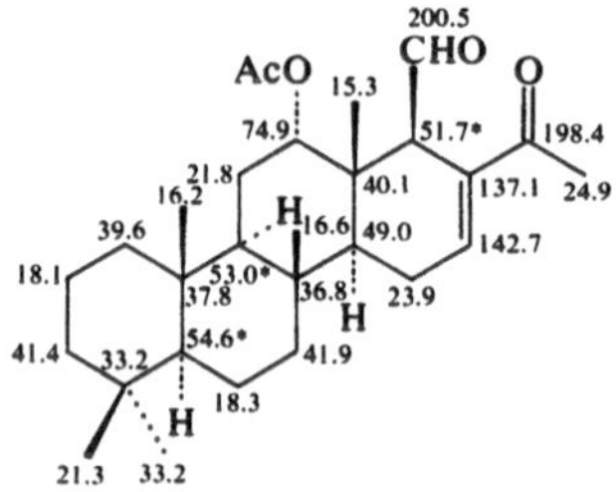

(119)[a]

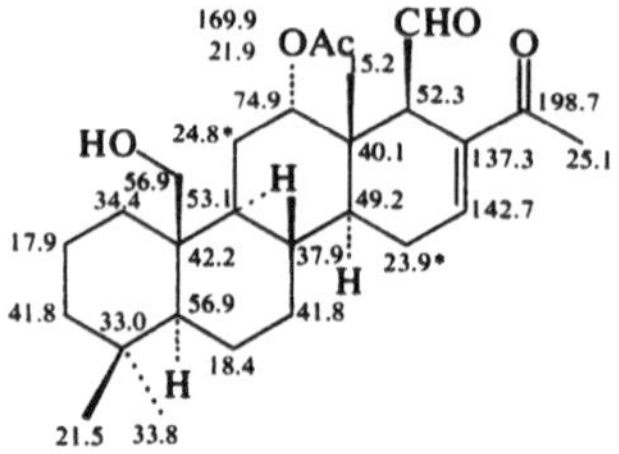

(120)[a]

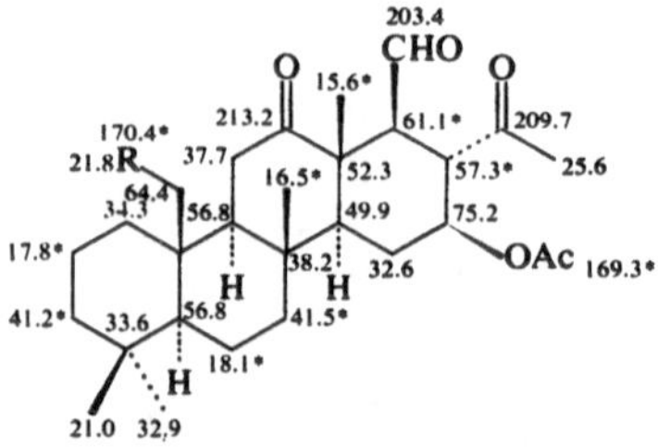

(121) R=H Tentative structure
(122)[a] R=OAc ^{13}C data shown

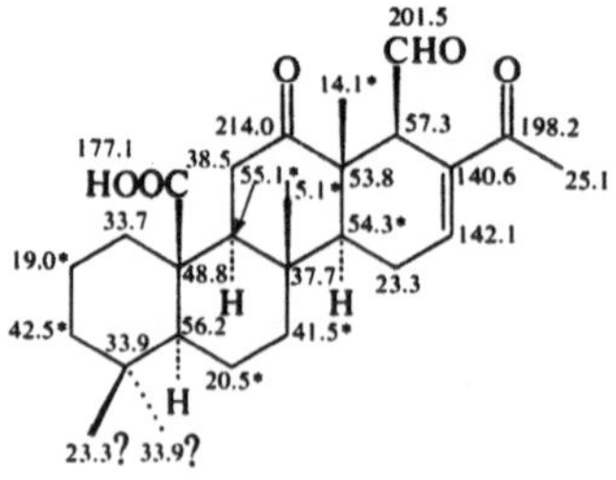

(123)[a]

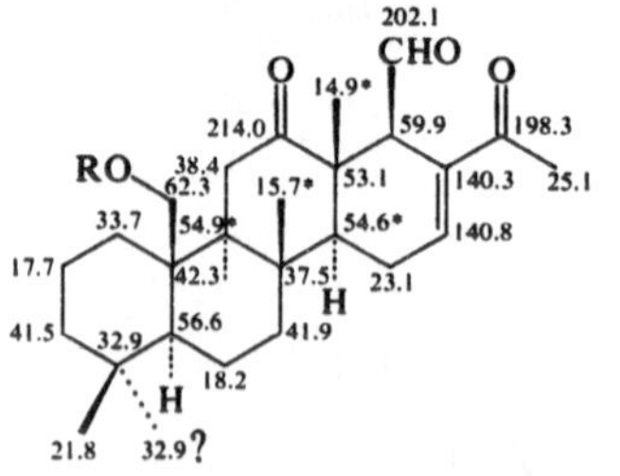

(124)[a] R=H ^{13}C data shown
[125] R=Ac

[a] Our ^{13}C-NMR assignments

Table 6 *(continued)*

E₂+.   Tetracyclic Homosesterterpenes (continued)

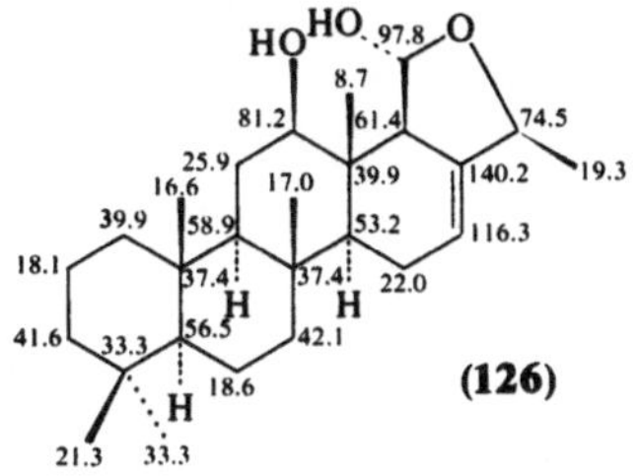

(126)

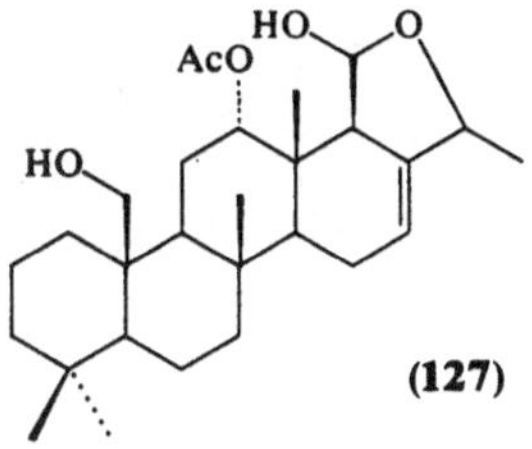

(127)

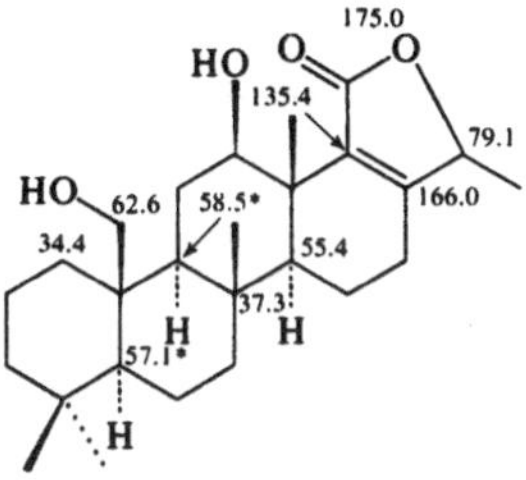

(128)[a] (Incomplete ¹³C assignment)

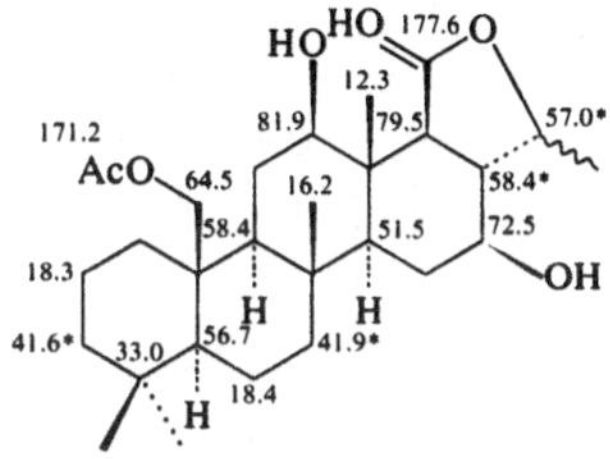

(129)[a] (Incomplete ¹³C assignment)

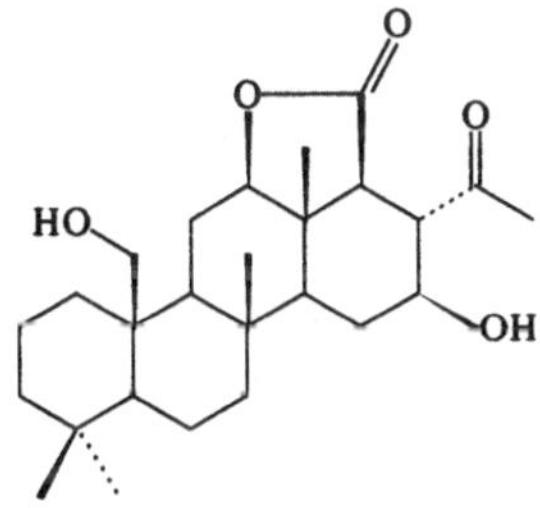

(130)

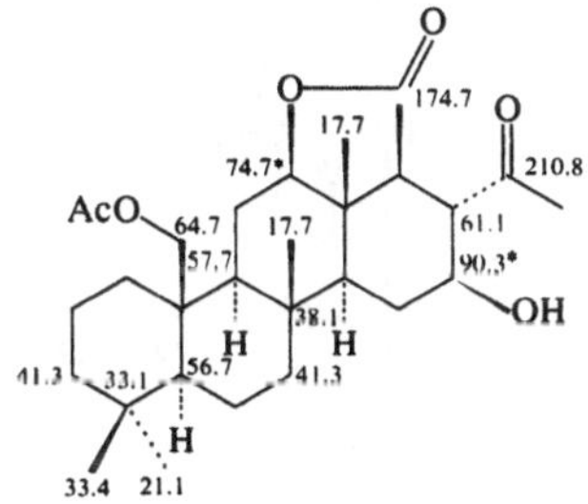

(131)[a] (Incomplete ¹³C assignment)

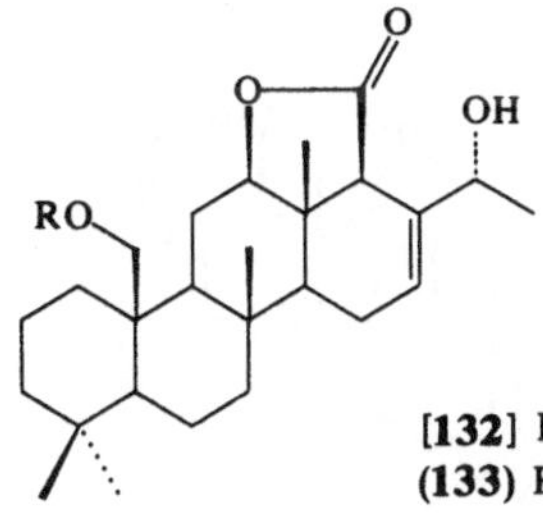

[132] R = H Sednolide
(133) R = Ac

[a] Our ¹³C-NMR assignments

Table 6 *(continued)*

E$_2$+'. Tetracyclic Homosesterterpenes

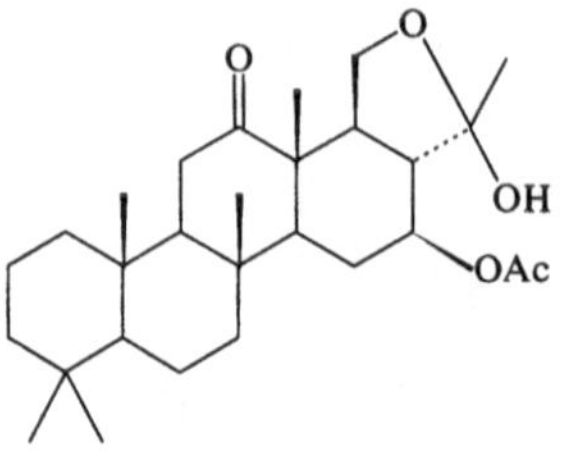

(134) Scalardysin-A

E$_2$+ +. Tetracyclic Homosesterterpenes

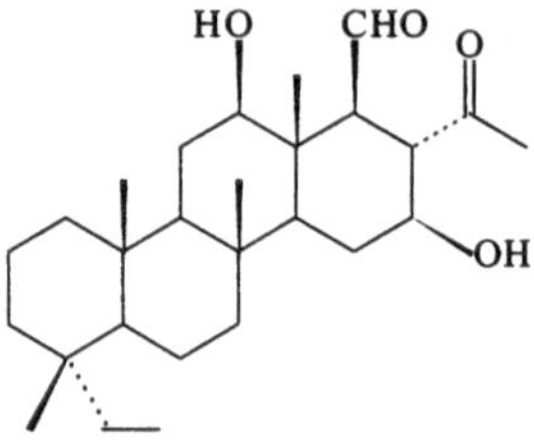

(135) Tentative structure

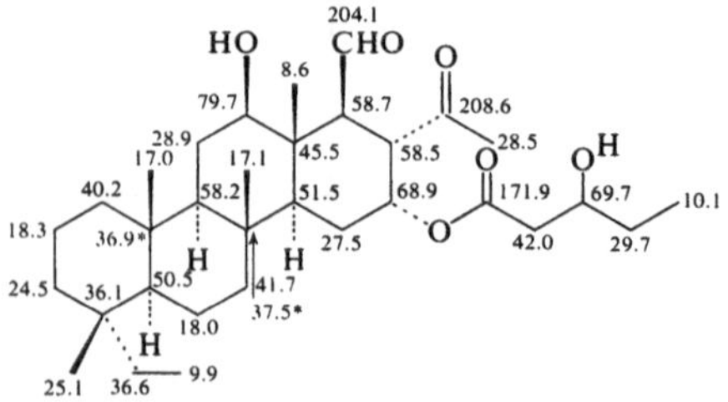

(136) Foliaspongin

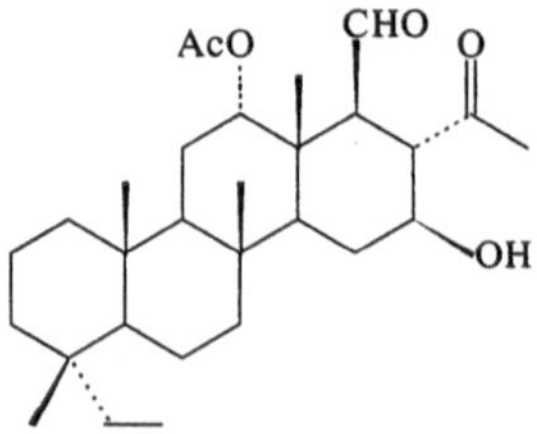

(137) Tentative structure

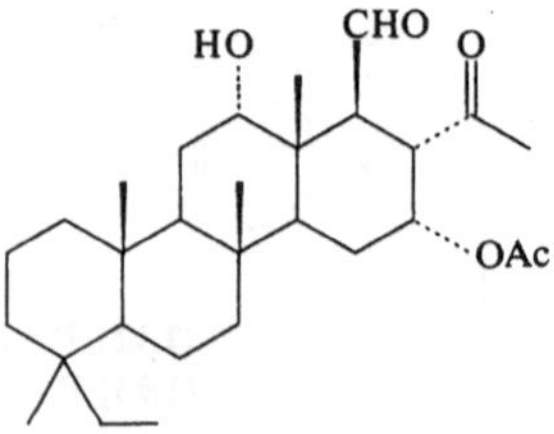

(138) Scalarherbacin-B

References, pp. 262—269

Table 6 *(continued)*

E$_2$ + +. Tetracyclic Homosesterterpenes (continued)

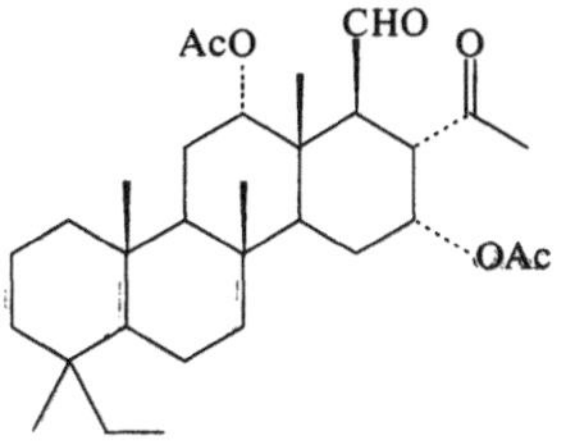

(139) Scalarherbacin-B acetate

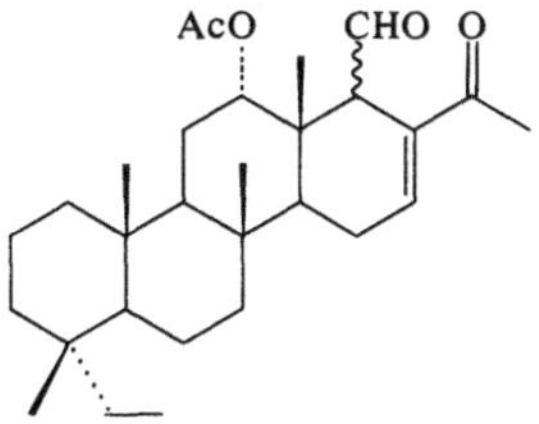

(140) Tentative structure
(CHO at C$_{18}$ missing in
literature drawing)

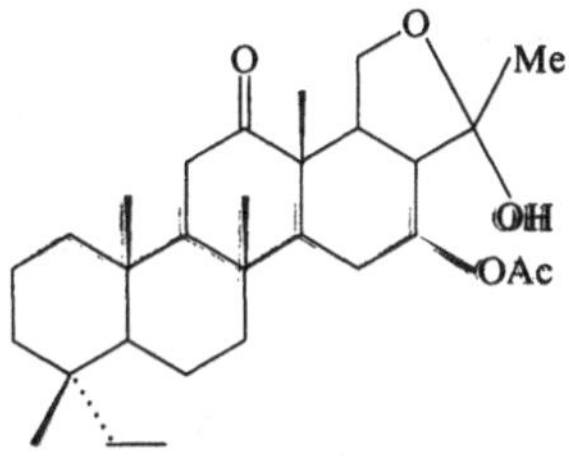

(141) Scalardysin-B

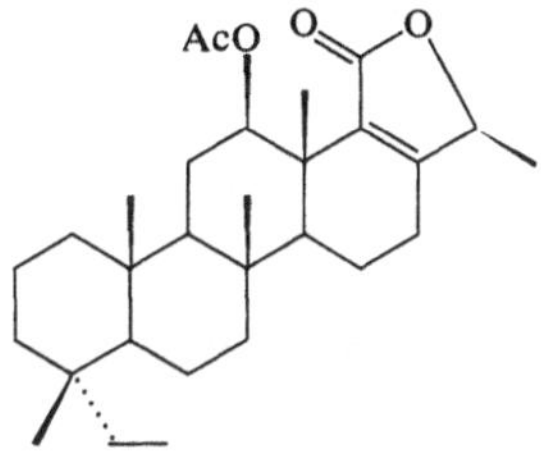

(142) Tentative structure

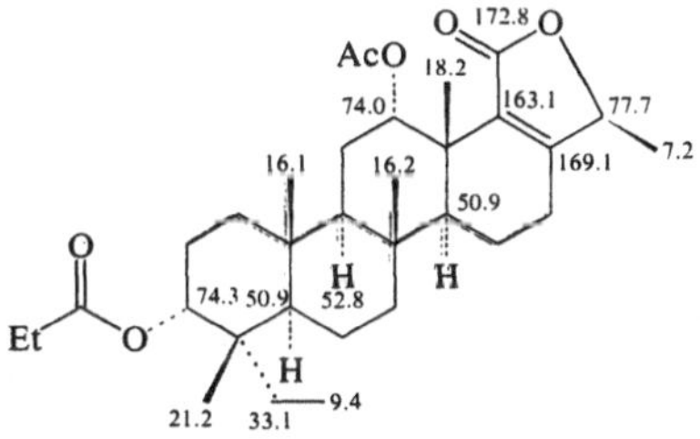

(143)

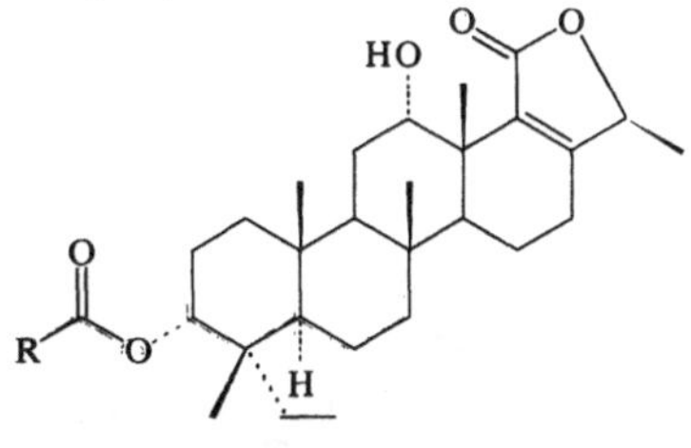

(144) R = Me
(145) R = Et
(146) R = Pr
All tentative structures

Table 6 *(continued)*

$E_2 + +$. Tetracyclic Homosesterterpenes (continued)

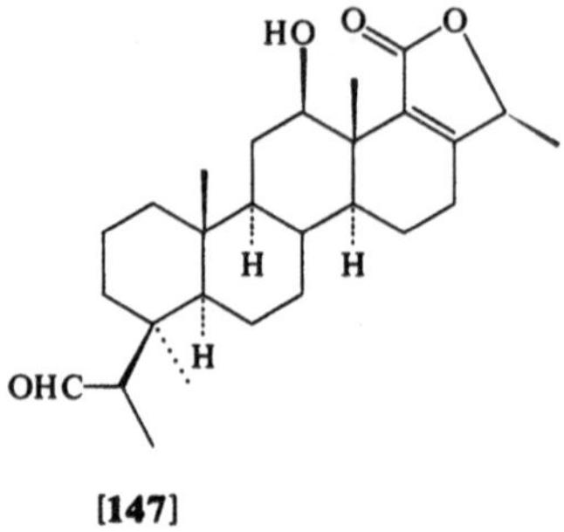

[147]

$E_2\#$.   Derivatives of Tetracyclic E_1 Sesterterpenes with Mixed Biogenesis

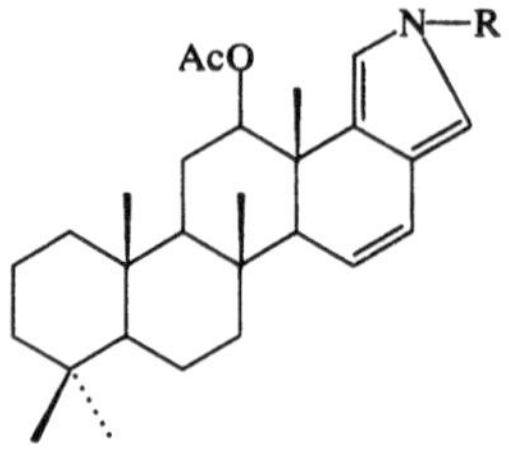

(148) $R = CH_2$	Molliorin-a
(149) $R = CH_2$	Molliorin-c
(150) $R = CH_2$	Molliorin-d
(151) $R = CH_3$	Molliorin-e

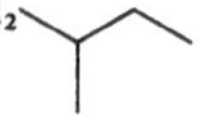

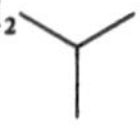

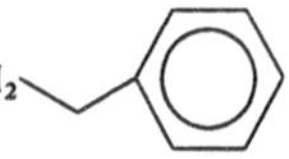

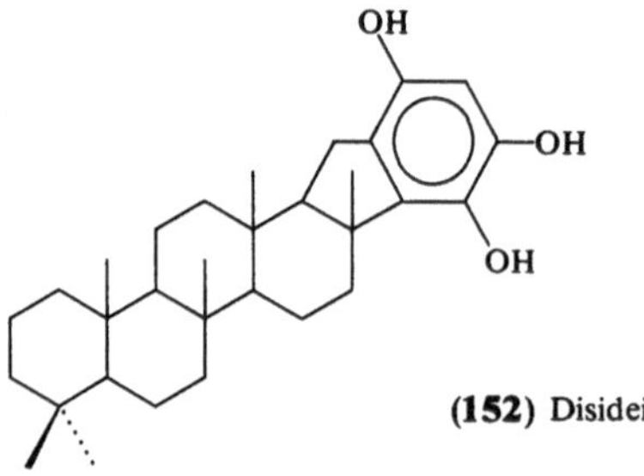

(152) Disidein

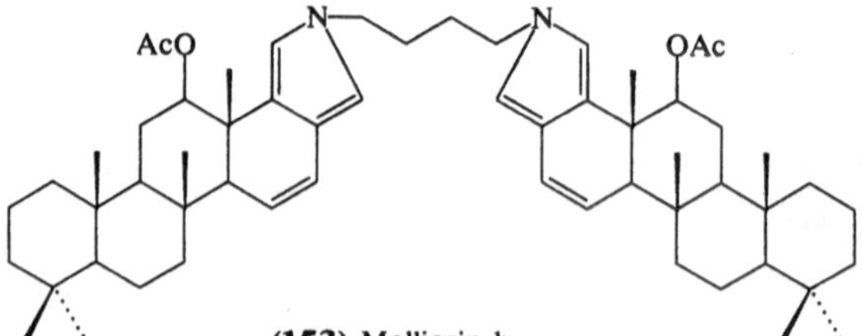

(153) Molliorin-b

Table 6 *(continued)*

E₃.   Tetracyclic Sesterterpenes

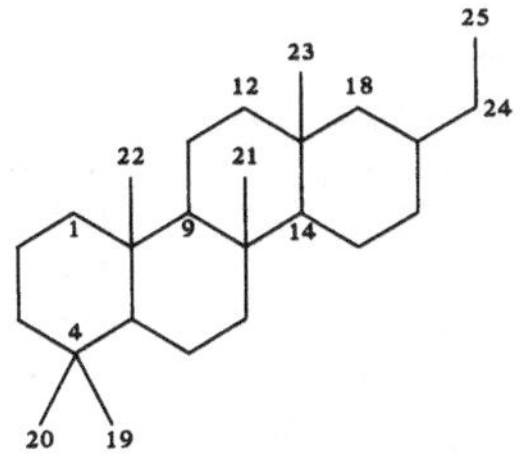

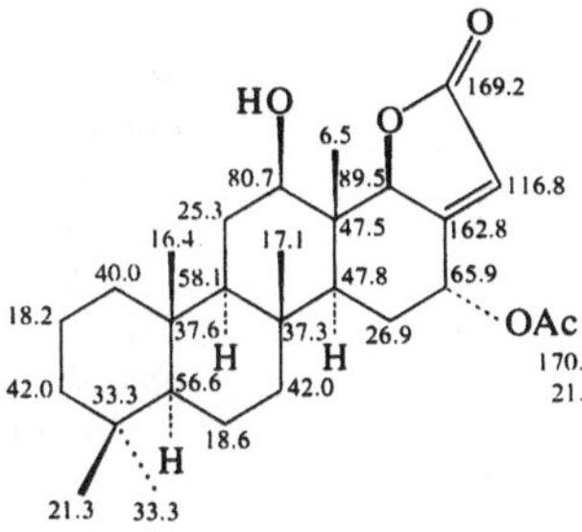

(154) Scalarolbutenolide

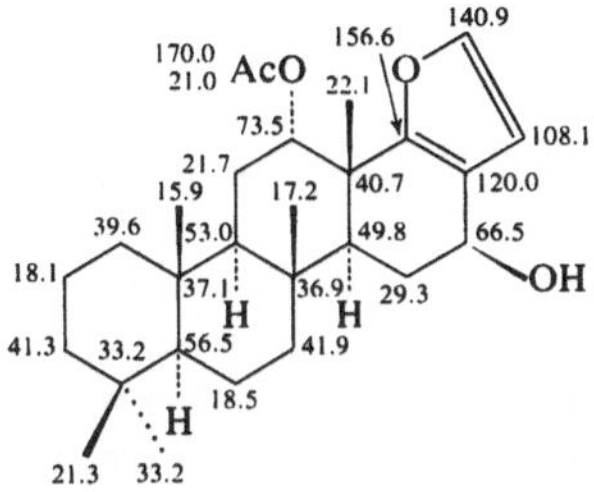

(155) Furoscalarol

E₄#.   Derivatives of Tetracyclic E₄ Sesterterpenes with Mixed Biogenesis

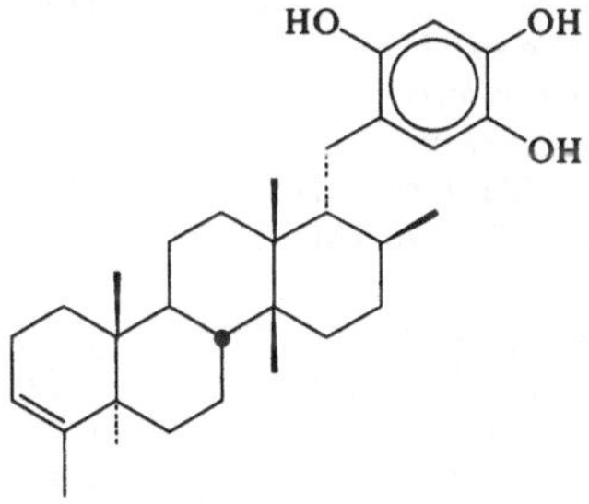

(156) Toxislyide-A

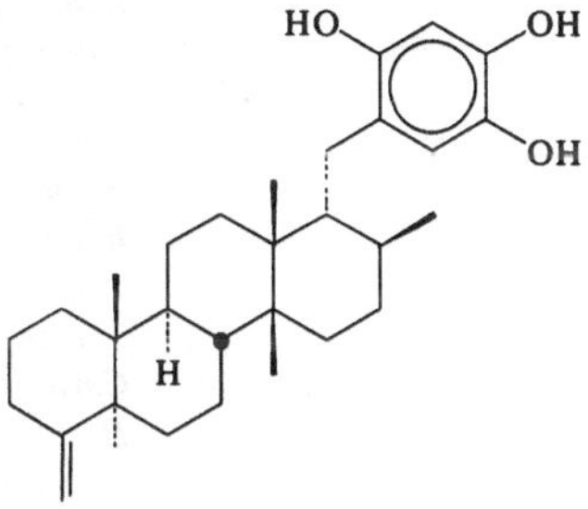

[157] Toxislyide-B

F₁.   Pentacyclic Sesterterpenes

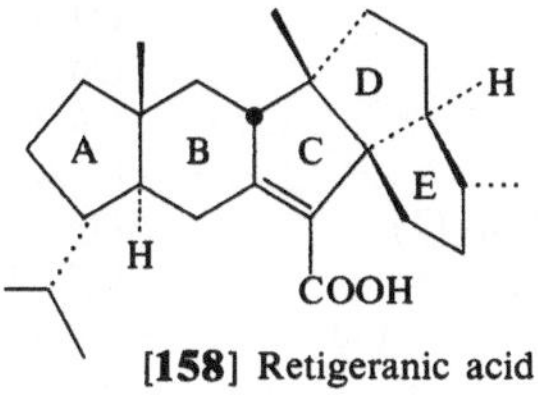

[158] Retigeranic acid

Table 7. *Physical Properties*

Compound	MP	[α]	Organism source	References
(1)	oil		*Solanum tuberosum* (T)	*(102)*
(2)			*Cochliobolus heterostrophus* (T)	*(7)*
(3)			*Ceroplastes albolineatus* (T)	*(83, 84)*
			Fasciospongia fovea (M)	*(103)*
(4)	oil		*Ceroplastes albolineatus* (T)	*(84)*
(5)	oil		*Ircinia spinosula* (M)	*(85)*
			Spongia sp. (M), *Ircinia* sp. (M),	
			Carteriospongia sp. (M), *Fascio-*	
			spongia fovea (M)	*(21, 24)*
			Spongia idia	*(67)*
(6)		−6.6	*S. idia, Cadlina marginata* (M)	*(67)*
(7), (8)	a mixture		*Spongia officinalis* (M)	*(104)*
(9)			*I. strobilina* (M)	*(161)*
(10)	oil	−15.6	*Ircinia fasciculata* (M)	*(105)*
			I. variabilis (M)	*(106)*
(11)			*I. variabilis* (M)	*(86)*
			I. strobilina (M)	*(87)*
			Fasciospongia sp. (M)	*(21)*
			I. dendroides (M)	*(28)*
(12)	as acetate		*I. dendroides* (M)	*(28)*
(13)	as acetate		*I. dendroides* (M)	*(28)*
(14)			*I. strobilina* (M)	*(87)*
(15)	oil	+45.3	*I. variabilis* (M)	*(106)*
(16)			*I. halmiformis* (M)	*(21, 24)*
(17)			*Ircinia* sp. (M)	*(107)*
(18)			*Fasciospongia* sp. (M)	*(24)*
(19)		−34.7	*Cacospongia scalaris*	*(108)*
(20)			*Cacospongia scalaris*	*(108)*
(21)			*Ircinia oros* (M), *I.* sp. (M)	*(107, 109)*
(22)			*Ircinia oros* (M), *I.* sp. (M)	*(107, 109)*
(23)	oil		*Thorecta marginalis* (M)	*(103)*
(24)	oil		*T. marginalis* (M)	*(103)*
(25)	oil		*Carteriospongia dendyi* (M)	*(110)*
(26)	oil		*Dysidea pallescens* (M)	*(111)*
(27)			*Ircinia ramosa* (M)	*(21)*
(28)	glass	−25.9	*Luaffariella variabilis* (M)	*(112)*
(29)	glass	−27.8	*L. variabilis* (M)	*(112)*
(30)	glass	+16.2	*L. variabilis* (M)	*(112)*
(31)	amorphous solid		*L. variabilis* (M)	*(113)*
(32)	oil	+31.6	*Prianos* sp. (M)	*(68, 91, 114)*
(33)	oil		*Prianos* sp. (M)	*(107)*
(34)			*Prianos* sp. (M)	*(107)*
(35)		−35.7	*Ceroplastes ceriferus* (T)	*(49, 90)*
(36)		−28.1	*C. ceriferus* (T)	*(49, 78, 115)*
(37)			*Ceroplastes albolineatus* (T)	*(49, 116, 117)*
(38)	as ester		*C. ceriferus* (T)	*(49, 51, 77)*
(39)		−69.4	*C. ceriferus* (T)	*(49, 78, 115)*

Table 7 (continued)

Compound	MP	[α]	Organism source	References
(40)		−69.5	C. ceriferus (T)	(49, 115)
(41)		−71.6	C. ceriferus (T)	(49, 115)
(42)		−83.5	C. ceriferus (T)	(49, 51)
(43)		−30.6	C. ceriferus (T)	(49, 78, 115)
(44)		−116.2	C. ceriferus (T)	(49, 115)
(45)	as ester	−77.5	C. ceriferus (T)	(49, 90)
(46)	oil	−83	C. ceriferus (T)	(49)
(47)	as ester	−93.5	C. ceriferus (T)	(49)
(48)	oil	−32.5	C. ceriferus (T)	(90)
(49)		−48.6	C. ceriferus (T)	(90)
(50)	165	+25	Salvia hypoleuca (T)	(118)
(51)	gum	+147	S. hypoleuca (T)	(118)
(52)	oil	+1.5	Unidentified sponge (M)	(119)
(53)	as ester		Sigmosceptrella laevis (M)	(65)
[54]	oil	+54	S. laevis (M)	(66)
[55]	as ester		S. laevis (M)	(65, 92)
			Prianos sp. (M)	(91)
(56)	oil	−61.2	S. laevis (M)	(65)
(57)	as ester		S. laevis (M)	(65)
[58]	165-7	−232	Ircinia wistarii (M)	(40, 120)
(59)	oil	+130	I. wistarii (M)	(120)
(60)	114-5	−23.4	Ceroplastes albolineatus (T)	(41, 121)
(61)	182-3	+30.4	Cheilanthes farinosa (T)	(122, 123)
(62)	218	+9.5	Ircinia sp. (M)	(38)
[63]	as a DNPH derivative		Cadlina luteomarginata (M)	(70)
(64)			Athabissca oil sand bithumen (T)	(54, 55)
(65)	177	+11.7	Cheilanthes farinosa (T)	(124)
(66)		+70	Ceroplastes albolineatus (T)	(125)
(67)		+47	C. albolineatus (T)	(126)
(68)			C. albolineatus (T)	(121, 127)
(69)		+139	C. albolineatus (T)	(128)
[70]			Ceroplastes albolineatus (T)	(76, 121)
(71)		+87	C. albolineatus (T)	(76)
(72)	80-1	+23	Cochliobolus heterostrophus (T)	(7)
(73)	121	+363	C. heterostrophus, H. zizaniae (T)	(129, 130)
(74)	174	+273	Helminthosporium zizaniae (T)	(130)
[75]	175	+300	H. oryzae (T)	(131)
			Cochliobolus heterostrophus (T)	(129, 130)
	182	+270	Helminthosporium oryzae (T)	(130, 131)
			H. pania (T), H. miliacei (T), H. zizaniae (T), Cochliobolus heterostrophus (T)	(9)
			Cochliobolus miyabeanus (T)	(3, 129, 133)
(76)	136		Ophiobolis miyabeanus (T)	(130)
	136-7			(132)
[77]	139	+76.2	Cephalosporium caerulens (T)	(72, 129)
(78)	as ester		Ceroplastes albolineatus (T)	(125)
(79)	as ester		C. albolineatus (T)	(134)

Table 7 *(continued)*

Compound	MP	[α]	Organism source	References
(80)	120-2	+ 143.7	*C. albolineatus* (T)	*(135)*
(81)	as ester		*C. rubens* (T)	*(136)*
(82)	oil	− 98.7	*C. rubens* (T)	*(136)*
(83)	oil	− 45	*C. rubens* (T)	*(136)*
(84)	oil	− 67.8	*C. rubens* (T)	*(136)*
[85]	123-4		*Gascardia madagascariensis* (T)	*(2, 6)*
[86]	224	+ 13.5	*Aspergillus stellatus* (T)	*(42)*
(87)		− 130.9	*Ceroplastes floridensis* (T)	*(43)*
(88)	149-51	− 203	*C. floridensis* (T)	*(43)*
[89]	62-3	− 81.2	*C. floridensis* (T)	*(43)*
(90)			*C. floridensis* (T)	*(43)*
[91]	123-6		*Gossypium hirsutum* (T)	*(46, 75)*
(92)	128-34		*G. hirsutum* (T)	*(137)*
(93)			*G. hirsutum* (T)	*(75)*
(94)	.		*G. hirsutum* (T)	*(75)*
(95)	as ester		*Ceroplastes rubens* (T)	*(136)*
(96)			*Hyrtios erecta* (M)	*(80)*
(97)	188-90	+ 36.5	*Spongia nitens* (M)	*(138)*
			Hyrtios erecta (M)	*(80)*
(98)	199-201		*Cacospongia scalaris* (M)	*(139)*
(99)	111-3	+ 47.3	*Cacospongia mollior* (M)	*(140)*
			C. scalaris (M)	*(139)*
			Spongia virgultosa (M)	*(141)*
			Hyrtios erecta (M)	*(138, 142)*
(100)	216-8	− 129	*Spongia idia* (M)	*(67)*
(101)	210-3	− 149	*Spongia nitens* (M)	*(138)*
			Hyrtios erecta (M)	*(80)*
(102)			*Dysidea herbacea* (M)	*(142)*
(103)	192-4	+ 13.9	*S. nitens, S. idia* (M)	*(37, 67, 141)*
(104)	166-8	+ 42.5	*Spongia officinalis* (M)	*(143)*
(105)	> 300	+ 24.9	*Spongia idia* (M)	*(67)*
(106)	176-7	− 104	*Hyrtios erecta* (M)	*(80, 144, 145)*
			C. scalaris (M)	*(139)*
			Spongia idia (M)	*(67)*
(107)	182	− 19.8	*Spongia idia* (M)	*(67)*
(108)			*Hyrtios erecta* (M)	*(80)*
(109)			*Hyrtios erecta* (M)	*(80)*
(110)			*Hyrtios erecta* (M)	*(80)*
(111)	236-8	− 57	*S. nitens* (M)	*(141)*
			Hyrtios erecta (M)	*(80)*
(112)	133-5	+ 43.2	*Cacospongia scalaris* (M)	*(146)*
			Spongia virgultosa (M)	*(141)*
(113)			*Hyrtios erecta* (M)	*(69)*
(114)	amorphous solid	+ 82.4	*Carteriospongia* sp. (M)	*(147)*
(115)			*Dysidea herbacea* (M)	*(142)*
(116)			*Dysidea herbacea* (M)	*(142)*
(117)	227-8	+ 124.6	*Carteriospongia radiata* (M)	*(22, 147)*

Table 7 *(continued)*

Compound	MP	[α]	Organism source	References
(118)	glass	+12	*C. dendyi* (M)	*(22, 147)*
(119)	foam	+32	*Carteriospongia dendyi* (M)	*(22, 147)*
(120)	foam		*Lendenfeldia* sp. (M)	*(149)*
(121)			*Carteriospongia* sp. (M)	*(22)*
(122)	107-8	+132.2	*Carteriospongia* sp. (M)	*(147)*
(123)	251-2	+33.5	*Lendenfeldia* sp. (M)	*(149)*
(124)	204	+110	*Carteriospongia* sp. (M)	*(147)*
[125]	193-4	+95.4	*Carteriospongia* sp. (M)	*(147)*
(126)	glass	+4.2	*Chromodoris sedna* (M)	*(37)*
(127)			*C. sedna* (M)	*(37)*
(128)	278-9		*C. sedna* (M)	*(37)*
(129)	270-1		*Lendenfeldia* sp. (M)	*(149)*
(130)	279-80	+46.8	*Lendenfeldia* sp. (M)	*(149)*
(131)	244-6	+27.7	*Lendenfeldia* sp. (M)	*(149)*
[132]	268-72		*Chromodoris sedna* (M)	*(37)*
(133)	oil		*C. sedna* (M)	*(37)*
(134)			*Dysidea herbacea* (M)	*(142)*
(135)			*Carteriospongia radiata* (M)	*(22)*
(136)	186-9	+44	*Carteriospongia foliascens* (M)	*(148, 150)*
(137)			*Carteriospongia* sp. (M)	*(22)*
(138)			*Dysidea herbacea* (M)	*(142)*
(139)			*Dysidea herbacea* (M)	*(142)*
(140)			*Carteriospongia foliascens* (M)	*(22)*
(141)			*Dysidea herbacea* (M)	*(142)*
(142)			*C. foliascens* (M)	*(22)*
(143)			*C. foliascens* (M)	*(22)*
(144)			*Carteriospongia radiata* (M)	*(22)*
(145)			*C. radiata* (M)	*(22)*
(146)			*C. radiata* (M)	*(22)*
[147]	252-4	+20	*Carteriospongia* sp. (M)	*(151)*
(148)	102-5	+51.7	*Cacospongia mollior* (M)	*(152)*
(149)		−46.9	*C. mollior* (M)	*(153)*
(150)		+11.6	*C. mollior* (M)	*(154)*
(151)		+6.5	*C. mollior* (M)	*(154)*
(152)	>260	+24	*Dysidea herbacea* (M)	*(111)*
(153)	173-4	+14.6	*Cacospongia mollior* (M)	*(155)*
(154)	220-2	+1.9	*Spongia nitens* (M)	*(150, 156)*
(155)	181-3	+14.7	*Cacospongia mollior* (M)	*(157, 158)*
(156)	as acetate		*Microciona toxistyla* (M)	*(159)*
[157]	as acetate		*M. toxistyla* (M)	*(159)*
[158]	221-2	−99.2	*Lovaria retigera* (T)	*(44)*
			L. isidiosa, L. subretigera (T)	*(44)*
(158)	218-21	−59	*L. retigera, L. subretigera* (T)	*(160)*

Acknowledgement

Support for the research reviewed here from the University of California, Santa Cruz was under a grant (to P. C.) from NOAA, National Sea Grant Program, Department of Commerce, under U. C. Project Number R/MP-33. We appreciated assistance from Ms. Judith Tillson and Ms. Melba Wallace who skillfully drew most of the structures in the original manuscript. We also thank Prof. Werner Herz for his many useful suggestions about this manuscript.

References

1. Orsenigo, M.: Toxin Production by *Helminthosporium oryzae*. II. Influence of Nutrition, pH, Temperature, and Age of Culture. Ann. Sper. Agrar. (Rome) **10**, 1809 (1965) [Chem. Abstr. **51**, 6791d (1957)].

2. Brochere-Ferreol, G., and J. Polonsky: Structure of a New Alicyclic Acid. Gascardic Acid, Isolated from Gum Lac of the Chochineal, *Gascardia madagascariensis*. Bull. Soc. Chim. France 993 (1960).

3. Nozoe, S., M. Morisaki, K. Tsuda, Y. Iitaka, N. Takahashi, S. Tamura, K. Ishibashi, and M. Shirasaka: The Structure of Ophiobolin, a C-25 Terpenoid Having a Novel Skeleton. J. Am. Chem. Soc. **87**, 4968 (1965).

4. Arigoni, D.: In: Chemical Society Autumn Meeting, Nottingham, 1965.

5. Scartazzini, R.: Ph. D. Dissertation, Nr. 3889, ETH, Zurich (1966).

6. Boeckman, R. K., Jr., D. M. Blum, E. V. Arnold, and J. Clardy: Structure of Gascardic Acid from an X-ray Diffraction Study. Tetrahedron Lett. 4609 (1979).

7. Nozoe, S., M. Morisaki, K. Fukushima, and S. Okuda: The Isolation of an Acyclic C-25 Isoprenoid Alcohol, Geranylnerolidol, and a New Ophiobolin. Tetrahedron Lett. 4457 (1968).

8. Devon, T. K., and A. I. Scott: Handbook of Naturally Occurring Compounds. Vol. II. New York: Academic Press. 1972.

9. Cordell, G. A.: The Sesterterpenes, a Rare Group of Natural Products. Prog. Phytochem. **4**, 209 (1977).

10. Glasby, J. S.: Encyclopedia of the Terpenoids. Vols. 1 and 2. New York: Wiley. 1982.

11. Baker, J. T., and V. Murphy: Handbook of Marine Science. Compounds from Marine Organisms. Vol. 1. Cleveland: CRC Press. 1976.

12. Turner, W. B.: Fungal Metabolites, Ch. 6, p. 248 – 252. New York: Academic Press. 1971.

13. Hanson, J. R.: Chemistry of Terpenes and Terpenoids, Ch. 2, Part 2, pp. 200 – 206 (Newman, A. A., ed.). New York: Academic Press. 1972.

14. Ito, S.: Natural Products Chemistry, Ch. 5, pp. 317 – 322 (Nakanishi, K., T. Goto, S. Ito, S. Natori, S. Nozoe, eds.). New York: Academic Press. 1974.

15. Klyne, W., and J. Buckingham: Atlas of Stereochemistry. Vol. II. London: Chapmann and Hall. 1978.

16. Wehrli, F. W., and T. Nishida: The Use of Carbon-13 Nuclear Magnetic Resonance Spectroscopy in Natural Products Chemistry. Prog. Chem. Org. Natur. Prod. **36**, 1. Wien-New York: Springer. 1979.

17. Cordell, G. A.: Occurrence, Structure Elucidation, and Biosynthesis of the Sesterterpenes. Phytochem. **13**, 2343 (1974).

18. Scheuer, P. J.: Chemistry of Marine Organisms, p. 22. New York: Academic Press. 1973.

19. Faulkner, D. J.: Interesting Aspects of Marine Natural Products Chemistry. Tetrahedron **33**, 1421 (1977).

20. — Antibiotics from Marine Organisms. In: Topics in Antibiotic Chemistry, Vol. 2, p. 20–29 (SAMMES, P. G., ed.). Ellis Horwood, 1978.
21. BAKER, J. T.: Some Metabolites From Australian Marine Organisms Pure Appl. Chem. **48,** 35 (1976).
22. WELLS, R. J.: New Metabolites from Australian Marine Species. Pure Appl. Chem. **51,** 1829 (1979).
23. KASHMAN, Y., A. GROWEISS, S. CARMELY, Z. KINANOM, D. CZARKIE, and M. ROTEM: Recent Research in Marine Natural Products from the Red Sea. Pure Appl. Chem. **54,** 1995 (1982).
24. MINALE, L., G. CIMINO, S. DE STEFANO, and G. SODANO: Natural Products from Porifera. Progr. Chem. Org. Natur. Prod. **33,** 1. Wien-New York: Springer. 1976.
25. MINALE, L.: In: Marine Natural Products Chemical and Biological Perspectives, Vol. 1, Ch. 4 (SCHEUER, P. J., ed.). New York: Academic Press. 1978.
26. BERGQUIST, P. R.: In: Sponges, p. 213. Berkeley: University of California Press. 1978.
27. BERGQUIST, P. R., and R. J. WELLS: In: Marine Natural Products Chemical and Biological Perspectives, Vol. V, Ch. 1 (SCHEUER, P. J., ed.). New York: Academic Press. 1983.
28. GONZALEZ, A. G., M. L. RODRIGUEZ, and A. S. M. BARRIOENTOS: On the Stereochemistry and Biogenesis of C-21 Linear Furanoterpenes in *Ircinia* Sp. J. Nat. Prod. **46,** 256 (1983).
29. DUNN, A. W., R. A. W. JOHNSTONE, T. J. KING, and B. SKLARZ: Isolation of C-25 Polyisoprenoids from *Aspergillus* sp.: Crystal Structure of Andibenin. J. Chem. Soc. Chem. Commun. 270 (1976).
30. SIMPSON, T. J.: Biosynthesis of Highly Modified Meroterpenoids in *Aspergillus variecolor*. Incorporation of Carbon-13 Labeled Acetates and Methionine into Anditomin and Andilesin C. Tetrahedron Lett. **22,** 3785 (1981).
31. SIMPSON, T. J., D. J. STENZEL, R. N. MOORE, L. A. TRIMBLE, and J. C. VEDERAS: Biosynthesis of the Meroterpenoid, Austin, by *Aspergillus ustus*: Incorporation of $^{18}O_2$ Soodium [1-^{13}C, $^{18}O_2$]-Acetate, and [Me-^{13}C, $^2H_3$]-Methionine. J. Chem. Soc. Chem. Commun. 1242 (1984).
32. POLONSKY, J., Z. VARON, T. PRANGE, C. PASCARD, and C. MORETTI: Structures of Simarinolide and Guanepolide (X-ray Analysis), New Quassinoid from *Simaba* cf. *orinocensis*. Tetrahedron Lett. **22,** 3605 (1981).
33. GRIECO, P. A., Y. MASAKI, and D. BOXLER: Sesterterpenes. I. Stereospecific Total Synthesis of Moenocinol. J. Org. Chem. **40,** 2261 (1975).
34. MARCH, J.: In: Advanced Organic Chemistry, pp. 878–983. New York: McGraw-Hill. 1977.
35. FUJIWARA, S., M. AOKI, T. UYEHARA, and T. KATO: Confirmation of the Structure of Ceriferol and Its Related Sesterterpenes by the Total Synthesis. Tetrahedron Lett. **25,** 3003 (1984).
36. NAYLOR, S., F. J. HANKE, L. V. MANES, and P. CREWS: Chemical and Biological Aspects of Marine Monoterpenes. Prog. Org. Chem. Org. Natur. Prod. **44,** 190. Wien-New York: Springer 1983.
37. HOCHLOWSKI, J. E., D. J. FAULKNER, L. S. BASS, and J. CLARDY: Metabolites of the Dorid Nudibranch *Chromodoris sedna*. J. Org. Chem. **48,** 1738 (1983).
38. MANES, L. V., S. NAYLOR, P. CREWS, and G. BAKUS: Suvanine, a Novel Sesterterpene from an *Ircinia* Marine Sponge. J. Org. Chem., **50,** 284 (1985).
39. COATES, R. M.: Biogenetic-Type Rearrangements of Terpenes. Prog. Org. Chem. Org. Natur. Prod. **33,** 156–168. Wien-New York: Springer. 1976.
40. HOFHEINZ, W., and P. SCHONHOLZER: Ircinianin, a Novel Sesterterpene from a Marine Sponge. Helv. Chim. Acta **60,** 1367 (1977).

41. Rios, T., L. Quijano, and J. Calderon: Albolineol, a Sesterterpene with a Novel Bicyclic Skeleton. J. Chem. Soc. Chem. Comm. 728 (1974).
42. Qureshi, I. H., S. A. Husain, R. Noorani, N. Murtaza, Y. Iitaka, S. Iwasaki, and S. Okuda: Stellatic Acid: A New Class of Sesterterpenoid; X-ray Crystal Structure. Tetrahedron Lett. 21, 1961 (1980).
43. Naya, Y., K. Yoshihara, T. Iwashita, H. Komura, K. Nakanishi, and Y. Hata: Unusual Sesterterpenoids from the Secretion of Ceroplastes floridensis (Coccidae), an Orchard Pest. Application of the Allylic Benzoate Method for Determination of Absolute Configuration. J. Am. Chem. Soc. 103, 7009 (1981).
44. Kaneda, K., R. Takahashi, Y. Iitaka, and S. Shibata: Retigeranic acid, A Novel Sesterterpene Isolated from the Lichens of Lobaria retigera group. Tetrahedron Lett. 4609 (1972).
45. Herz, W., and J. S. Prasad: Biogenetic Type Synthesis of Scalaranes. J. Org. Chem. 47, 4171 (1982).
46. Stipanovic, R. D., A. A. Bell, D. H. O'Brien, and M. J. Lukefahr: Heliocide H-2: An Insecticidal Sesterterpene from Cotton. Tetrahedron Lett. 567 (1977).
47. Boeckman, R. K., Jr., D. M. Blum, and S. D. Arthur: A Total Synthesis of Gascardic Acid. J. Amer. Chem. Soc. 101, 5060 (1979).
48. Vig, O. P., S. D. Sharma, S. S. Bari, and S. S. Rana: Terpenoids. Part CXXXVI. Synthesis of Geranylfarnesol. Indian J. Chem. 17 B, 31 (1979).
49. Pawlak, J. K., M. S. Tempesta, T. Iwashita, K. Nakanishi, and Y. Naya: Structures of Sesterterpenoids from the Scale Insect Ceroplastes Ceriferus. Revision of the 14-Membered Ceriferene Skeleton from 2-t/6-c/10-t to 2-c/6-t/10-t. Chem. Lett. (Japan) 1069 (1983).
50. Ikeda, Y., M. Aoki, T. Uyehara, T. Kato, and T. Yokoyama: Synthesis of dl-(2E,6Z,10E)-Cericerene for the Elucidation of the Proposed Structure of Cericerol-I. Chem. Lett. 1073 (1983).
51. Naya, Y., F. Miyamoto, K. Kishida, T. Kusumi, H. Kakisawa, and K. Nakanishi: Revised Structure of Ceriferic Acid. Chem. Lett. (Japan) 883 (1980).
52. Moiseenkov, A. M., E. V. Polunin, and A. V. Semenovsky: Synthesis of (2Z,6Z,10Z,14E,18E)-farnesylfarnesol. Tetrahedron Lett. 22, 3309 (1981).
53. Akio, S., N. Kenji, K. Noriaki, T. Yoshimasa, K. Shizumasa, A. Shinya, and Y. Kouzi: Brit. U.K. Patent Appl. GB. 2,068,370. Chem. Abs. 96, 199940s (1982).
54. Sierra, M. G., R. M. Cavero, M. A. Laburde, and E. A. Ruveda: Stereoselective Synthesis of (.+−.)-18,19-dinor-13.beta.H,14-alpha-H-cheilanthane, the Most Abundant Tricyclic Compound from Petroleums and Sediments. J. Chem. Soc. Chem. Commun. 417 (1984).
55. Ekweozor, C. M., and O. P. Strausz: 18,19-bisnor-13-14-cheilanthane: A Novel Degraded tricyclic Sesterterpenoid Type Hydrocarbon from the Athabasca Oil Sands. Tetrahedron Lett. 23, 2711 (1982).
56. Venkateswaran, R. V., D. Mukherjee, and P. C. Dutta: Synthetic Studies on Terpenoids. Part 21. Synthesis of a Few Perhydrophenanthrene Derivatives Related to Cheilanthatriol and Its Degradation Products. J. Chem. Soc. Perkin I, 1603 (1981).
57. Boeckman, R. K., Jr., J. B. Bershas, J. Clardy, and B. Solheim: Sesterterpenes. 1. Stereospecific Construction of the Ceroplastol and Ophiobolin Ring Systems via a Common Bicyclic Intermediate. J. Org. Chem. 42, 3630 (1977).
58. Dauben, W. G., and D. J. Hart: A Synthesis of the Ophiobolin Nucleus. J. Org. Chem. 42, 922 (1977).
59. Coates, R. M., and P. D. Senter: Annelative Ring Expansion Via Intramolecular (2+2) Photocycloaddition of Alpha,beta-unsaturated Gamma-lactones and Reductive Cleavage — Synthesis of Hydrocyclopentacyclooctene-5-carboxylates. J. Org. Chem. 47, 3597 (1982).

60. WEE, S.-H. L.: Studies Related to the Synthesis of Ophiobolin. Ph. D. Thesis University of California, Berkeley, 1982. Diss. Abs. **44**, 177-B (1983).

61. DAS, T. K., P. C. DUTTA, G. KARTHA, and J. M. BERNASSAU: Synthetic Studies on Terpenoids. Part 19. Synthesis of 3-beta, 10-alpha, 14-beta, trimethyl-1-beta-H, 11-beta-H, tricyclo(9.3.0.0) tetradec-6-en-5 one, a Tricyclic Ketone related to the Ophiobolins. J. Chem. Soc. Perkin I, 1287 (1977).

62. DAS, T. K., and P. C. DUTTA: Synthetic Studies on Ophiobolins. Synthesis of 1-beta(H)-3,7-alpha,11-beta,trimetyhl-cis-bicyclo (6.3.0)undecan-4 one. Synth. Commun. **6**, 253 (1976).

63. HUDLICKY, T., and R. P. SHORT: Terpenic Acids by Cyclopentene Annulation of Exocyclic Dienes. J. Org. Chem. **47**, 1522 (1982).

64. DAUBEN, W. G., and D. J. HART: A New Entry to the Tricyclo : 6.3.0.04,8 : Undecane Ring System. J. Org. Chem. **42**, 3783 (1977).

65. ALBERICCI, M., J. C. BRAEKMAN, D. DALOZE, and B. TURSCH: Chemical Studies of Marine Invertebrates. XIV. The Chemistry of Three Norsesterterpene Peroxides from the Sponge *Sigmosceptrella laevis*. Tetrahedron **38**, 1881 (1982).

66. ALBERICCI, M., M. COLLART-LEMPEREUR, J. C. BRAEKMAN, D. DALOZE, B. TURSCH, J. P. DECLERCQ, G. GERMAIN, and M. VAN MEERSSCHE: Chemical Studies of Marine Invertebrates. XII. Sigmosceptrellin-A Methyl Ester, a Nor-sesterterpenoid Peroxide from the Sponge *Sigmosceptrella laevis*. Tetrahedron Lett. 2687 (1979).

67. WALKER, R. P., J. E. THOMPSON, and D. J. FAULKNER: Sesterterpenes from *Spongia idia*. J. Org. Chem. **45**, 4976 (1980).

68. MANES, L. V., G. BAKUS, and P. CREWS: Bioactive Marine Sponge Norditerpene and Norsesterterpene Peroxides. Tetrahedron Lett. **25**, 931 (1984).

69. CREWS, P., P. BESCANSA, and G. BAKUS: A Non-Peroxide Norsesterterpene from a Marine Sponge. Experientia, **41**, 690 (1985).

70. HELLOU, J., R. J. ANDERSEN, S. RAFII, E. ARNOLD, and J. CLARDY: Luteone, a Twenty Three Carbon Terpenoid from the Dorid Nudibranch *Cadlina luteomarginata*. Tetrahedron Lett. **22**, 4173 (1981).

71. ISHIBASHI, K.: Studies on Antibiotics from *Helminthosporium* Sp. Fungi. V. J. Antibiotic **115**, 88 (1962).

72. ITAI, A., S. NOZOE, K. TUSUDA, and S. OKUDA: The Structure of Cephalonic Acid, a Pentaprenyl Terpenoid. Tetrahedron Lett. 4111 (1967).

73. TUSUDA, K., S. NOZOE, M. MORISAKI, K. HIRAI, A. ITAI, S. OKUDA, L. CANONICA, A. FIECCHI, M. G. KIENLE, and A. SCALA: Nomenclature of Ophiobolins. Tetrahedron Lett. 3369 (1967).

74. ISHIBASHI, K.: Studies on Antibiotics from *Helminthosporium* Sp. Fungi. Part III. Ophibolin Production by *Helminthosporium tuarcicum*. J. Agr. Chem. Soc. Japan **35**, 323 (1961), **36**, 323 (1961).

75. O'BRIEN, D. H., and R. D. STIPANOVICS: Carbon-13 Magnetic Resonance of Cotton Terpenoids: Carbon-Proton Long Range Couplings. J. Org. Chem. **43**, 1105 (1978).

76. IITAKA, Y., I. WATANABE, I. T. HARRISON, and S. HARRISON: The Structure of Ceroplasteric Acid and Ceroplastol I. Sesterterpenes from an Insect Wax. J. Amer. Chem. Soc. **90**, 1092 (1968).

77. KUSUMI, T., T. KINOSHITA, K. FUJITA, and H. KAKISAWA: A New Macrocyclic Sesterterpene Acid from *Ceroplastes ceriferus*. Chem. Lett. (Japan) 1129 (1979).

78. NAYA, T.: The Secretion of a Scale Insect, *Ceroplastes ceriferus*. Rev. Latinoamer. Quim. **10**, 186 (1979).

79. CREWS, P., S. NAYLOR, B. W. MYERS, J. LOO, and W. V. MANES: Residually Coupled Attached Proton Test in the ^{13}C NMR Assignment of Natural Products. Mag. Reson. in Chem., in press (1985).

80. CREWS, P., and P. BESCANSA: Sesterterpenes from a Common Marine Sponge, *Hyrtios erecta.* submitted.
81. BATES, R. B., D. M. GAYLE and B. J. GRUNER: The Stereoisomeric Farnesols. J. Org. Chem. **28,** 1080 (1963).
82. YASUYUKI, T., H. SATO, and A. KAGEYU: Structural Characterization of Polyprenols by 13C-N.M.R. Spectroscopy: Signal Assignments of Polyprenol Homologues. Polymer. **23,** 1987 (1982).
83. RIOS, T., and S. PEREZ: Geranylfarnesol, a New Acyclic C25 Isoprenoid Alcohol Isolated from Insect Wax. J. Chem. Soc. Chem. Comm. 214 (1969).
84. QUIJANO, L., J. S. CALDERON, and T. RIOS: Omega-Hydroxygeranylfarnesol, a New C-25 Isoprenoid Alcohol, Isolated from Insect Wax. Chem. Lett. (Japan) 1387 (1979).
85. CIMINO, G., S. DE STEFANO, and L. MINALE: Pentaprenyl Derivatives from the Sponge *Ircinia spinosula.* Tetrahedron **28,** 1315 (1972).
86. FAULKNER, D. J.: Variabilin, an Antibiotic from the Sponge, *Ircinia variabilis.* Tetrahedron Lett. 3821 (1973).
87. ROTHBERG, L., and P. SHUBIAK: Structure of Antibiotics from the Sponge *Ircinia strobilina.* Tetrahedron Lett. 769 (1975).
88. IKEDA, Y., M. AUKI, T. UYEHARA, T. KATO, and T. YOKOYAMA: Cyclization of Polyenes. 37. Synthesis of dl-(2E,6Z,10E)-cericerene for Elucidation of the Proposed Structure of Cericerol-I. Chem. Lett. 1073 (1983).
89. CALDERON, J., L. QUIJANO, M. GUZMAN, and T. RIOS: Nonenzymatic Cyclization of a Sesterterpene, Geranylfarnesol. Rev. Latinoam. Quim. **11,** 102 (1980).
90. MIYAMOTO, F., H. NAOKI, Y. NAYA, and K. NAKANISHI: Study of the Secretion from a Scale Insect *(Ceroplastes ceriferus).* Tetrahedron **36,** 3481 (1980).
91. SOKOLOFF, S., S. HALEVY, V. USIELI, A. COLORNI, and S. SAREL: Prianicin A and B, Norsesterterpenoid Peroxide Antibiotics from Red Sea Sponges. Experientia **38,** 337 (1982).
92. PICCINNI-LEOPARDI, C., G. GERMAIN, M. VAN MEERSSCHE, M. ALBERICCI, J. C. BRAEKMAN, D. DALOZE, and B. TURSCH: Chemical Studies of Marine Invertebrates. Part 46. Confirmation of the Molecular Structure of Sigmosceptrellin-B by X-ray Diffraction Analysis. J. Chem. Soc. Perkin **II,** 1523 (1982).
93. CREWS, P., and E. KHO-WISEMANN: Stereochemical Assignments in Marine Natural Products by C-13 Gamma Effects. Tetrahedron Lett. 2483 (1978).
94. SCMITZ, F. J., V. LAKSHMI, D. R. POWELL, and D. VAN DER HELM: Arenarol and Arenarone: Sesquiterpenoids with Rearranged Drimane Skeletons from the Marine Sponge *Dysidea arenaria.* J. Org. Chem. **49,** 241 (1984).
95. LUIBRAND, R. T., T. R. ERDMANN, J. J. VOLLMER, P. J. SCHEUER, J. FINER, and J. CLARDY: Ilimaquinone, A Sesquiterpenoid Quinone from a Marine Sponge. Tetrahedron **35,** 609 (1979).
96. LETEIJN, J. M., A. VAN VELDHUIZEN, and A. DE GROOT: The Assignment of C-13 NMR Shift Data in Clerodanes and Related Structures. Org. Magn. Reson. **19,** 95 (1982).
97. ELIEL, E. L.: Stereochemistry of Carbon Compounds, pp. 95—97 (1962).
98. MARTINEZ-RIPOLL, M., and J. FAVOS: On the Absolute Configuration Determination by X-ray Diffraction Data. Z. Kristallogr. **152,** 189 (1980).
99. SHIRAHATA, K., and N. HIRAYAMA: Revised Absolute Configuration of Mitomycin C. X-ray Analysis of 1-N-(-p-Bromobenzoyl)mitomycin C. J. Am. Chem. Soc. **105,** 7199 (1983).
100. TSUJI, N., T. KAMIGAUCHI, H. NAKAI, and M. SHIRO: X-Ray Analysis of Dibromogriseusin A. Revised Absolute Configuration of Griseusins. Tetrahedron Lett. **24,** 389 (1980).
101. HARADA, N., J. IWABUCHI, Y. YOKOTA, and H. UDA: A Chiroptical Method for Determining the Absolute Configuration of Allylic Alcohols. J. Am. Chem. Soc. **103,** 5590 (1981), and references therein.

102. TOYODA, M., M. ASAHINA, H. FUKAWA, and T. SHIMIZU: Isolation of New Acyclic C-25-Isoprenyl Alcohol from Potato Leaves. Tetrahedron Lett. 4879 (1969).

103. KAZLAUSKAS, R., P. T. MURPHY, R. J. QUINN, and R. J. WELLS: Two New Sesterterpene Lactones from a Sponge. Tetrahedron Lett. 2635 (1976).

104. CIMINO, G., S. DE STEFANO, and L. MINALE: Further Linear Furanoterpenes from Marine Sponges. Tetrahedron **28,** 5983 (1972).

105. CAFIERI, F., E. FATTORUSSO, C. SANTACROCE, and L. MINALE: Fasciculatin, A Novel Sesterterpene from the Sponge *Ircinia Fasciculata.* Tetrahedron **28,** 1579 (1972).

106. ALFANO, G., G. CIMINO, and S. DE STEFANO: Palinurin, a New Linear Sesterterpene from a Marine Sponge. Experientia **35,** 1136 (1979).

107. MANES, L. V., and P. CREWS: Unpublished results.

108. FUSETANI, N., Y. KATO, S. MATSUNAGA, and K. HASHIMOTO: Bioactive Marine Metabolites V. Two New Furanosesterterpenes, Inhibitors of Cell Division of the Fertilized Starfish Eggs, from the Marine Sponge *Cacospongia scalaris.* Tetrahedron Lett. **25,** 4941 (1984).

109. CIMINO, G., S. DE STEFANO, L. MINALE, and E. FATTORUSSO: Ircinin-1 and -2, Linear Sesterterpenes from the Marine Sponge *Ircinia oros.* Tetrahedron **28,** 333 (1972).

110. KAZLAUSKAS, R., P. T. MURPHY, and R. J. WELLS: Furodendin, A C22 Degraded Terpene from the Sponge *Phyllospongia dendyi.* Experientia **36,** 814 (1980).

111. CIMINO, G., P. DE LUCA, S. DE STEFANO, and L. MINALE: Disidein, a Pentacyclic Sesterterpene Condensed with an Hydroxyhydroquinone Moiety, from the Sponge *Disidea pallascens.* Tetrahedron **31,** 271 (1975).

112. SILVA DE, E. D., and P. J. SCHEUER: Three New Sesterterpenoid Antibiotics from the Marine Sponge *Luffariella variabilis* (Polejaff). Tetrahedron Lett. **22,** 3147 (1981).

113. — — Manoalide, an Antibiotic Sesterterpenoid from the Marine Sponge *Luffariella variabilis* (Polejaeff). Tetrahedron Lett. **21,** 1611 (1980).

114. KASHMAN, Y., and M. ROTEM: Muqubilin, a New C24-isoprenoid from a Marine Sponge. Tetrahedron Lett. 1707 (1979).

115. MIYAMOTO, F., H. NAOKI, T. TAKEMOTO, and Y. NAYA: New Macrocyclic Sesterterpenoids from a Scale Insect (Ceroplastes Ceriferus). Tetrahedron **35,** 1913 (1979).

116. QUIJANO, L., R. VELOZ, J. S. CALDERON, and T. RIOS: Isolation and Determination of the Structure of Albocerol, a New Macrocyclic Sesterterpene. Rev. Latinoamer. Quim. **6,** 196 (1975).

117. VELOZ, R., L. QUIJANO, L. S. CALDERON, and T. RIOS: Albocerol, a new Macrocyclic Sesterterpene. J. C. S. Chem. Comm. 191 (1975).

118. RUSTAIYAN, A., A. NIKNEJAD, L. NAZARIANS, J. JAKUPOVIC, and F. BOHLMANN: Naturally Occurring Terpene Derivatives. Part 403. Sesterterpenes from *Salvia hypoleuca.* Phytochem. **21,** 1812 (1982).

119. SULLIVAN, B., and D. J. FAULKNER: An Antimicrobial Sesterterpene from a Palauan Sponge. Tetrahedron Lett. **23,** 907 (1982).

120. GREGSON, R. P., and D. OUVRIER: Wistarin, a Tetracyclic Furanosesterterpene from the Marine Sponge *Ircinia wistarii.* J. Nat. Prod. **45,** 412 (1982).

121. RIOS, T., and F. COLUNGA: Three New Alcohols from Insect Wax. Ceroplastol I, II and Albolineal. Chem. Ind. 1184 (1965).

122. KHAN, H., A. ZAMAN, G. L. CHETTY, A. S. GUPTA, and S. DEV: Cheilanthatriol, a New Fundamental Type in Sesterterpenes. Tetrahedron Lett. 4443 (1971).

123. GUPTA, A. S., S. DEV, M. SANGARE, B. SEPTE, and G. LUKACS: Sur la Structure du Cheilanthatriol. Une Etude par la RMN du C-13. Bull. Soc. Chim. France 1879 (1976).

124. THANU-IYER, R., K. N. N. AYENGAR, and S. RANGASWAMI; Structure of Cheilarinosin, a New Sesterterpene from *Cheilanthes farinosa.* Indian J. Chem. **10,** 482 (1972).

125. Calderon, J. S., L. Quijano, and T. Rios: Ceralbic Acids I and II: Two New Sesterterpenic Acids Isolated from Insects' Wax. Chem. Ind. 584 (1978).

126. — — — Ceralbol, a New Sesterterpenic Alcohol Isolated from Insect Wax. Experientia 34, 421 (1978).

127. Rios, T., and L. Quijano: The Structure of Ceroplastol II, a Sesterterpene Alcohol Isolated from Insects Wax. Tetrahedron Lett. 1317 (1969).

128. Rios, T., and F. Gomez: Albolic Acid, A New Sesterterpene Acid Isolated from Insect Wax. Tetrahedron Lett. 2929 (1969).

129. Radics, L., M. Kajtar-Peredy, S. Nozoe, and H. Kobayashi: Carbon-13 Nuclear Magnetic Resonance Spectra of Ophiobolins. Tetrahedron Lett. 4415 (1975).

130. Nozoe, S., K. Hirai, and K. Tsuda: The Structure of Zizanin-A and -B, C25-Terpenoids isolated from Helminthosporium zizaniae. Tetrahedron Lett. 2211 (1966).

131. Canonica, L., A. Fiecchi, M. Galli Kienle, and A. Scala: Isolation and Constitution of Cochliobolin B. Tetrahedron Lett. 1329 (1966).

132. Nozoe, S., M. Morisaki, K. Tsuda, and S. Okuda: Biogenesis of Ophiobolins. The Origin of the Oxygen Atoms in the Ophiobolins. Tetrahedron Lett. 3365 (1967).

133. Canonica, L., A. Fiecchi, M. G. Kienle, B. M. Ranzi, and A. Scala: The Biosynthesis of Cochliobolins A and B. Tetrahedron Lett. 3035 (1966).

134. Quijano, L., J. S. Calderon, and T. Rios: Structure of Ceroplastolic Acid: A New Sesterterpene Isolated from Insect Wax. Chem. Ind. 592 (1979).

135. — — — The Structure of Ceroplastodiol, a New Tricyclic Sesterterpene Isolated from Insect Wax. Experientia 37, 542 (1981).

136. Tempesta, M. S., T. Iwashita, F. Miyamoto, K. Yoshihara, and Y. Naya: New Class of Sesterterpenoids from the Secretion of Ceroplastes rubens (Coccidae). J. C. S. Chem. Comm. 1182 (1983).

137. Stipanovic, R. D., A. A. Bell, D. H. O'Brien, and M. J. Lukefahr: Heliocide H-3, An Insecticidal Terpenoid from Gossypium hirsutum. Phytochem. 17, 151 (1978).

138. Cimino, G., S. De Stefano, and A. Di Luccia: Further Sesterterpenes from the Sponge Spongia nitens: 12-Epi-scalaradial and 12,18-Diepi-scalaradial. Experientia 35, 1277 (1979).

139. Yasuda, F., and H. Tada: Desacetylscalaradial, a Cytotoxic Metabolite from the Sponge Cacospongia scalaris. Experientia 37, 110 (1981).

140. Cimino, G., S. De Stefano, and L. Minale: Scalaradial, a Third Sesquiterpene with the Tetracarbocyclic Skeleton of Scalarin, from the Sponge Cacospongia mollior. Experientia 30, 846 (1974).

141. Cimino, G., S. De Stefano, L. Minale, and E. Trivellone: 12-Epi-scalarin and 12-Epi-deoxoscalarin, Sesterterpenes from the Sponge Spongia nitens. J. Chem. Soc. Perkin I, 1587 (1977).

142. Kashman, Y., and M. Zviely: New Alkylated Scalarins from the Sponge Dysidea herbacea. Tetrahedron Lett. 3879 (1979).

143. Cimino, G., S. De Stefano, and L. Minale: Deoxyscalarin, a Further Sesterterpene with the Unusual Tetracyclic Carbon Skeleton of Scalarin, from Spongia officinalis. Experientia 29, 934 (1973).

144. Kazlauskas, R., P. T. Murphy, R. J. Quinn, and R. J. Wells: Heteronemin, a new Scalarin Type Sesterterpene from the Sponge Heteronema erecta. Tetrahedron Lett. 2631 (1976).

145. Kashman, Y., and A. Rudi: The Carbon-13 Nmr Spectrum and Stereochemistry of Heteronemin. Tetrahedron 33, 2997 (1977).

146. Fattorusso, E., S. Magno, C. Santacroce, and D. Sica: Scalarin, a New Pentycyclic C-25 Terpenoid from the Sponge Cacospongia scalaris. Tetrahedron 28, 5993 (1972).

147. Kazlauskas, R., P. T. Murphy, R. J. Wells, and J. J. Daly: Terpenoid Constituents from Two Phyllospongia Sp. Aust. J. Chem. 33, 1783 (1980).

148. KIKUCHI, H., Y. TSUKITANI, I. SHIMIZU, M. KOBAYASHI, and I. KITAGAWA: Foliaspongin, an Antiinflammatory Bishomosesterterpene from the Marine Sponge *Phyllospongia foliascens* (Pallas). Chem. Pharm. Bull. **29**, 1492 (1981).

149. KAZLAUSKAS, R., P. T. MURPHY, and R. J. WELLS: Five New C26 Tetracyclic Terpenes from a Sponge (Lendenfeldia sp.). Aust. J. Chem. **35**, 51 (1982).

150. KIKUCHI, H., Y. TSUKITANI, I. SHIMIZU, M. KOBAYASHI, and I. KITAGAWA: Marine Natural Products. XI. An Antiinflammatory Scalarane-Type Bishomosesterterpene, Foliaspongin, from the Okinawan Marine Sponge *Phyllospongia foliascens* (Pallas). Chem. Pharm. Bull. **31**, 552 (1983).

151. CROFT, K. D., E. L. GHISALBERTI, B. W. SKELTON, and A. H. WHITE: Structural Study of a New Dialkylated Scalarane from a *Carteriospongia* sp. J. Chem. Soc. Perkin I, 155 (1983).

152. CAFIERI, F., L. DE NAPOLI, E. FATTORUSSO, C. SANTACROCE, and D. SICA: Molliorin-A: A Unique Scalarin-Like Pyrroloterpene from the Sponge *Cacospongia mollior*. Tetrahedron Lett. 477 (1977).

153. CAFIERI, F., L. DE NAPOLI, A. IENGO, and C. SANTACROCE: Molliorin-C: A Further Pyrroloterpene Present in the Sponge *Cacospongia mollior*. Experientia **34**, 300 (1978).

154. — — — — Minor Pyrroloterpenoids from the Marine Sponge *Cacospongia mollior*. Experientia **35**, 157 (1979).

155. CAFIERI, F., L. DE NAPOLI, E. FATTORUSSO, and C. SANTACROCE: Molliorin-B: A Second Scalarin-Like Pyrroloterpene from the Sponge *Cacospongia mollior*. Experientia **33**, 994 (1977).

156. CIMINO, G., S. DE ROSA, and S. DE STEFANO: Scalarolbutenolide, a New Sesterterpenoid from the Marine sponge *Spongia Nitens*. Experientia **37**, 214 (1981).

157. CIMINO, G., F. CAFIERI, L. DE NAPOLI, and E. FATTORUSSO: Carbon-13 Nmr Spectrum and Absolute Stereochemistry of Furoscalarol. Tetrahedron Lett. 2041 (1978).

158. CAFIERI, F., L. DE NAPOLI, E. FATTORUSSO, C. SANTACROCE, and D. SICA: Furoscalarol, Scalarin-Like Furanosesterterpenoid from the Marine Sponge *Cacospongia mollior* Gazz. Chim. Ital. **107**, 71 (1977).

159. CIMINO, G., S. DE STEFANO, L. MINALE, R. RICCIO, K. HIRTOSU, and J. CLARDY: Two Novel Sesterterpene Hydroxyquinols from the Sponge *Microciona toxistyla*. Tetrahedron Lett. 3619 (1979).

160. RAO, P. S., K. B. SARMA, and T. S. SESHADRI: Chemical Components of *Lobaria subisidiosa, L. retigera,* and *L. subertigera* from the Western Himalayas. Curr. Sci. **35**, 147 (1966).

161. SHUBIAK, P.: The Isolation and Structure Elucidation of Novel Sesterterpenes with Antibiotic Properties from the Sponge *Ircinia strobilina*. Ph. D. Thesis, Diss. Abstr. Int. B 1978, **38**, 4775.

162. HEISSLER, D., R. OCAMPO, P. ALBRECHT, J. J. RIEHL, and G. OVRISSON: Identification of Long Chain Tricyclic terpene Hydrocarbons ($C_{21} - C_{30}$) in Geological Samples. J. Chem. Soc. Chem. Commun. 496 (1984).

163. FREITAS DE, J. C., L. A. BLANKEMEIER, and R. S. FACOBS: In vitro Inactivation of the Neurotoxic Action of Beta-bungarotoxin by the Marine Natural Product, Manoalide. Experientia **40**, 864 (1984).

164. WOODWARD, R. W., L. MASCARO, JR., R. HORHAMMER, S. EISENSTEIN, and H. G. FLOSS: Stereochemistry of Indolmycin Biosynthesis. Steric Course of C- and N-methylation Reactions. J. Am. Chem. Soc. **102**, 6314 (1980).

(Received December 29, 1984)

Author Index

Page numbers printed in *italics* refer to References

Subject Index

By

R. Berner, Wien

Fortschritte der Chemie organischer Naturstoffe

Progress in the Chemistry of Organic Natural Products

Volume 47:

1985. 16 figures. VIII, 290 pages.
Cloth DM 198,—. ISBN 3-211-81864-2

Contents: R. SOUTHGATE and S. ELSON: Naturally Occurring β-Lactams. — I. HOWE and M. JARMAN: New Techniques for the Mass Spectrometry of Natural Products. — P. G. McDOUGAL and N. R. SCHMUFF: Chemical Synthesis of the Trichothecenes. — J. POLONSKY: Quassinoid Bitter Principles II.

Volume 46:

1984. 7 figures. IX, 253 pages.
Cloth DM 178,—. ISBN 3-211-81804-9

Contents: O. TANAKA and R. KASAI: Saponins of Ginseng and Related Plants. — E. FUJITA, M. NODE: Diterpenoids of *Rabdosia* Species. — S. JOHNE: The Quinazoline Alkaloids.

Volume 45:

1984. 2 figures. VIII, 288 pages.
Cloth DM 194,—. ISBN 3-211-81755-7

Contents: D. A. H. TAYLOR: The Chemistry of the Limonoids from Meliaceae. — J. A. ELIX, A. A. WHITTON, and M. V. SARGENT: Recent Progress in the Chemistry of Lichen Substances. — Y. SHIMIZU: Paralytic Shellfish Poisons.

Volume 44:

1983. 72 partly coloured figures. IX, 326 pages.
Cloth DM 208,–. ISBN 3-211-81754-9

Contents: F. J. EVANS and S. E. TAYLOR: Pro-Inflammatory, Tumour-Pro-
moting and Anti-Tumour Diterpenes of the Plant Families Euphorbiaceae
and Thymelaeaceae. – A. MONDON and B. EPE: Bitter Principles of Cneo-
raceae. – S. NAYLOR, F. J. HANKE, L. V. MANES, and P. CREWS: Chemical
and Biological Aspects of Marine Monoterpenes. – J. G. BUCHANAN: The
C-Nucleoside Antibiotics.

Volume 43:

1983. VIII, 383 pages.
Cloth DM 208,–. ISBN 3-211-81741-7

Contents: J. L. INGHAM: Naturally Occurring Isoflavonoids (1855–1981). –
A. KOSKINEN and M. LOUNASMAA: The Sarpagine-Ajmaline Group of
Indole Alkaloids.

Volume 42:

1982. VII, 323 pages.
Cloth DM 164,–. ISBN 3-211-81706-9

Contents: Y. ASAKAWA: Chemical Constituents of the Hepaticae. –
M. HEIDELBERGER: Cross-Reactions of Plant Polysaccharides in Anti-
pneumococcal and Other Antisera, an Update.

All Volumes and Cumulative Index 1–20 available

Price reduction for subscribers: 10%

**Special reduced price (20% reduction) for the complete Series
Vols. 1–48 incl. the Cumulative Index to Vols. 1–20**

Springer-Verlag Wien New York